WITHDRAWN
from the RIT Libraries

NMR Quantum Information Processing

NMR Quantum Information Processing

Ivan S. Oliveira
Brazilian Center for Research in Physics, Rio de Janeiro, Brazil

Tito J. Bonagamba
*Institute of Physics of São Carlos, University of São Paulo,
São Carlos, Brazil*

Roberto S. Sarthour
Brazilian Center for Research in Physics, Rio de Janeiro, Brazil

Jair C.C. Freitas
Federal University of Espírito Santo, Vitória, Brazil

Eduardo R. deAzevedo
*Institute of Physics of São Carlos, University of São Paulo,
São Carlos, Brazil*

ELSEVIER

AMSTERDAM • BOSTON • HEIDELBERG • LONDON • NEW YORK • OXFORD
PARIS • SAN DIEGO • SAN FRANCISCO • SINGAPORE • SYDNEY • TOKYO

Elsevier
Radarweg 29, PO Box 211, 1000 AE Amsterdam, The Netherlands
Linacre House, Jordan Hill, Oxford OX2 8DP, UK

First edition 2007

Copyright © 2007 Elsevier B.V. All rights reserved

No part of this publication may be reproduced, stored in a retrieval system or transmitted in any form or by any means electronic, mechanical, photocopying, recording or otherwise without the prior written permission of the publisher

Permissions may be sought directly from Elsevier's Science & Technology Rights Department in Oxford, UK: phone (+44) (0) 1865 843830; fax (+44) (0) 1865 853333; email: permissions@elsevier.com. Alternatively you can submit your request online by visiting the Elsevier web site at http://elsevier.com/locate/permissions, and selecting *Obtaining permission to use Elsevier material*

Notice
No responsibility is assumed by the publisher for any injury and/or damage to persons or property as a matter of products liability, negligence or otherwise, or from any use or operation of any methods, products, instructions or ideas contained in the material herein. Because of rapid advances in the medical sciences, in particular, independent verification of diagnoses and drug dosages should be made

Library of Congress Cataloging-in-Publication Data
A catalog record for this book is available from the Library of Congress

British Library Cataloguing in Publication Data
A catalogue record for this book is available from the British Library

ISBN: 978-0-444-52782-0

For information on all Elsevier publications
visit our website at books.elsevier.com

Printed and bound in The Netherlands
07 08 09 10 11   10 9 8 7 6 5 4 3 2 1

**Working together to grow
libraries in developing countries**

www.elsevier.com | www.bookaid.org | www.sabre.org

ELSEVIER   BOOK AID International   Sabre Foundation

To the people who give
support to our lives

Preface

Quantum Computation and Quantum Information (generally referred as QIP) deals with the identification and use of quantum resources for information processing. This includes three main branches of investigation: quantum algorithm design, quantum simulation and quantum communication, including quantum cryptography. Along the past few years, QIP has become one of the most active areas of research in both theoretical and experimental physics, attracting young students and researchers fascinated, not only by the potential practical applications of quantum computers, but also by the possibility of studying fundamental physics at the deepest level of quantum phenomena.

From a practical viewpoint, any experimental technique candidate to implement QIP in large scale, must satisfy the following basic demands: (i) to have a good physical representation for the quantum unit of information: the qubit; (ii) to be able to generate a complete set of universal quantum gates, and (iii) to be applicable to a scalable physical system. Nuclear Magnetic Resonance (NMR) perfectly satisfies the first two demands. Indeed, nuclear spins are nearly ideal qubits, and radiofrequency pulses correctly implement unitary transformations which can easily build a complete set of universal quantum logic gates. Since 1997, after the discovery of the so-called pseudo-pure states, every single quantum algorithm has been demonstrated by the use of liquid-state NMR. In this approach, qubits are represented by nuclear spins in molecules of a liquid. The main advantage of this approach is the straight use in QIP of a highly advanced technique, definitely established in science and technology by more than 50 years of development! However, it has also a main drawback: it is not scalable. That basically means that liquid-state NMR is an excellent technique to study the fundamentals of QIP, but not to build a large-scale quantum computer. However, this also means that if we want to take the advantages of NMR technology to build large-scale quantum computers, one must develop alternatives to liquid-state samples. And this is quickly developing in different fronts. In one front, techniques of atom-by-atom manipulation became a reality which will allow in the near future the construction of solid-state qubit arrays for large-scale QIP. In another front, Magnetic Resonance Force Microscopy (MRFM) has raised as a main breakthrough, capable of increasing NMR sensitivity from the current 10^{14} to a single spin! This technique can be used to implement the main steps necessary to practical implementation of NMR QIP.

This book describes the fundamentals of NMR QIP, and the main developments which can lead to a large-scale quantum processor. It is aimed at senior undergraduate students and graduates entering this area of research. It can also be used as a reference book in advanced quantum mechanics courses. It is our wish that the book will be useful as a reference for researches in the area of QIP, and other correlated areas. The text starts with a general chapter on the interesting topic of the physics of computation. The very first ideas which sparkled the development of QIP came from basic considerations of the physical processes underlying computational actions. In Chapter 2 an introduction it is made to NMR, including the hardware and other experimental aspects of the technique. In Chapter 3 we revise the fundamentals of Quantum Computation and Quantum Information.

The chapter is very much based on the extraordinary book of Michael A. Nielsen and Isaac L. Chuang (Cambridge, 2002), with an upgrade containing some of the latest developments, such as QIP in phase space. Chapter 4 describes how NMR generates quantum logic gates from radiofrequency pulses, upon which quantum protocols are built. It also describes the important technique of Quantum State Tomography for both quadrupole and spin 1/2 nuclei. Chapter 5 describes some of the main experiments of quantum algorithm implementation by NMR, quantum simulation and QIP in phase space. The important issue of (pseudo-)entanglement in NMR QIP experiments is discussed in Chapter 6. This has been a particularly exciting topic in the literature. The chapter contains a discussion on the theoretical aspects of NMR entanglement, as well as some of the main experiments where this phenomenon is reported. Finally, Chapter 7 is an attempt to address the future of NMR QIP, based on very recent developments in nanofabrication and single-spin detection experiments. Each chapter is followed by a number of problems, all with detailed solutions, which confers to the whole text a didactic character and allows it to be used as text-book in undergraduate or graduate courses. It is therefore our wish that this book will be useful for researches in the area of QIP and other correlated areas, as well as for general readers interested in the applications of quantum mechanics.

Acknowledgments

We would like to express our gratitude to the following people who contributed to the final form of the manuscript: Professors Alfredo M.O. de Almeida, Rubem L. Sommer, Alberto P. Guimarães, and Dr. Raúl O. Vallejos, from the Brazilian Center for Research in Physics (CBPF), in Rio de Janeiro; to Dr. Renato Portugal, from the National Laboratory of Scientific Computation (LNCC), in Petrópolis; to Dr. Alviclér Magalhães and Dr. Edsom L.G. Vidoto, from the University of São Paulo at São Carlos (USP-São Carlos). To our students, Alexandre M. de Souza, Carolina Cronemberger, Suenne R. Machado, Walter Lima Jr., André G. Viana, Ruben A. Estrada, João Teles C. Neto, Diogo O.S. Pinto, Felipe O.S. Pinto, André A. de Souza, André L.B.S. Bathista, Arthur G.A. Ferreira, Carlos A. Brasil, Gregório C. Faria and Roberto Tozoni. To Mr. Marcio Paranhos, for the EPS figures. Finally, we acknowledge the support received from the Brazilian Millennium Institute of Quantum Information.

Contents

Preface vii

Acknowledgments ix

Brief Historical Survey and Perspectives 1
    References 6

1 Physics, Information and Computation 9
 1.1 Turing Machines, logic gates and computers 9
 1.2 Knowledge, statistics and thermodynamics 15
 1.3 Reversible versus irreversible computation 18
 1.4 Landauer's principle and the Maxwell demon 20
 1.5 Natural phenomena as computing processes. The physical limits of computation 21
 1.6 Moore's law. Quantum computation 24
    Problems with solutions 27
    References 31

2 Basic Concepts on Nuclear Magnetic Resonance 33
 2.1 General principles 33
 2.2 Interaction with static magnetic fields 35
 2.3 Interaction with a radiofrequency field – the resonance phenomenon . . . 38
 2.4 Relaxation phenomena 41
 2.5 Density matrix formalism: populations, coherences, and NMR observables 44
 2.6 NMR of non-interacting spins 1/2 47
 2.7 Nuclear spin interactions 52
  2.7.1 Chemical shift 54
  2.7.2 Dipolar coupling 55
  2.7.3 *J*-coupling 56
  2.7.4 Quadrupolar coupling 57
  2.7.5 General form of the internal Hamiltonians 60
 2.8 NMR of two coupled spins 1/2 62
 2.9 NMR of quadrupolar nuclei 68
 2.10 Density matrix approach to nuclear spin relaxation 73
 2.11 Solid-state NMR 75
  2.11.1 Dipolar decoupling 76
  2.11.2 Magic-angle spinning (MAS) 76
  2.11.3 Cross-polarization (CP) 78
  2.11.4 The CP-MAS experiment 78

	2.12	The experimental setup	79
	2.13	Applications of NMR in science and technology	83
		Problems with solutions	83
		References	90
3	**Fundamentals of Quantum Computation and Quantum Information**		**93**
	3.1	Historical development	93
	3.2	The postulates of quantum mechanics	95
	3.3	Quantum bits	96
	3.4	Quantum logic gates	97
		3.4.1 Some examples of application of the postulates	98
		3.4.2 The controlled NOT – CNOT – gate	99
	3.5	Graphical representation of gates and quantum circuits	100
		3.5.1 The SWAP logic gate	101
		3.5.2 The Quantum Fourier Transform – QFT	102
	3.6	Quantum state tomography	104
		3.6.1 The density matrix	104
		3.6.2 Determining ρ	105
	3.7	Entanglement	106
		3.7.1 Some applications of entanglement	109
	3.8	Quantum algorithms	111
		3.8.1 The Deutsch's algorithm	112
		3.8.2 The quantum search algorithm	113
		3.8.3 The quantum factorizing algorithm	116
	3.9	Quantum simulations	124
	3.10	Quantum information in phase space	125
		3.10.1 The Wigner function	125
		3.10.2 Measuring the Wigner function	127
		3.10.3 Quantum states in phase space	127
	3.11	Determining eigenvalues and eigenvectors	130
		Problems with solutions	131
		References	135
4	**Introduction to NMR Quantum Computing**		**137**
	4.1	The NMR qubits	137
	4.2	Quantum logic gates generated by radiofrequency pulses	140
		4.2.1 Elementary single-qubit gates and their implementations using RF pulses	140
		4.2.2 Elementary two-qubit gates and their implementation in NMR	146
		4.2.3 Multi-qubit gates	150
		4.2.4 Use of strongly modulated RF pulses for quantum gate implementation in NMR QIP	151
	4.3	Production of pseudo-pure states	153
		4.3.1 Temporal averaging	154
		4.3.2 Spatial averaging	158
		4.3.3 State labeling	160

	4.4	Reconstruction of density matrices in NMR QIP: Quantum State Tomography	162
		4.4.1 NMR Quantum State Tomography	163
		4.4.2 NMR Quantum State Tomography in coupled spin 1/2 systems	163
		4.4.3 NMR Quantum State Tomography of quadrupole nuclei	165
	4.5	Evolution of Bloch vectors and other quantities obtained from tomographed density matrices	168
		Problems with solutions	171
		References	180
5	**Implementation of Quantum Algorithms by NMR**		183
	5.1	Numerical simulation of NMR spectra and density matrix calculation along an algorithm implementation	183
	5.2	NMR implementation of Deutsch and Deutsch–Jozsa algorithms	185
	5.3	Grover search tested by NMR	187
	5.4	Quantum Fourier Transform NMR implementation	189
	5.5	Shor factorization algorithm tested in a 7-qubit molecule	190
	5.6	Algorithm implementation in quadrupole systems	193
	5.7	Quantum simulations	193
	5.8	Measuring the discrete Wigner function	198
		Problems with solutions	201
		References	204
6	**Entanglement in Liquid-State NMR**		207
	6.1	The problem of liquid-state NMR entanglement	207
	6.2	The Peres criterium and bounds for NMR entanglement	209
	6.3	Some NMR experiments reporting pseudo-entanglement	211
		Problems with solutions	217
		References	220
7	**Perspectives for NMR Quantum Computation and Quantum Information**		221
	7.1	Silicon-based proposals: solution for the scaling problem	222
	7.2	NMR quantum information processing based on Magnetic Resonance Force Microscopy (MRFM)	226
	7.3	Single spin detection techniques: solution for the sensitivity problem	231
	7.4	NMR on a chip: towards the NMR quantum chip integration	234
		Problems with solutions	236
		References	241
Index			243

Brief Historical Survey and Perspectives

Various names are commonly associated to the invention and development of modern computing science. Among them, are George Boole (1815–1864), author of a work published in 1854 with the title: *An investigation into the laws of thought, on which are founded the mathematical theories of logic and probabilities*, which founded the nowadays called Boolean Algebra, and Claude Shannon (1916–2001) who, in 1938 on his MIT MSc Thesis, *A symbolic analysis of relay and switching circuits*, proposed a way for representing Boolean logic operators through relays and switches.

However, the Theory of Computation became an area of abstract mathematics only after the work of Alan Turing (1912–1954) and Alonzo Church (1903–1995). On his attempt to answer one of the challenges proposed by the great mathematician David Hilbert in 1928, the *entscheidungsproblem* or *decision problem*, Turing arrived to an abstract model of computation known as the *Turing Machine*. His idea was published in 1936 as a ground breaker paper entitled *On computable numbers, with an application to the entscheidungsproblem* [1]. A Turing Machine operates with a minimum number of symbols and instructions to perform logic operations: it is the embryo of all modern programmable computers.

Another breakthrough paper appeared twelve years afterwards, in 1948, again by Claude Shannon: *A mathematical theory of communication* [2]. On this paper, Shannon defined the unit of information, the *binary digit*, or *bit*,[1] and established the theory which tells us the amount of information (i.e., the number of bits) which can be sent per unit time through a communication channel, and how this information can be fully recovered, even in the presence of noise in the channel. This work founded the Theory of Information.

The computation and information technologies have developed very close to each other, in an astonishingly rapidly pace, for the last 50 years. Nowadays, a few square centimeters computer chip possesses hundreds of millions of electronic constituents, and a hairy thin optical fibre can transmit and maintain millions of conversations simultaneously!

On the side of pure Physics, the 20th Century also produced some "miracles", one of them – and possibly the most important of all – was *Quantum Mechanics*. The early development of this theory has attached to it a whole team of brilliant scientists: Max Planck, Niels Bohr, Albert Einstein, Louis de Broglie, Erwin Schrödinger, Wolfgang Pauli, Werner Heisenberg, only to name some of the best known. Quantum mechanics contains the rules of how to approach and solve problems involving particles such as electrons, protons, nuclei, atoms, molecules, and the interactions between these particles and radiation. Along the years, computers entered physics as a powerful ally for the analysis and development of physical models in particle and nuclear physics, condensed matter, gravitation, astrophysics, biological and ecological systems, and so on. In particular, the development of condensed matter magnetism and semiconductor physics resulted in important feedback to computer technology itself. This symbiotic relationship between physics and computers, deepened for decades until the point where computers themselves started to be seen by

---

[1] On his original paper, Shannon attributes the word *bit* to a suggestion made by J.W. Turkey.

the physicists, no longer as an auxiliary tool for the solution of complicated mathematical problems, but as physical systems, subject to the laws of physics, just like everything else! This insight led to a novel and exciting area of research in Physics: Quantum Computation and Quantum Information.

Quantum Information is the area of research in physics in which quantum resources are identified for the application in information processing, as well as the means to produce, store, send and recover information traveling through communication channels. One example of quantum resource for communication is *entanglement*, and one example of quantum information processing is *superdense coding*. To the more specific application of quantum resources to the development of quantum computer algorithms and quantum hardware, we call *Quantum Computation*. One example of quantum algorithm is the Shor factorization algorithm, and one example of quantum computing hardware are nuclear spins.

The "formal" beginning of the research field called Quantum Computation and Quantum Information can be attributed to a paper published in 1980 by Paul Benioff [3]: *The computer as a physical system: a microscopical quantum mechanical Hamiltonian model of computers as represented by Turing machines*. In this paper it is pointed out for the first time that unitary transformations undergone by quantum systems can be used to implement computing logical operations. However, the work of Benioff was inspired by an earlier paper, published in 1973 by the IBM physicist Charles Bennett [4]. In his paper, *Logical reversibility of computation*, Bennett showed that computation could be built entirely on the basis of reversible logic, although actual computers operate with irreversible processes. Indeed, computation is carried out in computers through the action of the so-called logic gates. One complete set of such gates are the NOT, AND and OR gates. Whereas NOT is a reversible gate (in the sense that the information at the input of the gate can be recovered applying the gate to the output), AND and OR are irreversible, in the sense that information is lost in their action, implying an increase of entropy equal to at least $k_B \ln 2$ for each bit which is lost.[2] On the other hand, quantum unitary transformations are reversible: from the knowledge of the state of a quantum system in time t_0, one can obtain the state in later time t: $|\psi(t)\rangle = U(t, t_0)|\psi(t_0)\rangle$, where $U(t, t_0)$ is a unitary propagator which satisfies the Schrödinger equation. However, since $UU^\dagger = \mathbf{1}$, where $\mathbf{1}$ is the identity matrix, one can recover $|\psi(t_0)\rangle$ from $|\psi(t)\rangle$ through the operation: $|\psi(t_0)\rangle = U^\dagger(t, t_0)|\psi(t)\rangle$. Of course, this is only valid for isolated systems. One of the major triumphs of Quantum Information Theory has been the development of tools which allow the treatment of non-isolated systems for quantum computation.

After Benioff, in the year of 1985, David Deutsch gave a decisively important step towards quantum computers presenting the first example of a quantum algorithm [6]. The Deutsch algorithm shows how quantum superposition can be used to speed up computational processes. Another influent name is Richard Feynman, who was involved about the same time in the discussions of the viability of quantum computers and their use for quantum systems simulations [7].

However, it was in 1994 that a main breakthrough happened, calling the attention of the scientific community for the potential practical importance of quantum computation and its possible consequences for modern society. Peter Shor discovered a quantum algorithm capable of factorizing large numbers in polynomial time [8]. Classical factorization is a kind of problem considered by computation scientists to be of exponential complexity.

---

[2]This result is due to R. Landauer (see Landauer 1961 [5]).

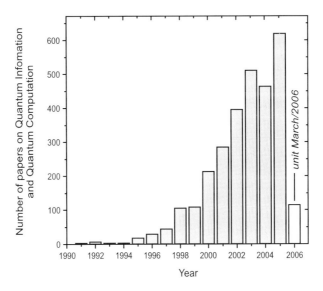

Figure 1 Number of papers published on quantum information and quantum computation in indexed scientific journals since 1990.

This basically means that the amount of time required to factorize a number N bits long, increases exponentially with N. In contrast, a quantum computer running Shor algorithm would require an amount of time which would be a polynomial function of N. This is a huge difference! To give an example, if $N = 1024$ bits, a classical algorithm would take about 100 thousand years to factorize the number, whereas Shor algorithm would accomplish the task in a few minutes!

Shor algorithm has not yet been tested in numbers that long, but its quantum working principles have already been demonstrated in laboratory, through the technique of nuclear magnetic resonance (NMR) [9]. The algorithm clearly raises important concerns about the security of cryptosystems based on the factorization of large numbers, such as the RSA protocol. Arthur Eckert captures the essence of the problem in the quote [10] "...*modern security systems are in a sense already insecure*...".

A few years after the discovery of Shor algorithm, in 1997, another important algorithm was discovered by Lov Grover [11]. The so-called Grover algorithm is a quantum search algorithm, which makes use of quantum superposition and quantum phase interference to find an item in a disordered list of N items with a squared speedup with respect to an equivalent classical algorithm. After the discoveries of Shor and Grover algorithms the interest in quantum computation and quantum information has grown dramatically along the years, as exemplified in Figure 1, which shows the number of refereed papers published in the subject from 1990 till nowadays.[3]

Quantum computation and quantum information, as much as their classical counterparts, depend upon the availability of natural resources, such as energy and entropy. However, if we think of classical phenomena as an approximation of the quantum world, one can expect the existence of quantum resources with no classical correspondence. One example of such a quantum resource is the quantum information unit, the *qubit*. One qubit can

---

[3]Database: *Web of Science*.

assume the classical values '0' or '1', but can also be put in *any* superposition of both '0' and '1'. However, possibly, the most counterintuitive and strange quantum resource is called *entanglement*. This property of some quantum superposition states implies non-local effects between qubits. It is interesting to note that entangled states are eigenstates of the so-called Heisenberg Hamiltonian [12], which is the basis of condensed matter models for magnetic phenomena in matter! For two particles, one example of such state is the so-called singlet spin state:

$$|\psi\rangle = \frac{|\uparrow\downarrow\rangle - |\downarrow\uparrow\rangle}{\sqrt{2}}$$

Quantum mechanics tells us that, before observation is made, both spins share – with equal weight – the states $|\uparrow\rangle$ and $|\downarrow\rangle$. Before a measurement, the probability of either spin to be found in either state is 50%. However, if one performs a measurement, say, in the *first spin*, the state of the *second spin* becomes determined, no matter the distance between them! For many years, this non-local property of entanglement has been perhaps the most controversial and debated aspect of quantum mechanics, since Einstein, Podolsky and Rosen pointed the problem out in a historical paper published in 1935 [13]. Since the EPR paper, as it became known, many decades were necessary until the discovery of a criterion to decide whether non-locality was a physical reality or just a mathematical property of the quantum formalism. This was a main contribution of John Bell, who in 1964 presented such a criterion [14]. The so-called *Bell inequality* is a statistical test for quantum non-locality. However, in 1964 there were no experimental conditions to implement such a test in a real physical system. This came about only in 1982 as a seminal work published by Aspect, Grangier and Roger [15], entitled *Experimental realization of Einstein–Podolsky–Rosen–Bohm gedankenexperiment: a new violation of Bell's inequalities*. This paper is considered – at least for the great majority of physicists – as the work where the non-locality, inherent to entangled states, is demonstrated to be definitely part of the physical world.

In the context of quantum computation and quantum information, entanglement is the natural resource which is behind the exponential speedup observed in algorithms such as Shor algorithm [16,17]. Furthermore, entanglement is at the basis of a number of novel applications in quantum computation and quantum information [18]: *superdense coding*, *quantum error correction codes*, *quantum cryptography*, and *quantum teleportation*. Every one of these applications has been demonstrated in successful experiments. Teleportation, in particular, was first implemented in 1997 by Bouwmeester and collaborators utilizing photons [19], by Nielsen, Knill and Laflamme in 1998 [20] utilizing NMR, and by Barret and collaborators [21] and Riebe and collaborators [22] in 2004 utilizing atomic traps.

In the year of 1997 NMR appeared in the context of quantum information and quantum computation as one of the most promising techniques candidate to be part of the quantum computing hardware. This was due to the discovery of the so-called *pseudo-pure states*, made by Gershenfeld and Chuang [23] and Cory, Fahmy and Havel [24]. Isolated nuclear spins were first pointed out by Seth Lloyd as possible good qubits, and radiofrequency pulses as good ways to implement the necessary unitary transformation for quantum information processing [25,26]. However, NMR deals not with isolated spins, but rather with statistical ensembles. Gershenfeld, Chuang and Cory showed how to produce non-equilibrium states of ensembles which effectively behave as pure quantum states, hence the name *pseudo-pure states*. Since these landmark works, every single quantum algorithm has been demonstrated by NMR, the first successful implementation being the Deutsch algorithm, done by Jones and Mosca, in 1998 [27].

However, in 1997, even before Jones' and Mosca's experiment, Warren raised important questions about the usefulness of liquid-state NMR for quantum computation [28], and in 1999 Braunstein and co-workers [29] presented a mathematical proof that NMR density matrices representing room temperature pseudopure states could always be written as product states, at least for the experiments reported until then, utilizing less than 12 qubits. The most important consequence of this result for liquid-state NMR quantum computing is the fact that no true entanglement can take place in such samples. In Ref. [29] no account is taken on the effects of unitary transformations implemented by radiofrequency pulses over the density matrices. This was considered afterwards by Linden and Popescu [17], in the context of the role of entanglement for quantum computation. These authors showed that entanglement is a necessary but not sufficient condition to produce an exponential gain in the processing speed of a quantum computer. It is also necessary that the *noise* to be below some threshold. The result is applicable to any n-qubits density matrix which can be written in the form

$$\rho_\epsilon = (1-\epsilon)\frac{\mathbf{1}}{2^n} + \epsilon \rho_1$$

where $\mathbf{1}$ is the $2^n \times 2^n$ identity matrix, and ϵ a parameter which measures the amount of "white noise" present in the system. ρ_1 is a density matrix representing a pure state.

In the case of NMR, ϵ goes with the so-called *scaling factor*, $1/2^n$, related to the amplitude of the NMR signal. The presence of such a factor means an exponential loss of intensity with the increase in the number of qubits, and it is intrinsic to conventional experiments made at room temperature. It tells us that, far beyond the entanglement problem, a *liquid-state sample at room temperature* will never be a useful large scale quantum computer! Yet, it is worth mentioning that very highly pure initial states have been achieved, as described by Anwar and collaborators in Ref. [30]. In such a highly polarized systems genuine entanglement could possibly take place. It is still worth mentioning the very recent results of Negrevergne and co-workers [31] reporting a NMR benchmark experiment in which a 12 qubit pseudo cat-state is created. The entanglement limits found by Braunstein et al., could be tested in such a system.

The question raised by Braunstein [29] and Linden and Popescu [17] concerns rather the *kind of samples* used in NMR quantum computing experiments (liquid solutions at room temperature), and *not* the *dynamics* implemented by radiofrequency pulses. NMR quantum computation takes place when the density matrix is transformed upon the unitary action of radiofrequency pulses which represent quantum logic operations. The technique called *quantum state tomography* [18] allows the measurement of every complex element of a density matrix. The application of this technique has been demonstrated in various experiments, from which it is possible to conclude that, under the action of radiofrequency pulses, density matrices indeed transform according to the quantum mechanics prescriptions. Therefore, the question is: if we could circumvent the scaling problem, would NMR quantum computing be viable? The answer is yes, and a number of theoretical proposals and impressive experiments that have appeared since 1998 encourage us to think of NMR as playing an important part in the future of quantum computing.

The first concrete proposal for a NMR scalable quantum computer was made by Kane in 1998 [32]. He showed that an array of ^{31}P atoms (nuclear spin 1/2) embedded in a Silicon lattice, with the hyperfine field and interaction between nuclei controlled by electric gates, could work as a scalable NMR quantum computer. Difficulties with Kane original approach were raised by Koiller and co-workers [33]. Afterwards, Skinner, Davenport and Kane [34] proposed an alternative scheme in which such difficulties could be circumvented.

A very interesting proposal using *Magnetic Resonance Force Microscopy* (MRFM) was made by Berman and co-workers in 2000 [35]. In that paper it is shown that through single-spin electron measurement and electron-nucleus hyperfine coupling, NMR quantum computation could be implemented, including the steps of initial state preparation, unitary transformations and final readout.

In 2002, Ladd and co-workers [36] proposed an architecture for a Silicon scalable quantum computer. In this scheme, arrays of ^{29}Si atoms (nuclear spin 1/2) lay on the steps of a ^{28}Si superlattice (nuclear spin zero). The NMR frequencies are determined by a magnetic field gradient generated by a Dy-based micromagnet, and spin-spin interactions by the dipole fields. Upon initial polarization beyond a threshold, the scheme becomes scalable and could be used in a NMR quantum computer.

On the experimental side, impressive advances on NMR technology and nanofabrication can lead to the implementation of the schemes similar to those described above, particularly the proposal of Berman et al. [35]. In 2004 Rugar and co-workers [37] reported the detection of one single electron spin in silicon dioxide. In the direction of chip integration, Yusa and collaborators [38] reported in 2005 the construction of part of a NMR spectrometer inside a single semiconductor chip! Finally, Kitchen et al. [39] report an Mn *atom-by-atom substitution* in GaAs using STM, and Savukov, Lee and Romalis describe the *Optical detection of liquid-state NMR* [40]. All these amazing works are important developments in the direction of quantum chip manufacturing and further increase the NMR resolution and sensitivity. For sure, they point to an optimistic future for NMR QIP.

REFERENCES

[1] A.M. Turing, On computable numbers, with an application to the entscheidungsproblem, *Proc. Lond. Math.* **42** (1936) 230.
[2] C.E. Shannon, A mathematical theory of communication, *Bell System Tech. J.* **27** (1948) 379.
[3] P. Benioff, The computer as a physical system: a microscopical quantum mechanical Hamiltonian model of computers as represented by Turing machines, *J. Stat. Phys.* **22** (1980) 563.
[4] C.H. Bennett, Logical reversibility of computation, *IBM J. Res. Develop.* **17** (1973) 525.
[5] R. Landauer, Irreversibility and heat generation in the computing process, *IBM J. Res. Develop.* **5** (1961) 183.
[6] D. Deutsch, Quantum theory, the Church–Turing Principle and the universal quantum computer, *Proc. R. Soc. Lond. A* **400** (1985) 97.
[7] R.P. Feymann, Simulating physics with computers, *Int. J. Theor. Phys.* **21** (1982) 4667.
[8] P.W. Shor, Algorithms for quantum computation: discrete logarithms and factoring, in: *Proc. 35th Annual Symp. Found. Comput. Sci.* (IEEE Press, Los Alamitos, 1994).
[9] L.M.K. Vandersypen, M. Steffan, G. Breyta, C.S. Yannoni, M.H. Sherwood, I.L. Chuang, Experimental realization of Shor's quantum factoring algorithm using nuclear magnetic resonance, *Nature* **414** (2001) 883.
[10] A. Eckert, in: D. Bouwmeester, A. Eckert, A. Zeilinger (Eds.), *The Physics of Quantum Information* (Springer, 2001), Chapter 2.
[11] L.K. Grover, Quantum mechanics helps in searching a needle in a haystack, *Phys. Rev. Lett.* **79** (1997) 325.

[12] A.P. Guimarães, *Magnetism and Magnetic Resonance in Solids* (John Wiley & Sons, New York, 1998).
[13] A. Einstein, B. Podolsky, N. Rosen, Can quantum mechanical description of physical reality be considered complete?, *Phys. Rev.* **47** (1935) 777.
[14] J.S. Bell, On the Einstein–Podolsky–Rosen paradox, *Physics* **1** (1964) 195.
[15] A. Aspect, P. Grangier, G. Roger, Experimental realization of Einstein–Podolsky–Bohm *gedankenexperiment*: a new violation of Bell's inequalities, *Phys. Rev. Lett.* **49** (1982) 91.
[16] A. Steane, Quantum computing, *Rep. Prog. Phys.* **61** (1998) 117.
[17] N. Linden, S. Popescu, Good dynamics versus bad kinematics: is entanglement needed for quantum computation?, *Phys. Rev. Lett.* **87** (2001) 047901-1.
[18] M.A. Nielsen, I.L. Chuang, *Quantum Computation and Quantum Information* (Cambridge, 2002).
[19] D. Bouwmeester, J-P Wan, K. Mattle, M. Eibl, H. Weinfurter, A. Zeilinger, Experimental quantum teleportation, *Nature* **390** (1997) 575.
[20] M.A. Nielsen, E. Knill, R. Laflamme, Complete quantum teleportation using nuclear magnetic resonance, *Nature* **396** (1998) 52.
[21] M.D. Barret, J. Chiavefni, T. Schaetz, J. Britton, W.M. Itano, J.D. Jost, E. Knill, C. Langer, D. Leibfried, R. Ozerf, D.J. Wineland, Deterministic quantum teleportation of atomic qubits, *Nature* **428** (2004) 737.
[22] M. Riebe, H. Häffner, C.F. Roos, W. Hänsel, J. Benhelm, G.P.T. Lancaster, T.W. Körber, C. Becher, F. Schmidt-Kaler, D.F.V. James, R. Blatt, Deterministic quantum teleportation with atoms, *Nature* **428** (2004) 734.
[23] N. Gershenfeld, I.L. Chuang, Bulk spin resonance quantum computation, *Science* **275** (1997) 350.
[24] D.G. Cory, A.F. Fahmy, T.F. Havel, Ensemble quantum computing by NMR spectroscopy, *Proc. Natl. Acad. Sci. USA* **94** (1997) 1634.
[25] S. Lloyd, A potentially realizable quantum computer, *Science* **261** (1993) 1569.
[26] D.P. DiVincenzo, Two-bit gates are universal for quantum computation, *Phys. Rev. A* **51** (1995) 1015.
[27] J.A. Jones, M. Mosca, Implementation of a quantum algorithm on a nuclear magnetic resonance quantum computer, *J. Chem. Phys.* **109** (1998) 1648.
[28] W.S. Warren, The usefulness of NMR quantum computing, *Science* **277** (1997) 1688.
[29] S.L. Braunstein, C.M. Caves, R. Jozsa, N. Linden, S. Popescu, R. Schack, Separability of very noisy mixed states and implications for NMR quantum computing, *Phys. Rev. Lett.* **83** (1999) 1054.
[30] M.S. Anwar, D. Blazina, H.A. Carteret, S.B. Duckett, T.K. Halstead, J.A. Jones, C.M. Kozak, R.J.K. Taylor, Preparing highly pure initial states for nuclear magnetic resonance quantum computing, *Phys. Rev. Lett.* **93** (2004) 040501-1.
[31] C. Negrevergne, T.S. Mahesh, C.A. Ryan, M. Ditty, F. Cyr-Racine, W. Power, N. Boulant, T. Havel, D.G. Cory, R. Laflamme, Benchmarking quantum control methods on a 12-qubit system, *Phys. Rev. Lett.* **96** (2006) 170501.
[32] B.E. Kane, A Silicon-based nuclear spin quantum computer, *Nature* **393** (1998) 133.
[33] B. Koiller, X.D. Hu, S. Das Sarma, Exchange in Silicon-based quantum computer architecture, *Phys. Rev. Lett.* **88** (2002) 027903.
[34] A.J. Skinner, M.E. Davenport, B.E. Kane, Hydrogen spin quantum computer in Silicon: a digital approach, *Phys. Rev. Lett.* **90** (2003) 087901-1.
[35] G.P. Berman, G.D. Doolen, P.C. Hammel, V.Y. Tsifrinovich, Solid-state nuclear spin quantum computer based on magnetic resonance force microscopy, *Phys. Rev. B* **61** (2000) 14694.
[36] T.D. Ladd, J.R. Goldman, F. Yamaguchi, Y. Yamamoto, E. Abe, K.M. Itoh, All-silicon quantum computer, *Phys. Rev. Lett.* **89** (2002) 017901.
[37] D. Rugar, R. Budakian, H.J. Mamin, B.W. Chui, Single spin detection by magnetic resonance force microscopy, *Letters to Nature* **430** (2004) 329.
[38] G. Yusa, K. Muraki, K. Takashina, K. Hashimoto, Y. Hirayama, Controlled multiple quantum coherences of nuclear spins in a nanometer-scale device, *Letters to Nature* **434** (2005) 1001.
[39] D. Kitchen, A. Richardella, J.-Ming Tang, M.E. Flatté, A. Yazdani, Atom-by-atom substitution of Mn in GaAs and visualization of their hole-mediated interactions, *Letters to Nature* **442** (2006) 436.
[40] I.M. Savukov, S.-K. Lee, M.V. Romalis, Optical detection of liquid-state NMR, *Letters to Nature* **442** (2006) 1021.

– 1 –

Physics, Information and Computation

The Universe is not just a giant computer; it is a giant quantum computer – S. Lloyd and Y. Jack Ng [Sci. Am. November 2004, pp. 53–61]

1.1 TURING MACHINES, LOGIC GATES AND COMPUTERS

What is a computer? May be, for most computer users, computers are just those nice "black boxes" which connect us to the world through the Internet; or machines which entertain kids (and adults!) with fancy games; or auxiliary tools to help us planing our domestic budgets, etc. For engineers and technologists, may be, computers are essential tools without which would be impossible to safely couple a spacecraft to the International Space Station, or to land a robot in Mars, or yet to build the giant European A380 Airbus. For mathematicians and computer scientists computers may be viewed as a physical realization of a Turing Machine.

For scientists in general, for physicists in particular, computers have been a valuable tool in helping them with their research work and teaching. This help comes basically in three kinds of use: (i) solving complicated mathematical problems, (ii) controlling experiments and data acquisition in laboratories and, (iii) reviewing the literature through the Internet, preparing lectures, writing papers, theses, books, etc. Computers have become so inextricably tied to the scientific activity that we can hardly regard them as a simple chunk of matter, which is subject to very same Nature laws they help to unreveal! Yet, this seems to becoming the prevailing vision about computers, at least among the physicists. As put by David Deutsch [1],

> Computers are physical objects, and computations are physical processes. What computers can or cannot compute is determined by the laws of physics alone, and not by pure mathematics.

Besides using computers to help in their research, what are the possible interests of physicists in computers and computational processes? Even this question could lead to different routes. Researchers could, for instance, attack on the material science side, studying the physical properties of bulk semiconductor materials, the basic stuff from which chips are made of, or studying the magnetic materials, the basic stuff hard-discs are built from. One could take the route of the so-called nanoscience and nanotechnology and exploit the ultimate limits of miniaturization of computer components, down to the molecular size. Yet, we can take an entirely different route, and ask for very fundamental questions about computers and about computation. One could ask, for instance, what is the minimum amount of energy and time necessary to flip a bit of information, or whether it is possible to perform computation without any energy expenditure at all. Or still, what is the limit imposed by thermodynamics laws for the amount of information which can be stored in a computer memory. On this direction, Roger Penrose put together computer science and the laws of physics to raise instigating questions about the capabilities of computers to reproduce mental phenomena exclusive (so far!) of the human brain [2].

The approach towards computers and computation on the side of fundamental reasoning has led to deep insights about the nature of computing processes. One example is the solution of a century-old problem concerning the second law of thermodynamics. The problem is known as the Maxwell demon, and it was proposed by the great Scottish physicists, James Clerk Maxwell, in 1871 [3]. The solution to this problem, as it will be discussed below, came only in the eighties, thanks to the investigation of Rolf Landauer in 1961, about the energy requirements of computational processes. On the other hand, Seth Lloyd [4,5] has considered the "ultimate physical limits of computation", and has adopted the very appealing view that not less than the entire universe is a gigantic (quantum) computer. According to him, every single natural process can be interpreted as a computational action of the universe, whose output is everything we observe in Nature!

So, what is a computer? There seems to be no single answer to this question. The answer you give depends on the way you look at a computer, or actually, the way you see the world. Quantum computation (QC) and quantum information processing (QIP) appeared from considerations about the very basic physical processes of computation.

However interesting may be, QC and QIP would be restricted to a bunch of mathematical results if there was no way to implement them in the physical world, as much as a Turing Machine (see below) would be a mere theoretical curiosity without the existence of computers! This book deals with a particular way to implement QC and QIP: it is called Nuclear Magnetic Resonance, or simply NMR. There are excellent books in the subjects of quantum computation and quantum information [6,7], in NMR [8] and in (classical) computation [9]. This book exploits elements of these three different fields, and put them together in order we can understand NMR-QIP. In this chapter we will introduce the basic elements of computation, and will discuss the physics of computational processes. Chapters 2 and 3 introduce the necessary background of NMR and quantum computation theories, in order we can exploit the realizations of NMR-QIP in the subsequent chapters.

We will start with a very basic model of computation called Turing Machine [10]. This name is in honor of the great British mathematician Alan Turing (1912–1954). A Turing Machine is not a real computer, made of chips, printed boards and wires, but a mathematical idea which captures the essence of a computing action. What is most interesting and most important about Turing Machines is the fact that there is not known computation which can be proved to be carried out by an actual computer, but which cannot be carried out by a Turing Machine. It is in this sense that a real computer (made of chips and wires) is a physical realization of a Turing Machine.

Any Turing Machine is composed by the following basic ingredients:

- A tape, divided in cells;
- A tape read/write head;
- A set of symbols which can be written in the tape, called the *alphabet*;
- A set of very simple instructions. Each instruction is associated to an internal state of the machine, that tells her which action to take.

The tape is the analogous of a computer memory, and the set of instructions is the equivalent of a computer program. The tape has a starting cell, but is infinite in length, which is of course, an abstraction. In fact, only with this assumption a Turing Machine is capable of computing *any* function.

The head moves along the tape according to a specific instruction. It can move right or left, and it can write any symbol taken from the alphabet in a tape cell, if so instructed.

The way it works is as following: the head starts from the leftmost part of the tape, reads the symbol which is written in the first cell, takes an action, and moves to the next cell. The procedure is repeated until the machine eventually concludes the computation and halts. Let us give a very simple illustrative example: a Turing Machine to perform the addition operation $3 + 5$. In order to do so, we must first define the symbols to represent these numbers on the tape. We will adopt the following representation: $3 \equiv ***$ and $5 \equiv *****$. So, the expected result of our calculation is $8 = ********$. The input state is simply the initial two blocks of '*'s separated by a blank, which we will represent by a small box: $\square$. So, our alphabet has only two symbols: $\{*, \square\}$.

Next, we have to define the actions the machine must take at each step. Each action is labeled by an internal state of the machine. The table below shows the states and actions necessary to accomplish the task we want:

Machine state labels	Action to be taken if the head reads '*'	Action to be taken if the head reads '$\square$'
1	move to the right; remain in 1	erase; write '*'; move to the right; go to state 2
2	move to the right; remain in 2	move to the left; go to state 3
3	erase; stop	

The machine begins in the state 1, with its tape written with the symbols sequence on its leftmost cells:

$$*** \square ***** \square \square \cdots$$

The head starts moving to the right and reads '*' in the first cell. Therefore, it is instructed to stay in state 1, and move right until the $\square$ cell is reached. Then, it replaces '$\square$' by '*' and jump to state 2. It keeps moving to the right until the last symbol '*' is reached. After this, the head reads '$\square$', and the machine is instructed to move one cell left and jump to state 3. Then, it erases the last '*' symbol and stops. The final sequence in the tape is therefore:

$$********$$

which is the desired result. This example is oversimplified, but it captures the essence of the working principle of a Turing Machine. The most important fact is that with enough time, and the appropriate alphabet and instructions, *any* computation which can be performed in the most powerful computer in World can also be computed in a specific Turing Machine!

A rather more practical way to approach computers is through the basic elements of any computer action: the logic gates. Logic gates are logical operators which act on bits, taking them from an initial state to a final one. There is not a single choice for a complete set of logic gates, as much as there is not a single choice for the vector basis to represent a 3-dimensional vector: one can, for instance write a vector **A** in rectangular coordinates,

$$\mathbf{A} = A_x \mathbf{i} + A_y \mathbf{j} + A_z \mathbf{k}$$

or in spherical coordinates:

$$\mathbf{A} = A_\rho \mathbf{e}_\rho + A_\theta \mathbf{e}_\theta + A_\phi \mathbf{e}_\phi$$

The vector is the same, but the basis is different. In a similar fashion, computational actions can be written as combinations of different sets of elementary logic gates.

One example of logic gates formed by three elements is the set of gates AND, OR and NOT. It is an amazing fact that *any* computational operation can be decomposed in a specific action of a combination of these three gates.

The action of a logic gate can be characterized through the so-called *truth table*. The truth table returns the output of the gate, given its input. For instance, the action of the gate NOT is simply to invert the value of the input bit: if the input is '0', NOT transforms it into '1', and if the input is '1', NOT returns '0'. As simple as that! Table 1.1 is the truth table of NOT.

The gates AND and OR are a little bit more complicated than that. They take two bits at the input and return only one bit at the output. The simple fact that one bit is lost, or erased, in the action of these gates, has deep thermodynamics consequences. It is related to irreversibly, entropy growth and energy consumption, as we will see in the following sections. The truth tables for the gates AND and OR are shown in Tables 1.2 and 1.3, respectively.

Besides truth tables, logic gates have *circuit representations*. This is a very useful way to visualize the action of the gates in diagrammatic logic circuits. Figure 1.1 shows the symbols of AND, OR and NOT gates.

The gates AND, OR and NOT can be combined to produce more complicated logical operations, which can in turn be considered new gates. Two of such combinations are NOR (NOT-OR) and NAND (NOT-AND). These gates are simply the NOT operation following OR and AND operations, respectively. Figure 1.2 shows their circuit symbols.

One very important combination of gates is shown in Figure 1.3. It is called Exclusive-OR gate, or simply XOR. It is a *conditional* NOT operation: the first bit acts as

Table 1.1.

Truth table of logic gate NOT

Input	Output
0	1
1	0

1.1. Turing Machines, logic gates and computers

Table 1.2.

Truth table of logic gate AND

Inputs	(AND)	Output
0	0	0
0	1	0
1	0	0
1	1	1

Table 1.3.

Truth table of logic gate OR

Inputs	(OR)	Output
0	0	0
0	1	1
1	0	1
1	1	1

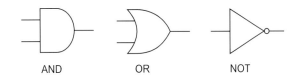

    AND   OR   NOT

Figure 1.1 Circuit representation of AND, OR and NOT logic gates.

    NOR   NAND

Figure 1.2 Circuit representation of NOR and NAND logic gates.

a control bit over the state of the second bit. For this reason it is called *control bit*, whereas the second bit is called *target bit*. The target bit flips only if the control bit is set to '1'. The truth table of XOR is shown in Table 1.4. Notice that gate implements the addition of two bits, except for the fact that there is no *carry* bit.

As we said before, the set AND, OR and NOT is only one possible choice for the elementary logic gates. It is possible, for instance, to generate an AND gate from a combination of only NAND or NOR gates. Actually, any logical operation can be built from NAND or NOR. Therefore, any of these gates form a basic set on their own. These gates can now be combined to implement different logical operations. Figure 1.4 is an example of combination

14 1. Physics, Information and Computation

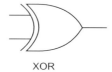

XOR

Figure 1.3 Representation of Exclusive-OR, or XOR gate.

Table 1.4.

Truth table of logic gate XOR

Inputs	(XOR)	Output
0	0	0
0	1	1
1	0	1
1	1	0

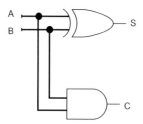

Figure 1.4 Half-adder circuit. A and B are input bits, S the sum and C the carry bit, which is not used in this circuit.

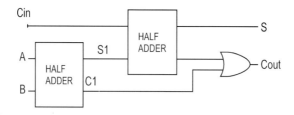

Figure 1.5 The full adder circuit is built from two half adders.

of NOR and AND gates to produce a half-adder circuit. A full adder circuit, built from two half-adders and an OR gate, is shown in Figure 1.5.

But what is the actual physical stuff a logic gate is made of? The description so far has been only symbolic and does not tell us anything about how those actions can be implemented in physical systems. Of course this has to do with the complex computer technology, and a detailed description of how conventional computer elements work is beyond the scope of this book. There are excellent books on the subject (see, for instance, [11]) explaining how logic gates can be built from very basic electronic circuit elements, such as diodes and resistors, and how these gates are integrated into chips containing millions of them.

This is all we will see about classical logic gates. Quantum computation is also constructed upon the action of (quantum) logic gates (Chapter 3). Besides, quantum logic gates can reproduce the action of the classical ones, but the opposite is not true. This should not be surprising, since classical phenomena is only a particular case of quantum phenomena. On the following sections we will discuss some interesting aspects of the physics of computation.

1.2 KNOWLEDGE, STATISTICS AND THERMODYNAMICS

Both, computation and communication deal with the processing of information. Information, on its turn, can assume different forms: it can be an image, a text, a set of numbers, a sound, etc. Besides, sometimes we want information to be publicly displayed, as in an advertisement, and sometimes we want it to be secret, as when we buy something through the Internet using our credit cards. Whatever the situation, information can always be converted into a bunch of 0's and 1's. But to make something useful with information, we must have the means to represent the 0s and 1s in the physical world, in order we can process information. In doing so, information becomes subject to the laws of physics.

In computers, information is represented by bits in electronic circuits. But any physical object with two clearly stable distinguishable states can represent a bit of information. A collection of such objects is a physical system where information can be stored and processed. Let us take a very simple object to work as a bit: a coin. The head and tail "states" of a regular coin are very stable, and they can be associated to the usual logic labels of a bit: '0' and '1'. How much information can be stored in a collection of coins? For the sake of argument let us considerer 4 coins. Heads will be represented by an empty circle, and tails by a full circle. Let us associate the logic label '0' to heads, and '1' to tails.

It is intuitive that if we could not change the state of a coin, our capability of representing information would be dramatically reduced in this system. For instance, if we glue all four coins on the desk, tails up, we would have the bit sequence '1111', which may represent, for instance, the number 15.[1] This would be all we could do! In order to represent other numbers, one must be able to change the state of each coin. If we allow each coin to have its state changed, there would be 16 different configuration in this system. This number turns out to be equal 2^4. In general, if we had n coins, there would be 2^n different configurations. So, the number of different configurations grows *exponentially* with the number of elements in the system.

Let us take a closer look in the possible configurations of 4 coins. This is shown in Figure 1.6. Notice that there are 6 configurations for the case of two heads and two tails. Then, there are four configurations for either three heads and one tail or three tails and one heads. Finally, there is only one configuration for either four heads or four tails.

Now, let us imagine that instead of choosing one particular configuration among the 16 available ones, we simply adopt a *statistical* procedure: we throw each coin and simply

---

[1] Notice that $15 = 1 \times 2^0 + 1 \times 2^1 + 1 \times 2^2 + 1 \times 2^3$.

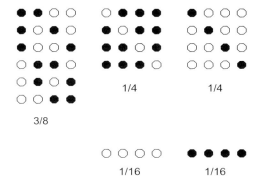

Figure 1.6 Possible configurations of heads and tails of 4 coins. Numbers are the probabilities for each configuration, obtained dividing the multiplicity of the configuration by the total number of possibilities.

look the resulting configuration. What is the one most likely to occur? Common sense tells us that since the configuration "half–half" (two heads and two tails) is the most numerous one, this must be the most likely to occur. Proving this is simple: the probability for heads or tails of each coin is $1/2$. Thus, the probability of *any* particular configuration of the four coins[2] is $1/2^4 = 1/16$. However, since we do not distinguish the coins and the configuration of two heads and two tails has 6 different possibilities, the total probability for this configuration is $6 \times 1/16 = 3/8$. The probabilities for the other configurations are $4 \times 1/16 = 1/4$ and $1 \times 1/16$. Notice that the total probability is:

$$\frac{3}{8} + 2 \times \frac{1}{4} + 2 \times \frac{1}{16} = 1$$

The fact that the configuration with two heads and two tails is the most likely one has nothing to do with physics; it is purely statistical. It reflects the fact that if we take a single coin and throw it a large number of times, at the end we will have about half heads and half tails.

But why this statistical game is so important to the physics of computation? Simply because when we deal with physical systems containing a large number of components,[3] we cannot follow the behavior of each individual, and must make use of statistical methods! That means we must talk about probabilities, instead of certainties, and have a way to quantify our lack of information about the state of the system. For instance, suppose you close your eyes and throw each of the four coins. Keeping the eyes closed, what would be your guess for the result? According to the above, the best you can say is that the "half–half" configuration is the most likely to occur. Of course, when you open your eyes, you can verify the actual result, which can be far different from your guess. However, the point is that if you *repeat the procedure a large number of times*, the "half–half" configuration will occur more times than the others, in the exact proportion of the probabilities calculated above. In one single throw, what changes between the point your eyes are shut to when they are open, is the *information* you have about the system. It goes from many possibilities (eyes shut) to a single one (eyes opened)! In physics the quantity which is associated to

---

[2] The other way to see this is simply observing that, since the total number of configurations is 16, the probability to get a particular one is $1/16$.

[3] Imagine that instead of only 4 coins, we had to deal with 10^{23} coins!

1.2. Knowledge, statistics and thermodynamics

knowledge and information is the *entropy*. There are different definitions of entropy. One of them is the so-called *Shannon* or *information* entropy:[4]

$$S = -\sum_k p_k \log(p_k) \tag{1.2.1}$$

where p_k is the probability associated to the occurrence of an event. The logarithm is taken in base two, and the sum is over all possible configurations. The Shannon entropy for the case of our 4 coins game is:

$$S = -\left[\frac{3}{8}\log\left(\frac{3}{8}\right) + 2 \times \frac{1}{4}\log\left(\frac{1}{4}\right) + 2 \times \frac{1}{16}\log\left(\frac{1}{16}\right)\right] \approx 2.03$$

This number somehow quantifies your lack of information before you open your eyes! At the moment you open them, you become aware of the actual configuration, and all p_k collapse to zero, except a specific one, which will be equal to 1. In this situation, the entropy is $S = 0$.

Of course, a physical system such as 4 coins is useless for any practical purpose concerning computation and information processing. We are interested in systems containing a very large number of components, something like 10^{23}, such as spins in a solid or in a liquid, in the presence of a magnetic field. In such systems, the spins occupy the energy levels according to some probability distribution, but of course in this case we cannot "open our eyes" and see which spin is in which state!

Entropy is also a key concept in thermodynamics.[5] For instance, if you apply a static magnetic field in an initially demagnetized paramagnetic system in contact with a thermal reservoir,[6] there will appear a net magnetization in the system. Before the field is applied, the magnetic moments are at random and magnetization is zero; after the field is applied, they point, on average, to the direction of the field, giving rise to a net magnetization. This change from a disordered situation (moments at random) to an ordered one (moments aligned with the field) corresponds to a decrease in the entropy of the magnetic system. The second law of thermodynamics tells us that to compensate the decrease of entropy in the magnetic system, there must be a heat flow to the bath. If the field is removed, the magnetic moments become disordered again, and entropy increases.

Whenever disorder occurs, the entropy increases. Disorder, on its turn, can be caused by different agents: friction, magnetic hysteresis, electric resistance, corrosion, etc. When two gases of different species are mixed, the entropy increases. When a volume containing a gas is doubled, the entropy increases. And so on. Increase in entropy always corresponds to a loss of information. When the volume of a gas is doubled, there will be more positions for its constituents molecules to occupy and therefore a loss of information about the positions of the molecules. Of course, if the volume is reduced, keeping the temperature of the gas constant, we gain information about the positions of the molecules.[7] In this case, the entropy of the gas is reduced.

---

[4] We will use throughout the same symbol for entropy: S.
[5] Notice the fact that entropy is a fundamental concept for both, information theory and physics. This is no accident!
[6] For instance, a paramagnetic salt in contact with a liquid helium bath at 4.2 K.
[7] Consider the limit case in which the volume is reduced until all the molecules occupy a single point in space. We would know exactly their position!

Entropy appears in thermodynamic functions, such as the Helmholtz free-energy [12]:

$$A = U - TS \tag{1.2.2}$$

Here, U is the internal energy and T the temperature. On another hand, there is a definition of the Helmholtz free-energy from a statistical approach [12]. Given a system with an energy spectrum $\{E_k\}$, the Helmholtz free-energy is defined as:

$$A = -k_B T \ln \mathcal{Z} \tag{1.2.3}$$

where k_B is de Boltzmann constant, and $\mathcal{Z}$ the so-called *partition function* of the system:

$$\mathcal{Z} = \sum_k e^{-E_k/k_B T} \tag{1.2.4}$$

The bridge which connects the thermodynamic equation (1.2.2) to the statistical definition (1.2.3) is precisely the *statistical entropy*, defined as:

$$S = -k_B \sum_k p_k \ln(p_k) \tag{1.2.5}$$

This expression differ from the Shannon entropy in two basic aspects: first, it has the multiplying Boltzmann factor, a signature of thermodynamic phenomena, and second, the logarithm is taken in the natural basis. The probability p_k is the *occupation number* for the energy level E_k, or its *population*. At thermal equilibrium at temperature T, the probabilities can be obtained from the principle of maximization of the entropy [12]:

$$p_k = \frac{e^{-E_k/k_B T}}{\mathcal{Z}} \tag{1.2.6}$$

These numbers represent the chance the level k is occupied.[8]

1.3 REVERSIBLE VERSUS IRREVERSIBLE COMPUTATION

We saw in the last section that entropy is a measure of order, and therefore of the information we have about the configuration of a physical system. In particular, every time entropy increases, there is a loss of information. On another hand, computationally speaking, loss of information means loss of bits, or *erasure*. Consider, for instance, the action of the logic gate AND. This gate accepts two bits on its input, and returns only one bit at the output. The same happens in the action of the gate OR. Consequently, information is lost. On the contrary, the gate NOT conserves the number of bits and, therefore, conserves information. If we apply NOT to its output, what we obtain the input back. This property is called *reversibility*. The NOT gate is reversible whereas the AND and OR gates are *irreversible*, because we cannot obtain the input bits of either these gates, by applying them to their output. So, there is a connection between entropy, information and reversibility.

---

[8] Notice that $\sum_k p_k = \mathcal{Z}/\mathcal{Z} = 1$.

Table 1.5.

Truth table of Toffoli gate

Input			Output		
0	0	0	0	0	0
0	0	1	0	0	1
0	1	0	0	1	0
1	0	0	1	0	0
0	1	1	0	1	1
1	0	1	1	0	1
1	1	0	1	1	1
1	1	1	1	1	0

Whenever an irreversible process happens, entropy increases. This can be seen from a simple example taken from standard thermodynamics: consider two isolated objects, A and B, at different temperatures, T_A and T_B. Let us assume that $T_A > T_B$. Suppose the two objects are brought together, and an amount of heat ΔQ flows from A to B. This is clearly an example of irreversible process. Thermodynamics tells us that the entropy of A will *decrease* by an amount $\Delta S_A = \Delta Q / T_A$ and the entropy of B will *increase* by $\Delta S_B = \Delta Q / T_B$. But since $T_A > T_B$, the increase of the entropy in B will be larger than the decrease of the entropy in A. Consequently, the total entropy increases in the process.

Another nice example of loss of information caused by an irreversible process is given by Bennett and Landauer [13]: suppose a rubber ball is dropped from two meters of height from the ground. If there is no friction, the ball will bounce back to exactly two meters high. So, just by watching the bouncing height, we can deduce the height it was dropped in first place. However, if there is friction, energy will be lost at each bounce and the height will decrease in an irreversible manner. In this case, the information about the initial height is lost. Yet another example (from the same reference) of irreversible loss of information, this time in a simple maths operation: if we are presented with the expression $2 + 2$ we gain more information than if we are presented with the result 4. This is because there are infinite ways to obtain 4 from the addition of two integers, but there is only one unique result for $2 + 2$!

So, every time an AND and OR gate is executed an irreversible operation takes place, information is lost, and the entropy increases. However, a main breakthrough happened in 1973, when Charles Bennett[9] showed [15] that computation can be performed entirely on the basis of reversible logic![10] One way to see this is to consider a logic gate called the *Toffoli* gate. This gate has three input bits and three output bits. Two of the input bits are control bits and the third one is the target bit. Therefore, the Toffoli gate is similar to a XOR gate with two control bits. The operation of the Toffoli gate is simple: the target bit flips only if the two control bits are set 1; otherwise the target bit remains unchanged. The truth-table for the Toffoli gate is shown in Table 1.5. The first two bits on the left columns are control, and the last one the target.

---

[9]The first report on reversible Turing Machines was actually due to Y. Lecerf in 1963 [14].
[10]This discovery of Bennett lead to the idea of quantum computation by Paul Benioff almost a decade afterwards [16].

Table 1.6.

Truth table of NAND gate

Input		Output
0	0	1
0	1	1
1	0	1
1	1	0

It is obvious to see from Table 1.5 that the input bits can be recovered by applying the gate to the output! Therefore, the Toffoli gate is reversible. Now, if a basic set of gates could be built from Toffoli, then it will be demonstrated that computation can be made reversible. In fact, it is a simple matter to implement NAND from Toffoli. All we have to do is to set the target bit as 1 at the input, and Toffoli will work just as NAND. In Table 1.5, these correspond to the 2nd, 5th, 6th and 8th lines. The first two entries of these lines, plus the corresponding bits on the last column, is just NAND logic (see Table 1.6).

Therefore, computation can be made entirely from reversible logic gates![11]

Reversible computation has a very important consequence: until 1961 scientists believed that any computational action would result in an energy cost. But in 1961 Rolth Landauer showed [17] that what do cost energy is *erasure*. In other words, if no bit is lost during the computation, it can be made at energy-free cost! This discovery lead to the solution of a century-old problem in thermodynamics: the Maxwell demon problem.

1.4 LANDAUER'S PRINCIPLE AND THE MAXWELL DEMON

In 1961 Rolf Landauer [17], studying the thermodynamics of computational processes, discovered that the action of erasing one bit of information has an energy cost of at least $k_B T \ln 2$, and increases the entropy of the environment by an amount of at least $k_B \ln 2$. Computation does not cost energy, but erasure does.

Landauer discovery opened the path to the solution of a century old problem in thermodynamics: the Maxwell demon problem. An excellent discussion about this problem is made in Refs. [18,19]. The Maxwell demon problem concerns a hypothetical situation in which the second law of thermodynamics is violated. The situation conceived by Maxwell was the following: suppose we have a container with its volume divided in the middle by a blocking wall, which contains an ideal (frictionless) sliding window. A gas at a certain equilibrium temperature fills the volume of the container. By opening the sliding window, molecules can pass either way. Maxwell imagined an intelligent being, a demon, controlling the window (Figure 1.7). The demon job is to open the sliding window every time a fast molecule hits the window from one side, or a slow molecule hits it from the other side. By doing so, after a while, the most energetic molecules would be separated from the less energetic ones, and we end up with two gases at different temperatures. This violates the second law of thermodynamics because no work was expended to produce a difference of

---

[11] There is another gate which can be used to prove reversible computation. It is the *Fredkin* gate; it performs a *controlled swap* operation between two bits [6].

1.5. Natural phenomena as computing processes. The physical limits of computation

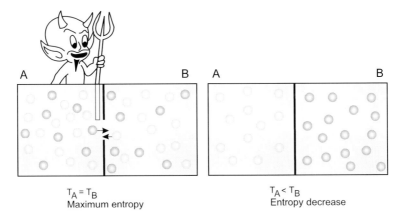

Figure 1.7 The Maxwell demon is capable of measuring the energy of each molecule in the gas, and let the slower (light ones) pass to the left, and the faster (dark ones) to the right, through the opening aperture he controls. After a while the left side of the container is at a temperature lower than the right side. Has the demon cheated the second law of thermodynamics?

temperature between the gases initially at equilibrium. If such a mechanism was possible, one would have a perpetual means to produce energy without expenditure![12]

Maxwell proposed this problem in 1871 [3]. Since then many attempts have been made to refute the argument, and save the second law of thermodynamics. A detailed discussion is made by Bennett in Refs. [18,13,19], who also arrived to the final solution of the problem. It has to do with Landauer's result: in order to configure a true thermodynamical cycle, every time the demon makes a measurement on the speed of a molecule, it has to forget the previous measurement.[13] But, according to Landauer's result, erasing information expends energy and increases entropy. Therefore, the demon cannot cheat the second law of thermodynamics!

1.5 NATURAL PHENOMENA AS COMPUTING PROCESSES. THE PHYSICAL LIMITS OF COMPUTATION

You look at a mirror and see your image reflected on it. What does this common act have to do with computers? Apparently, everything! You can regard light as carrying the input information to the mirror, the physical laws which govern light reflection by a body as a computer program and the image you see as the output. Any physical phenomenon can be interpreted as a process containing these three stages of a computer action: (1) information input, (2) logical processing, and (3) information output. David Deutsch [20] called the attention to the analogy between physical phenomena and computation; Table 1.7 resumes the parallel.

This view has been deepened by Seth Lloyd [21], who considers natural phenomena, not only as analog to computation, but as a *result* of computation itself! According to this view, the entire universe can be seen as a computer running a very peculiar program: the

---

[12] In a fridge, for instance, work must be done in order a difference of temperature can be maintained. Conversely, work can be extracted from systems in which a difference of temperature exists.

[13] Otherwise there would be no cycle, with the information growing forever on the demon's memory.

Table 1.7.

Comparison between physics and computation

Computation		Physics
Computer	⟷	Physical system
Computation	⟷	Experiment
Input	⟷	Initial state
Computer program	⟷	Physical laws
Output	⟷	Final state

laws of physics. The output is everything we observe in Nature! Since the laws of physics are fundamentally governed by quantum mechanics, the Universe can be view as a quantum computer. Within this expanded view of natural phenomena, it becomes natural to ask whether there are limits imposed by the physical laws to computation [4]. The answer is yes, there are natural limits for computation, for both, processing speed and memory capability. The speed is limited by the amount of energy available in the system, and the memory is limited by the entropy. To arrive to quantitative results, Lloyd [4] assumes a computer model of a mass of 1 kg occupying a volume of 1 litre. Since this is approximately the dimensions for a conventional laptop, the model is named the *ultimate laptop*.

The first step to calculate the ultimate computer speed, is to show that the minimum amount of time necessary to flip a bit is that given by the uncertainty principle. At this point we will advance some ideas which will be developed in detail in the subsequent chapters. We will consider a known quantum system, composed by a magnetic moment evolving under a magnetic field. The nuclear magnetic moment, μ, relates to the nuclear spin $\mathbf{I}$ through $\boldsymbol{\mu} = \gamma_n \hbar \mathbf{I}$, where γ_n is the nuclear gyromagnetic ratio. Let us represent the spin eigenstates of I_z by the vectors:

$$|\uparrow\rangle = \begin{pmatrix} 1 \\ 0 \end{pmatrix}; \qquad |\downarrow\rangle = \begin{pmatrix} 0 \\ 1 \end{pmatrix} \tag{1.5.1}$$

Since $I_z|\uparrow\rangle = +1/2|\uparrow\rangle$ and $I_z|\downarrow\rangle = -1/2|\downarrow\rangle$, the matrix for I_z is:

$$I_z = \frac{1}{2}\begin{pmatrix} 1 & 0 \\ 0 & -1 \end{pmatrix} \tag{1.5.2}$$

Suppose that at $t = t_0$ the spin is in the eigenstate $|\uparrow\rangle$. Then, a magnetic field of amplitude B is applied along the x-direction. This field will interact with the magnetic moment according to the Hamiltonian

$$\mathcal{H} = -\boldsymbol{\mu} \cdot \mathbf{Bi} = -\hbar \omega I_x = -\frac{1}{2}\hbar \omega \sigma_x \tag{1.5.3}$$

where $\omega \equiv \gamma_n B$ is a characteristic frequency, and σ_x is the x-component of Pauli matrices. Obviously, $\mathcal{H}$ is not diagonal in the basis of I_z:

$$\mathcal{H} = -\frac{1}{2}\hbar\omega\begin{pmatrix} 0 & 1 \\ 1 & 0 \end{pmatrix} \tag{1.5.4}$$

1.5. Natural phenomena as computing processes. The physical limits of computation

The energy eigenvalues of this Hamiltonian are $E_0 = -\hbar\omega/2$ and $E_1 = +\hbar\omega/2$. Therefore, an estimate for the spread in energy is simply $\Delta E \approx \hbar\omega$. Replacing in the time-energy expression to the uncertainty principle, $\Delta E \times \Delta t \approx h/2$, yields $\Delta t \approx h/\hbar\omega = \pi/\omega$. We will show now that this is the minimum time necessary to rotate the spin from the state $|\uparrow\rangle$ to $|\downarrow\rangle$.

The action of the field on the spin is to produce a torque that causes the spin to rotate about the field direction. Quantum mechanics tells us that the state evolution during an interval of time Δt is given by:[14]

$$|\psi(t)\rangle = U(t-t_0)|\psi(t_0)\rangle = e^{i\omega\Delta t \sigma_x/2}|\uparrow\rangle$$

$$= \left[\cos\left(\frac{\omega\Delta t}{2}\right)\mathbf{1} + i\sin\left(\frac{\omega\Delta t}{2}\right)\sigma_x\right]|\uparrow\rangle \quad (1.5.5)$$

where we have used the relation $\exp(\theta\sigma_x) = \cos(\theta)\mathbf{1} + i\sin(\theta)\sigma_x$. We are interested in the time the field takes to lead the spin from the state $|\uparrow\rangle$ to the state $|\downarrow\rangle$. As we will see in the subsequent chapters, this transformation can be viewed as the logic quantum NOT operation, and it is the simplest logic operation we can perform in the system. Remembering that $\sigma_x|\uparrow\rangle = |\downarrow\rangle$, all we have to do is to set $\omega\Delta t/2 = \pi/2$ in Equation (1.5.5) to find:

$$\Delta t = \frac{\pi}{\omega} \quad (1.5.6)$$

which matches the time given by the uncertainty principle. This result is no accident. In fact, the modern interpretation of the time-energy uncertainty principle [22,23] is that a system with average energy ΔE takes an interval of time of at least $\Delta t = \pi\hbar/(2\Delta E)$ to evolve from a quantum state to another one which is orthogonal to it, and therefore distinguishable. This is precisely the case of the states $|\uparrow\rangle$ and $|\downarrow\rangle$, since $\langle\uparrow|\downarrow\rangle = 0$.

We have shown that the minimum time to perform a logical operation is that given by the uncertainty principle. How can we use this fact to estimate the maximum speed of a computer? The maximum speed will be given by the inverse of the minimum time required to perform a logical operation. According to Special Relativity, the maximum energy available in a physical system with mass m is $E = mc^2$. Replacing $m = 1$ kg for the ultimate laptop, we find $E \approx 9 \times 10^{16}$ J, and therefore:

$$\frac{1}{\Delta t} \approx \frac{mc^2}{h} \approx \frac{9 \times 10^{16}}{6 \times 10^{-34}} \approx 10^{50} \text{ Hz}$$

An estimate for the maximum memory is not so straightforward. The memory is limited by the entropy of the system, and the number of operations per second depends also on the temperature. Lloyd [4] estimates the maximum memory space of the ultimate laptop as approximately 10^{31} bits and the maximum number of operations per bit per second it can perform is about 10^{19}.

Of course, usual computers will probably never achieve the performance of the ultimate laptop, but this is not the point. What is important here is that: first the recognition that natural phenomena can be interpreted as computing processes, and second that the laws

---
[14] We are assuming that in $t = t_0$ the spin was in the state $|\uparrow\rangle$, and that $\Delta t = t - t_0$.

of physics put a limit for the maximum processing capabilities of a computer. Seth Lloyd extends this discussion to the case of matter compressed to the limit of black holes [21]. Again, the discussion is about fundamental principles: if ordinary matter computes, would the compressed matter of a black hole compute too? The importance of this question lies in the fact until the 70s it was believed that nothing could escape the action of black holes. From the point of view of natural phenomena as computing process, this would put black hokes in a special category: it would have an input (whatever falls in it!), a processing action, but no output! Lloyd discuss this curious situation [21] in the light of Hawking radiation of black holes.

1.6 MOORE'S LAW. QUANTUM COMPUTATION

The transistor, invented in 1947 by John Bardeen, Walter Houser Brattain and Willian Bradford Shockley, is the main electronic component of computers. The first transistor was about 5 cm in length (Figure 1.8), and its miniaturization along the years lead to the present revolution in computation and communication. In the early 60s, Gordon Moore [24] observed that the number of transistors within computer chips was doubling at nearly every 18 months (Figure 1.9). This represents an exponential growth in the density of transistors inside computer chips. The discovery is known as the Moore's law, and is a consequence of the fast development of semiconductor technology. Table 1.8 shows the number of transistors in the various processors from 1971 to 2000.

The chip Pentium IV would cover an area of approximately 1 square kilometer if the size of the transistor was that of the first transistor!

Of course that if the number of transistors increases, but the size of the chips remain nearly the same, it means that the size of transistors decreases. From the point of view of physics, it is more interesting to express Moore's law in terms of the number of atoms necessary to represent a bit of information in computers. This is shown in Figure 1.9, taken from the excellent book of Williams and Clearwater [25], who adapted the figure from the

Figure 1.8 A photograph of the first transistor (Courtesy of The Porticus Centre).

1.6. Moore's law. Quantum computation

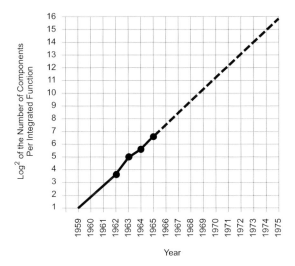

Figure 1.9 Reproduction of the original plot from Moore's paper published in the Electronics Magazine in 1965 [24].

Table 1.8.

Moore's law in number of transistors within computer processors until 2000

Processor model	Release year	Number of transistors
4004	1971	2 250
8008	1972	2 500
8080	1974	5 000
8086	1978	29 000
Intel 286	1982	120 000
Intel 386	1985	275 000
Intel 486	1989	1 180 000
Pentium	1993	3 100 000
Pentium II	1997	7 500 000
Pentium III	1999	24 000 000
Pentium IV	2000	42 000 000

work of Keyes [26]. Notice that the vertical scale is logarithmic. We see that at the time of Moore's observation, it was necessary about 10^{19} atoms to represent a bit of information in computers. In modern computers this number goes about a few thousand atoms.[15] If we believe that Moore's law will continue to hold for approximately two decades, we will come to the astonishing conclusion that by the year 2020 a bit of information will be represented by a single atom! On another hand, we know that the physics which govern the behavior of single atoms is quantum mechanics. Therefore, the observation made by Gordon Moore rises a problem that goes far beyond the "simple task" of manipulating smaller and smaller

---

[15] In August 2004, INTEL announced the manufacture of a SRAM chip containing over half-a-billion transistors. The dimension of a logic gate on these new generation circuits is of 35 nm!

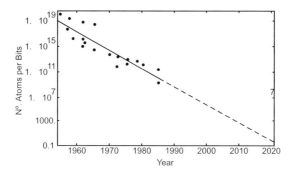

Figure 1.10 Decrease in the number of atoms necessary to represent one bit of information. Adapted with permission from Williams and Clearwater [25].

electronic components to fabricate computer chips. It is about something much deeper: a change in the current paradigm of computation and information processing, which is based in classical physics!

The proposal of implementing computer logic operations using the laws of quantum mechanics was first made in 1980 by Paul Benioff [16], who was building upon Bennett's paper of 1973 about reversible computation [15]. This proposal became to be known as *quantum computation*, and an amazing number of new ideas and results have appeared in this subject since the work of Benioff. Since quantum mechanics is such a nonintuitive theory, one can expect very strange behavior of quantum computers. At this point, there are three things we would like to say about strangeness in quantum mechanics, which will be detailed in the following chapters. First of all, the transformations in quantum computers are *reversible*. This is because the time evolution of *isolated* quantum states is governed by the Schrödinger equation: given an initial state $|\psi(0)\rangle$, quantum mechanics tells us that in a subsequent time t it will be $|\psi(t)\rangle = U(t)|\psi(0)\rangle$, where $U(t)$ is an unitary operator,[16] that is, $U(t)U^\dagger(t) = \mathbf{1}$. Reversibility means that the state $|\psi(0)\rangle$ can be recovered from $|\psi(t)\rangle$ through the operation $|\psi(0)\rangle = U^\dagger(t)|\psi(t)\rangle$.

Second, the superposition principle for quantum states means that logic states, which would be mutually excluding in a classical computer, in a quantum computer they can co-exist! To exemplify this, let us consider our previous example of a nuclear spin 1/2. There are two eigenstates of I_z: $|\uparrow\rangle$ and $|\downarrow\rangle$. Since these states are orthogonal to each other, they can be distinguished. As we will see in Chapter 3, spin states can be associated to the *quantum bit of information*, called the *qubit*.[17] One can make the following logical association to the spin eigenstates:

Physical state		Logical state
$|\uparrow\rangle$	$\longleftrightarrow$	$|0\rangle$
$|\downarrow\rangle$	$\longleftrightarrow$	$|1\rangle$

---

[16] This operator satisfies the Schrödinger equation.

[17] This is entirely similar to the example of coins given at the beginning of this chapter: a coin has two distinguishable states (heads and tails), which can be associated to the logic states of a classical bit.

The superposition principle of quantum states, says that one can construct qubit states like:

$$|\psi\rangle = \alpha|0\rangle + \beta|1\rangle \quad (1.6.1)$$

as long as $|\alpha|^2 + |\beta|^2 = 1$. Upon the action of an unitary transformation $U(t)$, the state $|\psi\rangle$ goes to:

$$U(t)|\psi\rangle = \alpha U(t)|0\rangle + \beta U(t)|1\rangle \quad (1.6.2)$$

This means that one can operate with '0's' and '1's' simultaneously in a quantum computer, something which is obviously impossible in a classical machine!

Finally, here it comes the really "weird" property. It concerns quantum states which can be produced in more-than-one qubit systems, like:

$$|\psi\rangle = \frac{|00\rangle + |11\rangle}{\sqrt{2}} \quad (1.6.3)$$

Such states are called *entangled*. This means that there are no individual qubit states $|\phi_1\rangle$ and $|\phi_2\rangle$ such that $|\psi\rangle = |\phi_1\rangle \otimes |\phi_2\rangle$, where the symbol '$\otimes$' denotes *tensor product*.[18]

Quantum mechanics tells us that if a measurement is made on *either* qubit in the state given in (1.6.3), there will be 50% of chance to find it on $|0\rangle$ and 50% to find it on $|1\rangle$. But, if we find one of the qubits in, for instance, $|0\rangle$, it means that, after the measurement, the second qubit will also be in $|0\rangle$, even if no measurement is made over it! In other words, the measurement of the state of one qubit in an entangled state, affects the state of the other qubit, independent on how distant[19] they can be from each other!

Perhaps the most striking feature about these strange properties of quantum states, is the fact that they all have been verified experimentally in laboratories! In other words, they are not just mathematical properties of the quantum formalism, but the very way Nature works at her deepest level. Furthermore, they are not only part of physical reality, but they are also the *natural resources* which make quantum computers so much powerful than the classical ones! If it was not so, quantum computation and quantum information would be nothing, but a mere mathematical curiosity.

PROBLEMS WITH SOLUTIONS

P1.1 - Work out the truth table for the half adder circuit of Figure 1.4.

Solution
From the figure it is easy to work out the following truth table:

A	B	S	C
0	0	0	0
0	1	1	0
1	0	1	0
1	1	0	1

---

[18] This notion, with the mathematical background necessary for this book, will be developed in Chapter 3.
[19] As long as the two qubits remain isolated from the environment.

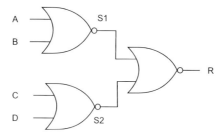

Figure 1.11 Three NOR gates are equivalent to an AND gate (Problem P1.2).

Notice that the output S is just the XOR output, which in turn is the addition of two bits, and the output C is the *carry* bit.

P1.2 - Show that the combination of three NOR gates of Figure 1.11 is equivalent to an AND gate.

Solution
It is instructive to write down all possible combinations of inputs and outputs for the circuit:

A	B	C	D	S1	S2	R
0	0	0	0	1	1	0
0	0	0	1	1	0	0
0	0	1	0	1	0	0
0	1	0	0	0	1	0
1	0	0	0	0	1	0
0	0	1	1	1	0	0
0	1	0	1	0	0	1
1	0	0	1	0	0	1
1	0	1	0	0	0	1
1	1	0	0	0	1	0
0	1	1	0	0	0	1
0	1	1	1	0	0	1
1	0	1	1	0	0	1
1	1	0	1	0	0	1
1	1	1	0	0	0	1
1	1	1	1	0	0	1

Notice that whenever A or B and C or D is '1', R is '1'. Otherwise R is '0'. That is AND logic.

P1.3 - What is the Shannon entropy associated with the throw of a fair coin? What happens to the entropy if there is a slight probability excess towards one of the faces?

Solution
For a fair coin, the probability of either output is $1/2$. Therefore,

$$S = -\frac{1}{2}\log\frac{1}{2} - \frac{1}{2}\log\frac{1}{2} = 1$$

This is the maximum entropy for one bit of information.

Suppose now there is an unbalance for the probabilities. For the sake of argument, let us write $p_1 = 1/2 + \delta$ and $p_2 = 1/2 - \delta$. Let us assume that $\delta \ll 1/2$. In this case,

$$S = -\left(\frac{1}{2}+\delta\right)\log\left(\frac{1}{2}+\delta\right) - \left(\frac{1}{2}-\delta\right)\log\left(\frac{1}{2}-\delta\right)$$

Using the approximation:

$$\log\left(\frac{1}{2}+\delta\right) \approx -1 + \frac{2\delta}{\ln 2}$$

we obtain:

$$S \approx 1 - \frac{2\delta^2}{\ln 2}$$

Therefore, the effect of favoring one of the outcomes is to slightly decrease the entropy.

P1.4 - Consider an ideal gas containing N molecules, which is isothermally compressed from an initial volume V_0 to a final volume V_1. Calculate the "information content" of the gas, defined as

$$\mathcal{I} = Nk_B \ln \frac{\Omega_0}{\Omega_1}$$

where Ω_0 and Ω_1 are the volumes in the phase space, before and after compression, respectively, and k_B the Boltzmann constant. Relate the result to the entropy variation in the gas.

Solution
Since the compression is made isothermally, there is no variation in the kinetic energy of the gas, and the ratio Ω_0/Ω_1 will be equal to the ratio V_0/V_1. Therefore,

$$\mathcal{I} = Nk_B \ln \frac{V_0}{V_1}$$

Now, the entropy associated to a volume Ω in phase space is

$$S = Nk_B \ln \Omega$$

The entropy difference when the volume changes from Ω_0 to Ω_1 is therefore,

$$S_0 - S_1 = Nk_B \ln \Omega_0 - Nk_B \ln \Omega_1 = Nk_B \ln \frac{\Omega_0}{\Omega_1}$$

which is precisely the expression for $\mathcal{I}$.

P1.5 - Calculate the Boltzmann entropy of N spins $1/2$ subject to a static magnetic field B_0, at equilibrium temperature T.

Solution
There are $N!$ different permutations of N spins $1/2$. In a magnetic field, there will be $N_\uparrow$ spins parallel to the field and $N_\downarrow$ contrary to the field. Obviously, $N = N_\uparrow + N_\downarrow$. The multiplicity of a given spin configuration[20] is:

$$\Omega = \frac{N!}{N_\uparrow! N_\downarrow!}$$

---

[20] In the example of the 4 coins discussed in the text, the multiplicity of the half–half configuration, for instance, is $4!/2!2! = 6$.

The entropy is

$$S = k_B \ln \Omega = k_B \ln \frac{N!}{N_\uparrow! N_\downarrow!}$$

For large N, one can make use of the Stirling's formula:

$$\ln(x!) \approx x \ln x - x,$$

to arrive at:

$$S = N k_B \ln N - N_\uparrow k_B \ln N_\uparrow - N_\downarrow k_B \ln N_\downarrow$$

Now, let $p_\uparrow$ and $p_\downarrow$ be the respective probabilities to find a spin in a state up and down. At a temperature T, they are given by

$$p_\uparrow = \frac{e^{-E_\uparrow/k_B T}}{\mathcal{Z}} \quad \text{and} \quad p_\downarrow = \frac{e^{-E_\downarrow/k_B T}}{\mathcal{Z}}$$

where $\mathcal{Z}$ is the partition function, and

$$E_\uparrow = -\frac{1}{2} \gamma_n \hbar B_0 \quad \text{and} \quad E_\downarrow = +\frac{1}{2} \gamma_n \hbar B_0$$

where γ_n is the nuclear gyromagnetic ratio.

Replacing $N_\uparrow = N p_\uparrow$ and $N_\downarrow = N p_\downarrow$ in the expression for S, and using the above expressions for the probabilities, we arrive at:

$$S = N k_B \ln \mathcal{Z} - \frac{N \gamma_n \hbar B_0}{2T} \coth\left(\frac{\gamma_n \hbar B_0}{2 k_B T}\right)$$

Notice that the first term also depends on the ratio B_0/T through $\mathcal{Z}$. However, in the limit of high temperature, $\ln \mathcal{Z} \approx \ln 2$, and $\coth(x) \to 0$. In this limit, $N k_B \ln 2$ is the maximum entropy. The application of a magnetic field reduces the entropy.

P1.6 - The Fredkin gate is a controlled-swap gate, which can be used to demonstrate reversible classical computation. The gate has three input bits; the first bit is the control, and the other two bits are target. The gate swaps the states of the target bits if the control bit is set '1'. Otherwise nothing happens. Work the truth table of a Fredkin gate out.

Solution

Let A be the control bit and B and C the input target bits. If $A = 1$, B and C are swapped. Let us call B' and C' the output target bits. The truth table is:

A	B	C	B'	C'
0	0	0	0	0
0	0	1	0	1
0	1	0	1	0
0	1	1	1	1
1	0	0	0	0
1	0	1	1	0
1	1	0	0	1
1	1	1	1	1

Obviously that B and C can be recovered by applying the gate to B' and C'. Therefore, the gate is reversible. Fredkin gate can be used to built an universal set of classical logic gates.

P1.7 - (a) Consider the result (1.5.5) for the evolution of the quantum state of a spin $1/2$ in a magnetic field. Calculate the time necessary to take the initial state $|\uparrow\rangle$ to the *non-orthogonal* state

$$|\phi\rangle = \frac{|\uparrow\rangle + i|\downarrow\rangle}{\sqrt{2}}$$

Why this state cannot be distinguished from the initial state?

(b) Calculate the expectation value for the magnetic moment, $\langle\psi(t)|\boldsymbol{\mu}|\psi(t)\rangle$, with $|\psi(t)\rangle$ given by (1.5.5). Interpret the result in the light of the Erhenfest theorem.

Solution

(a) To reach the state $|\phi\rangle$ from (1.5.5), take $\omega\Delta t/2 = \pi/4$, that is:

$$\Delta t = \frac{\pi}{2\omega}$$

and remember that $\sigma_x|\uparrow\rangle = |\downarrow\rangle$. Notice that this time is half of that required to evolve the initial state to an orthogonal one. The state $|\phi\rangle$ cannot be distinguished from the initial state, because in a measurement of $|\phi\rangle$ there is 50% of chance to find $|\uparrow\rangle$.

(b)

$$\langle\psi(t)|\boldsymbol{\mu}|\psi(t)\rangle = \gamma_n\hbar\langle\psi(t)|\mathbf{I}|\psi(t)\rangle$$

Let us calculate the expected value of the component z of the magnetic moment. The other components can be calculated in a straightforward way. Since $I_z|\uparrow\rangle = +1/2|\uparrow\rangle$ and $I_z|\downarrow\rangle = -1/2|\uparrow\rangle$, we obtain from (1.5.5):

$$I_z|\psi(t)\rangle = \frac{1}{2}\cos\left(\frac{\omega\Delta t}{2}\right)|\uparrow\rangle - \frac{1}{2}i\sin\left(\frac{\omega\Delta t}{2}\right)|\downarrow\rangle$$

Therefore,

$$\langle\psi(t)|I_z|\psi(t)\rangle = \frac{1}{2}\cos^2\left(\frac{\omega\Delta t}{2}\right) - \frac{1}{2}\sin^2\left(\frac{\omega\Delta t}{2}\right) = \frac{1}{2}\cos(\omega\Delta t)$$

and we have for the expected value of μ_z:

$$\langle\psi(t)|\mu_z|\psi(t)\rangle = \frac{\gamma_n\hbar}{2}\cos(\omega\Delta t)$$

It is easy to show that:

$$\langle\psi(t)|\mu_y|\psi(t)\rangle = \frac{\gamma_n\hbar}{2}\sin(\omega\Delta t)$$

$$\langle\psi(t)|\mu_x|\psi(t)\rangle = 0$$

This result tells us that, upon the application of a magnetic field in the x direction, the z component starts to precess about the field with angular frequency equal to ω. The same is true for μ_y. The component along the field, μ_x, remains constant, equal to zero. This is consistent with the classical torque equation, $d\boldsymbol{\mu}/dt = \gamma_n\boldsymbol{\mu}\times\mathbf{B}$.

REFERENCES

[1] D. Deutsch, in: D. Bouwmeester, A.K. Eckert, A. Zeilinger (Eds.), *The Physics of Quantum Information: Quantum Cryptography, Quantum Teleportation, Quantum Computation* (Springer, 2001) Chapter 4.

[2] R. Penrose, *The Emperor's New Mind: Concerning Computers, Minds and the Laws of Physics* (Oxford Press, 1989).
[3] H.S. Leff, A.F. Rex (Eds.), *Maxwell's Demon 2: Entropy, Classical and Quantum Information, Computing* (IOP Publishing, 2003).
[4] S. Lloyd, Ultimate physical limits to computation, *Nature* **406** (2000) 1047.
[5] S. Lloyd, *Programming the Universe. A Quantum Computer Scientist Takes on the Cosmos* (Knopf Publishing Group, 2006).
[6] M.A. Nielsen, I.L. Chuang, *Quantum Computation and Quantum Information* (Cambridge, 2002).
[7] J. Stolze, D. Suter, *Quantum Computing: A Short Course from Theory to Experiment* (WILEY-VCH Verlag GmbH & Co. KGaA, 2004).
[8] C.P. Slichter, *Principles of Magnetic Resonance*, third edition (Springer-Verlag, 1990).
[9] T.H. Cormen, C.E. Leiserson, R.L. Rivest, *Introduction to Algorithms* (MIT Press, 1990).
[10] A.M. Turing, On computable numbers, with an application to the entscheidungsproblem, *Proc. Lond. Math.* **42** (1936) 230.
[11] J.J. Brophy, *Basic Electronics for Scientists*, fifth edition (McGraw-Hill, 1990).
[12] L.E. Reichl, *A Modern Course in Statistical Physics* (University of Texas, 1980).
[13] C. Bennett, R. Landauer, The fundamental physical limits of computation, *Sci. Am.* **253** (1985) 48.
[14] Y. Lecerf, Machines de Turing reversibles, *Comptes Rendus* **257** (1963) 2597.
[15] C.H. Bennett, Logical reversibility of computation, *IBM J. Res. Develop.* **17** (1973) 525.
[16] P. Benioff, The computer as a physical system: a microscopical quantum mechanical Hamiltonian model of computers as represented by Turing Machines, *J. Stat. Phys.* **22** (1980) 563.
[17] R. Landauer, Irreversibility and heat generation in the computing process, *IBM J. Res. Develop.* **5** (1961) 183.
[18] C. Bennett, Demons, engines and the second law, *Sci. Am.* **257** (1987) 108.
[19] C. Bennett, Notes on Landauer's principle, reversible computation, and Maxwell's demon, *Studies in History and Philosophy of Modern Physics* **34** (2003) 501.
[20] See *Deutsch Lectures* at http://cam.qubit.org/video_lectures/lectures.php.
[21] S. Lloyd, Black hole computers, *Sci. Am.* (November 2004) 53.
[22] Y. Aharonov, D. Bohm, Time in the quantum theory and the uncertainty relation for the time-energy domain, *Phys. Rev.* **122** (1961) 1649.
[23] N. Morgolus, L.B. Levitin, The maximum speed of dynamical evolution, *Physica D* **120** (1998) 188.
[24] G.E. Moore, Cramming more components onto integrated circuits, *Electronics Magazine* **38** (1965).
[25] C.P. Williams, S.H. Clearwater, *Explorations in Quantum Computing* (Springer, 1998).
[26] R.W. Keyes, Miniaturization of electronics and its limits, *IBM J. Res. Develop.* **32** (2000) 84.

– 2 –

Basic Concepts on Nuclear Magnetic Resonance

The world of the nuclear spins is a true paradise for theoretical and experimental physicists. It supplies, for example, most simple test systems for demonstrating the basic concepts of quantum mechanics and quantum statistics, and numerous textbook-like examples have emerged. On the other hand, the ease of handling nuclear spin systems predestinates them for testing novel experimental concepts. Indeed, the universal procedures of coherent spectroscopy have been developed predominantly within nuclear magnetic resonance (NMR) and have found widespread application in a variety of other fields. – Richard R. Ernst (Nobel Prize Lecture, 1992)

Resonance is a ubiquitous phenomenon in Nature. Every time a system with a natural frequency is excited by an external periodic perturbation of frequency close to that natural frequency, then a strong increase in the amplitude of vibration takes place. If a particle possessing magnetic dipole moment is simultaneously placed in the presence of a static magnetic field and an electromagnetic field oscillating with appropriate frequency, resonant absorption/emission can occur. This phenomenon, named magnetic resonance, is present in many closely related techniques such as electron spin resonance, nuclear magnetic resonance, ferromagnetic resonance, and nuclear quadrupole resonance, among others [1]. Nuclear magnetic resonance (NMR) is one of the most extensively studied and applied magnetic resonance techniques. Along the past five decades, NMR has found an astonishing increase in the number of subjects where it is employed, going from the now routine use in medicine to applications in biological, chemical, physical, petrophysical, and materials sciences. The aim of this chapter is to give a brief overview on the main foundations of NMR, with emphasis on the aspects relevant for quantum computing. The subject is excessively lengthy to be treated in detail in a moderate sized text, so the reader is referred to specialized books such as references [2,3] for a more comprehensive treatment of the theoretical and practical aspects of NMR.

2.1 GENERAL PRINCIPLES

The phenomenon of nuclear magnetic resonance can be generally observed for nuclei having non-vanishing total angular momentum. The nuclear total angular momentum is usually called nuclear spin. It is a vector operator, in the formalism of quantum mechanics, and is usually represented by $\hbar \mathbf{I}$, where $\mathbf{I}$ is a dimensionless operator representing the total angular momentum of the nucleus. The nuclear spin has the same general properties as any angular momentum operator in quantum mechanics; it is characterized by a quantum number I, called *the nuclear spin quantum number*. Incidentally, the name nuclear spin is commonly used to refer to the spin quantum number itself, as well as the spin operator, so the context should make it clear which meaning is being used for the term "nuclear spin". The quantum characteristics of the spin nuclear operator are given by the eigenvalues and eigenvectors of its square modulus (denoted by $\mathbf{I}^2$) and its z-component (denoted by I_z) [2]:

$$\mathbf{I}^2|I,m\rangle = I(I+1)|I,m\rangle \qquad (2.1.1)$$

$$I_z|I,m\rangle = m|I,m\rangle \qquad (2.1.2)$$

The state vectors denoted by $|I,m\rangle$ correspond to the common eigenvectors of $\mathbf{I}^2$ and I_z, being specified by the quantum numbers I and m, with $m = -I, -I+1, \ldots, I-1, I$. Other spin operators essential for the understanding of magnetic resonance experiments are the raising and lowering operators, which are defined from the transverse components of the spin operator respectively by $I_+ = I_x + iI_y$ and $I_- = I_x - iI_y$. The actions of such operators on the $|I,m\rangle$ vectors are given by [2]:

$$I_+|I,m\rangle = \sqrt{I(I+1) - m(m+1)}\,|I,m+1\rangle \qquad (2.1.3)$$

$$I_-|I,m\rangle = \sqrt{I(I+1) - m(m-1)}\,|I,m-1\rangle \qquad (2.1.4)$$

In most usual NMR experiments, the magnitude of the energy involved is much smaller than the spacing between the ground and excited nuclear energy levels (typically meV for the former as compared to keV for the latter). Therefore, one can consider that the nucleus is permanently in its ground state and all states of interest are contained in the vector subspace spanned by the vectors $|I,m\rangle$, with I fixed. In this case one says that the nucleus is in a state of well-defined total angular momentum or that the nuclear spin is a constant of motion, and we can omit the label I in the state, and represent it simply by $|m\rangle$. The energy of a nucleus in such situations is therefore determined only by the quantum number m, which in a semiclassical picture specifies the orientation of the nuclear total angular momentum with respect to external electromagnetic fields.

The total angular momentum of an atomic nucleus is due to the contributions of all orbital and intrinsic angular momenta of the protons and neutrons constituting the nucleus (the so-called *nucleons*). Therefore, in spite of the name, the nuclear spin is actually the result of the addition of orbital and spin angular momenta of all nucleons. The detailed way in which these angular momenta couple to form the total nuclear angular momentum is in general complex, depending on the characteristics of the interactions (nuclear plus electromagnetic) between the nucleons [3]. In certain cases there are general rules, similar to the ones present in Atomic Physics, which allow the prediction of the nuclear spin from the number of protons and neutrons in a given nucleus. For instance, if the number of protons and the number of neutrons are both even, the nuclear spin is zero, indicating a compensation mechanism of angular momenta for pairs of nucleons. If there is only an unpaired nucleon, then the nuclear spin is equal to the total angular momentum of that single nucleon. These rules allow the understanding of the nuclear spin values for some common nuclei such as ^{1}H ($I = 1/2$), ^{12}C ($I = 0$), ^{13}C ($I = 1/2$), and so on.

All atomic nuclei having non-zero nuclear spin also posses a *magnetic dipole moment* (represented by $\boldsymbol{\mu}$). Likewise the angular momentum, the nuclear magnetic dipole moment is also the result of the composition of the magnetic dipole moments of all nucleons.

There is a general result from angular momentum theory in quantum mechanics, known as the Wigner–Eckart theorem [4], which allows the magnetic dipole moment to be directly related to the nuclear spin, according to:

$$\mu = \gamma_n \hbar \mathbf{I} \tag{2.1.5}$$

where γ_n is called the *gyromagnetic ratio* of the nucleus. This parameter is characteristic of each nuclear species. The result (2.1.5) can be understood based on a simple model that depicts the nucleus as a rapid rotating body with axis defined by the nuclear spin vector. Therefore, the magnetic dipole moment, although not strictly parallel to the vector $\mathbf{I}$, is on average given by its projection on the axis defined by $\mathbf{I}$. It is this average that is effectively involved in usual situations where the nucleus is kept in its ground state and the total angular momentum of the nucleus is kept constant, as is the case of NMR experiments. Therefore, in these situations the Wigner–Eckart theorem can be applied and the result (2.1.5) is obtained.

Besides the magnetic dipole moment, nuclei with spin higher than 1/2 also possesses an *electric quadrupole moment*.[1] In a semiclassical picture, the nuclear electric quadrupole moment informs about the deviation of the nuclear charge distribution from a spherical symmetry. Nuclei with spin 0 or 1/2 are therefore said to be spherical, with zero electric quadrupole moment. Quadrupolar nuclei, on the other hand, are not spherical, assuming cylindrically symmetrical shapes around the symmetry axis defined by the nuclear spin. Within the subspace $|I, m\rangle$, the nuclear electric quadrupole moment operator is a traceless tensor operator of second rank, with Cartesian components written is terms of the nuclear spin:

$$\mathbf{Q}_{\alpha\beta} = \frac{eQ}{I(2I-1)} \left[\frac{3}{2} (\mathbf{I}_\alpha \mathbf{I}_\beta + \mathbf{I}_\beta \mathbf{I}_\alpha) - \delta_{\alpha\beta} \mathbf{I}^2 \right] \tag{2.1.6}$$

where α and β indicate Cartesian coordinates, $\delta_{\alpha\beta}$ is the Kronecker delta (equal to 1 if $\alpha = \beta$ and zero otherwise), e is the elemental electric charge, and Q is called the electric quadrupole moment of the nucleus (measured in units of square length, usually in barns, with 1 barn = 10^{-24} cm^2). The value of Q, similarly to the gyromagnetic ratio γ_n, is also a property of each nuclear species, its magnitude informing about the degree of deviation of the nuclear charge distribution from spherical symmetry.

2.2 INTERACTION WITH STATIC MAGNETIC FIELDS

Atomic nuclei with non-zero total angular momentum interact with the electromagnetic fields present in their environment through the nuclear magnetic dipole moment and, in the case of nuclei with $I > 1/2$, the nuclear electric quadrupole moment. The basic interaction necessary to understand NMR is the so-called Zeeman interaction, occurring between a magnetic dipole moment and the magnetic fields (applied plus local fields) existing at the nuclear site. In this section one is concerned with this magnetic interaction, which constitutes the basis of all NMR experiments. The discussion about the details of the interaction

---

[1] Electric dipole moments are zero for all nuclei because of symmetry requirements on the nuclear wave function, which must have defined parity [3,4].

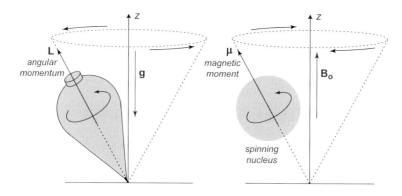

Figure 2.1 Analogy between a spinning top in a gravitational field and a magnetic moment in a magnetic field.

between the nuclear electric quadrupole moment of quadrupolar nuclei and local electric field gradients will be deferred to another section.

The Zeeman interaction between the nuclear magnetic dipole and an external static magnetic field gives rise to a manifold of energy levels for the nucleus depending on its orientation with respect to the axis defined by the magnetic field. The absorption and irradiation of energy associated with transitions between these levels constitute the physical phenomena observed in an experiment of magnetic resonance.

The classic interaction between a body with magnetic dipole moment $\boldsymbol{\mu}$ and an external static magnetic field $\mathbf{B}_0$ is described by an orientation-dependent potential energy $-\boldsymbol{\mu} \cdot \mathbf{B}_0$ and an associated torque $\boldsymbol{\mu} \times \mathbf{B}_0$. If the magnetic dipole moment is parallel and proportional to the angular momentum $\mathbf{L}$ of the body, as it is the case for example of spinning charged bodies, then this torque will cause the precession of the body around the axis of $\mathbf{B}_0$, in complete analogy with the motion of a child's spinning top acted on by the gravitational force (Figure 2.1). The resultant motion is called *Larmor precession* and the frequency of precession (named *Larmor frequency*) is easily obtained by solving the dynamic equation of motion. It is given by (see Problem P2.1):

$$\boldsymbol{\omega}_L = -\gamma_n \mathbf{B}_0 \tag{2.2.1}$$

This precession of the magnetic moment around the field is exactly what occurs for atomic nuclei with non-zero nuclear spin when in the presence of a static magnetic field. When an atomic nucleus with magnetic dipole moment given by (2.1.5) is placed in an external static magnetic field $\mathbf{B}_0$, the nuclear states $|m\rangle$ assume different energy values depending on the orientation of the nuclear spin with respect to the direction of $\mathbf{B}_0$. This splitting is known as nuclear Zeeman effect and the orientation-dependent interaction is described by a Hamiltonian of the form (named *Zeeman Hamiltonian*):

$$\mathcal{H}_Z = -\boldsymbol{\mu} \cdot \mathbf{B}_0 = -\mu_z B_0 = -\gamma_n \hbar B_0 I_z = -\hbar \omega_L I_z \tag{2.2.2}$$

where ω_L is a positive number (for the more usual case of $\gamma_n > 0$ [5]) that yields the magnitude of the Larmor frequency. The z-direction corresponds to the axis defined by the magnetic field $\mathbf{B}_0$ and all quantum operators act in the subspace spanned by $|m\rangle$ where $m = -I, -I+1, \ldots, I-1, I$. Under the action of such Hamiltonian, the expectation values of

2.2. Interaction with static magnetic fields

the Cartesian components of the nuclear spin operator in the plane perpendicular to the z-direction (i.e., $\langle I_x \rangle$ and $\langle I_y \rangle$) show an oscillatory behavior with time, with a frequency given by (2.2.1), whereas $\langle I_z \rangle$ is stationary (see Problem P2.2). Therefore, one can regard the nucleus as performing a precession motion around $\mathbf{B}_0$, in complete analogy with the Larmor precession of a classical magnetic dipole.

The eigenvalues of the Zeeman Hamiltonian (2.2.2), which are clearly proportional to the eigenvalues of the operator I_z, represent the energy levels of the nucleus, given by:

$$E_m = -m\hbar\omega_L \qquad (2.2.3)$$

Therefore, for a nucleus with spin I, there are $2I + 1$ energy levels equally spaced by the amount $\hbar\omega_L$. The lower energy states correspond to the higher (positive) m values. The ground state is thus the state with $m = I$, which means, in a semiclassical picture, that the nucleus is as aligned as possible with the direction of the field $\mathbf{B}_0$.

For an ensemble of identical nuclei in thermal equilibrium, the population of each energy level is given by the Boltzmann distribution [6]. In the case of $I = 1/2$, for example, one has a two-level system, with the populations n_- and n_+ of the $m = -1/2$ and $m = +1/2$ levels, respectively, related by the Boltzmann factor:

$$\frac{n_-}{n_+} = e^{-\hbar\omega_L/k_B T} \qquad (2.2.4)$$

where k_B is the Boltzmann constant and T is the absolute temperature of the ensemble. For protons (^{1}H nuclei) in a magnetic field of 5 T, $\hbar\omega_L$ is around 10^{-6} eV, whereas, at room temperature, $kT \cong 2.5 \times 10^{-2}$ eV, so the Boltzmann factor $e^{-\hbar\omega_L/kT}$ is very close to the unity. The fractional difference of populations in this case is about 1 part in 10^5, which shows the intrinsically low sensitivity of such experiments involving magnetic properties of nuclear populations (such as NMR).

The result (2.2.4) can be naively interpreted as meaning that there are more nuclei in the parallel than in the anti-parallel direction with respect to the magnetic field (see however the arguments given in [5] against this oversimplified point of view). This slight imbalance in the populations of the $m = -1/2$ and $m = +1/2$ levels is therefore the cause of appearance of a net equilibrium magnetization along the z-direction (parallel to the applied magnetic field). Semiclassically, one can visualize this net magnetization as the result of the Larmor precession of the nuclear spins around the direction of $\mathbf{B}_0$, resulting in a non-vanishing component of the magnetization along the z-direction and zero components in the transversal plane, due to the randomness of the motion of the spins around the precession cone (see Figure 2.2).

Therefore, the effect resulting of the application of a static magnetic field is the appearance of a nuclear magnetization parallel to that field. The thermal equilibrium magnetization for an ensemble of nuclei with $I = 1/2$, is given by [4,7]:

$$M_0 = \frac{n_0 \gamma_n^2 \hbar^2 B_0}{4 k_B T} \qquad (2.2.5)$$

where n_0 is the number of nuclei per unit volume. The dependence of the magnetization, increasing linearly with the field strength and inversely proportional to the temperature,

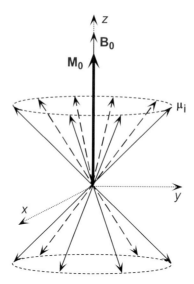

Figure 2.2 In the presence of a static magnetic field, there will be more spins precessing around the direction parallel to the field than against it. This inbalance creates a macroscopic magnetization which points to the direction of the field.

is characteristic of the *nuclear paramagnetism*, which is analogous to the electronic paramagnetism, but of magnitude much smaller. This means that the *static* magnetic properties of atomic nuclei are completely overwhelmed by electronic magnetism even in the case of diamagnetic substances, and such effects are never observed under ordinary circumstances (see however [8] for a discussion of exotic cases where these effects can be observed).

2.3 INTERACTION WITH A RADIOFREQUENCY FIELD – THE RESONANCE PHENOMENON

Transitions between the energy levels defined by the Hamiltonian (2.2.2) can be induced by the application of oscillating magnetic fields with the appropriate Larmor frequency, given by (2.2.1). For nuclear spins, the Larmor frequencies are of the order of MHz (for static fields of a few Tesla), so the excitation is achieved by a radiofrequency (RF) field. Incidentally, in the case of electron spin resonance, the Larmor frequency falls in the GHz range for similar magnetic field strengths, which means that the excitation electromagnetic field in such case is provided by microwaves.

The excitation of the nuclear spins system can be understood considering the effect of a second time-dependent magnetic field, $\mathbf{B}_1(t)$, applied perpendicularly to the static magnetic field $\mathbf{B}_0$, along the x-direction, for example. The Hamiltonian operator associated with this oscillating field, named *RF Hamiltonian* ($\mathcal{H}_{RF}$), is obtained in a similar way as the Zeeman Hamiltonian. Writing $\mathbf{B}_1(t) = 2B_1 \cos(\Omega t + \phi)\mathbf{i}$, where Ω and ϕ are respectively the frequency and the phase of the RF field and $\mathbf{i}$ is the unitary vector along the x-direction, then $\mathcal{H}_{RF}$ will be given by:

$$\mathcal{H}_{RF} = -\boldsymbol{\mu} \cdot \mathbf{B}_1(t) = -\gamma_n \hbar I_x \left[2B_1 \cos(\Omega t + \phi) \right] \quad (2.3.1)$$

2.3. Interaction with a radiofrequency field – the resonance phenomenon

The RF Hamiltonian can be treated as a perturbation to the main Zeeman Hamiltonian, considering that the magnitude of $\mathbf{B}_1(t)$, typically around a few Gauss (1 G = 10^{-4} T), is much smaller than that of $\mathbf{B}_0$. Therefore, the dominating role is still played by $\mathcal{H}_Z$ and the effect of $\mathcal{H}_{RF}$ can be determined using standard time-dependent perturbation theory [9]. Briefly, the result is that, when the frequency of the RF field is close to the Larmor frequency ($\Omega \cong \omega_L$), i.e., *on resonance*, transitions between the eigenstates of $\mathcal{H}_Z$ (specified by the quantum numbers m and n) are induced with a transition rate (or probability per unit time) given by the *Fermi golden rule*:

$$P_{m \longrightarrow n} = P_{n \longrightarrow m} \propto \gamma_n^2 \hbar^2 B_1^2 |\langle m|I_x|n\rangle|^2 \tag{2.3.2}$$

As one can see, this transition rate grows with the square of both, the gyromagnetic ratio of the nucleus and the magnitude of the RF magnetic field. Also, only magnetic fields $\mathbf{B}_1(t)$ perpendicular to $\mathbf{B}_0$ can induce such transitions, so as to give a non-vanishing value to the matrix element between the m, n states (as is the case of the operators I_x and I_y). The selection rule for the transitions is also obtained from the properties of the I_x (or I_y) operator: $\Delta m = \pm 1$ [2].

A semiclassical interpretation for the excitation of the nuclear spins is obtained by considering the linearly polarized magnetic field $\mathbf{B}_1(t)$ as composed of two circularly polarized fields, both with the same frequency and amplitude B_1, but precessing around the z-axis in opposite directions:

$$\mathbf{B}_1(t) = \mathbf{B}_1^+(t) + \mathbf{B}_1^-(t) \tag{2.3.3}$$

$$\mathbf{B}_1^+(t) = B_1[\cos(\Omega t + \phi)\mathbf{i} + \sin(\Omega t + \phi)\mathbf{j}] \tag{2.3.4}$$

$$\mathbf{B}_1^-(t) = B_1[\cos(\Omega t + \phi)\mathbf{i} - \sin(\Omega t + \phi)\mathbf{j}] \tag{2.3.5}$$

For $\Omega = \omega_L$, i.e., on resonance, the field $\mathbf{B}_1^-(t)$ rotates around the z-axis coherently with the nuclear Larmor precession described by Equation (2.2.1), whereas $\mathbf{B}_1^+(t)$ rotates in the opposite sense. In a coordinate frame rotating around the z-axis with frequency $\mathbf{\Omega} = -\Omega \mathbf{k}$ (named *rotating frame*), the field $\mathbf{B}_1^-(t)$ is stationary, as well as the nuclear spins, while $\mathbf{B}_1^+(t)$ rotates with twice the Larmor frequency. Therefore, only $\mathbf{B}_1^-(t)$ will have an effective influence on the nuclear spins, provided that both fields have magnitude much smaller than that of the static field $\mathbf{B}_0$.

The torque $\boldsymbol{\mu} \times \mathbf{B}_1^-$ will cause each magnetic moment $\boldsymbol{\mu}$, which is stationary in the rotating frame, to precess around the direction of the field $\mathbf{B}_1^-$. This direction is fixed in the rotating frame (let us call it the x'-direction) and rotates around the z-axis with the Larmor frequency in the laboratory reference frame. If the frequency of the RF field Ω is not equal to ω_L, i.e., in the off-resonance case, the precession of the magnetic moments in the rotating frame is around an axis defined by an effective magnetic field given by:

$$\mathbf{B}_{eff} = \left(B_0 - \frac{\Omega}{\gamma_n}\right)\mathbf{k} + B_1\mathbf{i}' \tag{2.3.6}$$

where $\mathbf{i}'$ is a unit vector along the x'-direction in the rotating frame, being it related to the unit vectors in the laboratory-fixed frame by $\mathbf{i}' = \cos(\Omega t + \phi)\mathbf{i} - \sin(\Omega t + \phi)\mathbf{j}$.

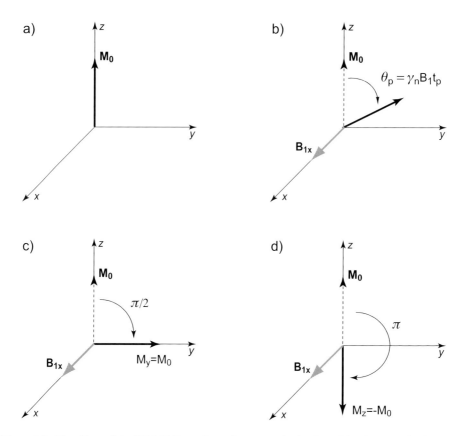

Figure 2.3 The effects of the RF field B_1 on the nuclear magnetization is to drive the magnetization vector away from the z-axis towards the transverse plane. Depending on the amplitude and duration of the RF pulse one can have a $\pi/2$ pulse, a π pulse, or, more generally, a pulse with a nutation angle θ_p.

The effect of the torque $\boldsymbol{\mu} \times \mathbf{B}_1^-$ on the collection of nuclear spins is to cause the net magnetization to deviate from the z-direction (Figure 2.3). In the on-resonance case, the magnetization $\mathbf{M}$ precesses, in the rotating frame, around the x'-direction with an angular frequency (named *nutation frequency*) with magnitude given by $\omega_1 = \gamma_n B_1$, in analogy with Equation (2.2.1). After the RF field is turned off, the magnetization points to a direction deviated from the z-axis by a nutation angle given by $\theta_p = \gamma_n B_1 t_p$, where t_p is the time interval during which the RF field was turned on. This transient RF field is named a *RF pulse*, and t_p is therefore the *pulse duration*. If $\theta_p = \pi/2$, the magnetization $\mathbf{M}$ immediately after the pulse lies in the plane transversal to $\mathbf{B}_0$; this is called a $\pi/2$ pulse. For a π pulse, on the other hand, the magnetization $\mathbf{M}$ is inverted at the end of the time of application of the pulse.

It should be remembered, however, that in the laboratory frame the magnetization is always precessing around the z-axis, with frequency corresponding to the Larmor frequency. Therefore, after a $\pi/2$ pulse, for example, there is an alternate electric signal induced by the precessing nuclear magnetization, which can be readily detected by a coil placed in the transversal plane. Basically this is the signal that is recorded in a conventional pulse NMR experiment. The signal, for reasons to be discussed in the next section, is not constant in amplitude. Instead, its amplitude decays with time after the RF pulse, typically in an exponential way. So the signal is commonly referred to as the *free induction decay* (FID), that is, a decaying signal detected in the absence of the excitation RF field.

The connection of such semiclassical description with the transitions between the quantum energy levels described by Equation (2.3.2) can be understood, for spin 1/2 systems in the simple case of $\pi/2$ and π pulses, as an equalization and an inversion of populations, respectively. This means that, after a $\pi/2$ or π pulse, the populations are not anymore given by the thermal equilibrium relation (2.2.4). The return to equilibrium needs the system give up some energy to the environment (generally named the *lattice*). This process is termed *relaxation* and is detailed in the next section.

2.4 RELAXATION PHENOMENA

After a single $\pi/2$ pulse, the collection of nuclear spins presents a resultant magnetization in the plane perpendicular to B_0 (named transverse plane) and precessing around this static field (as viewed from the laboratory-fixed coordinate system). This is clearly a non-equilibrium situation, since the only magnetic field existent now is the field B_0. If the spins were completely isolated from external influences and if there were no interactions between them, such non-equilibrium state would persist forever, with a sinusoidal electric signal induced in a coil placed in the transverse plane. However, the behavior observed in practice is completely different: the electric signal is not constant in amplitude; otherwise, it decays with time typically in an exponential way. Furthermore, after some time has elapsed, the magnetization returns completely to the initial z-direction, satisfying again the thermal equilibrium requirements (see Equations (2.2.4) and (2.2.5)). It is worth emphasizing that the exact meaning of the expression "some time" is largely dependent on the details of each particular nuclear spin system and its environment, ranging typically from microseconds to several hours. Two different processes, occurring simultaneously but (in general) independently, can be identified for this *relaxation* of the system of spins: the *transverse relaxation* and the *longitudinal relaxation*.

The transverse relaxation is the process that leads, after the end of the RF pulse, to the disappearance of the components of the nuclear magnetization **M** that are perpendicular to the field B_0. The origin of the transverse relaxation relies on the loss of coherence in the precession motion of the spins (or dephasing of the spins), caused by the existence of a spread in precession frequencies for the collection of nuclear spins. As shown before, the precession (or Larmor) frequency of each spin is dictated by the local magnetic field in the z-direction, which is affected by the external field and also by the internal fields that each nuclear spin creates on the position of other nuclei. This is the reason why the transverse relaxation is also known as *spin-spin* relaxation. However, this term can me misleading, since there are many other sources of local fields that can contribute to transverse relaxation, as will be discussed later. The spread in precession frequencies progressively results in a reduction of the resultant transverse components M_x and M_y of the magnetization, as depicted schematically in Figure 2.4. After some time, the spins distribute randomly in a precession cone around B_0 and the transverse magnetization is again zero, as it was before the application of the RF pulse.

In many simple cases, the decaying of the transverse components of the nuclear magnetization in the rotating frame can be described by a phenomenological differential equation

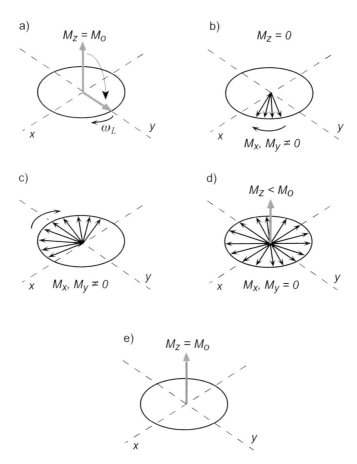

Figure 2.4 Sequence of the process of relaxation of nuclear spins: (a) application of a $\pi/2$ RF pulse; (b) to (e) time evolution of the transverse and longitudinal components of the magnetization. Note that the transverse relaxation has been concluded in (d), but the process of longitudinal relaxation continues up to (e), indicating the common situation of $T_1 > T_2$.

of the form:

$$\frac{dM_{x,y}}{dt} = -\frac{M_{x,y}}{T_2} \tag{2.4.1}$$

where T_2 is a parameter known as *transverse* or *spin-spin relaxation time*. The solution to Equation (2.4.1) is straightforward:

$$M_{x,y} = M_0 e^{-t/T_2} \tag{2.4.2}$$

where M_0 is the initial value of the transverse magnetization after the application of the RF pulse. This time-variation is superimposed in the laboratory frame to the oscillatory behavior of the transverse magnetization, which means that the decay expressed in Equation (2.4.2) is reflected in the amplitude of the signal detected in the coil. That is the reason

2.4. Relaxation phenomena

why the detected electric signal is termed free induction decay (FID). When the NMR spectrum is obtained by Fourier transform of the FID, this time-decay gives rise to a broadening of the resonance lines [5]. Therefore, materials with short T_2 values present generally broad resonance lines.

Simultaneously to the transverse magnetization, the longitudinal component of the nuclear spin magnetization also tends to recover its equilibrium value, after the end of the RF pulse. This process is physically distinct from the previously described transverse relaxation. The recovery of the M_z component of the magnetization is related to transitions between the nuclear spin levels mentioned in *Section 2.2*. When $M_z = 0$, just after the $\pi/2$ pulse, the populations of the $m = \pm 1/2$ levels (for a spin 1/2 nuclei system) are equalized. This state does not correspond to the thermal equilibrium described by Equation (2.2.4). Therefore, the natural tendency is the system gives up its excess of energy by effecting transitions preferably from the upper to the lower energy level, till the Boltzmann distribution is reestablished. This is exactly the same process that occurs when a non-magnetized sample is placed in the presence of the static magnetic field. When such transitions occur, energy is exchanged between the system of nuclear spins and its environment, which is generally named the *lattice*. That is why this process is also named *spin-lattice* relaxation. It involves necessarily the exchange of energy between the system of spins and the lattice, which is the main difference in comparison to the transverse relaxation case, which involves only the loss of coherence in the precession motion of the spins. It is important to stress that these transitions are not related to spontaneous emission of radiation. Otherwise, the transitions are of the induced type, being associated with time-fluctuating electromagnetic fields present in the material as consequence of the set of interactions involving the nuclear spins and their environment (to be described in detail later) [5].

Similarly to the transverse case, the longitudinal relaxation can also be described by a phenomenological equation of the form:

$$\frac{dM_z}{dt} = \frac{M_0 - M_z}{T_1} \qquad (2.4.3)$$

where T_1 is known as the *longitudinal* or *spin-lattice relaxation time* and M_0 is the thermal equilibrium magnetization. The solution of such equation is clearly:

$$M_z = M_0\left(1 - e^{-t/T_1}\right) \qquad (2.4.4)$$

The magnetization evolution described by Equation (2.4.4) cannot be directly detected, since it does not induce an electric signal in the coil placed in the transverse plane. The importance of such evolution in pulse NMR experiments is dictated by the fact that in most cases it is necessary to accumulate many FID's successively in order to attain a good signal-to-noise (S/N) ratio. Therefore, one has to wait the complete recovery of the longitudinal magnetization before the application of the next pulse in the loop, so as to avoid the loss of signal in the beginning of the next FID (an effect known as *saturation*). The materials presenting large values of T_1 are thus problematic from the point of view of acquiring NMR signals with suitable S/N ratio.

The relaxation times T_1 and T_2 are parameters characteristic of each particular system, whose magnitudes depend on factors such as temperature, physical state of the matter (solid or liquid), molecular mobility, magnitude of the external magnetic field, etc. [4,5].

It is universally found that $T_1 \geqslant T_2$, with the equality occurring mostly in liquids. On the other hand, in crystalline solids one has typically quite short values of T_2 and long values of T_1, which leads to broad resonance lines and poor sensitivity. This is one of the reasons why solid-state NMR is not so easy and informative as the liquid-state counterpart [5]. Moreover, it should be emphasized that the simple behavior described by Equations (2.4.2) and (2.4.4) is not actually observed in many practical cases, where there occurs a distribution of relaxation times leading to a multiexponential behavior for the decaying transverse magnetization and/or the recovery of the longitudinal magnetization.

2.5 DENSITY MATRIX FORMALISM: POPULATIONS, COHERENCES, AND NMR OBSERVABLES

The semiclassical description given in *Section 2.3*, using a vector model to describe the changes in the nuclear magnetization caused by the application of static and oscillating magnetic fields is oversimplified. There are many features of a NMR experiment that cannot be completely explained using this simple model. The most appropriate approach to describe NMR phenomena involves the use of the *density matrix formalism* from Quantum Statistical Mechanics. This approach is specifically appropriate for situations where a large number of particles is involved and one does not have access to the individual quantum states of the particles, only to macroscopic averages (or *ensemble averages*). This formalism is fully described in specialized textbooks (see References [11,12] for example). Only the general principles will be reviewed here, with emphasis on the most relevant aspects for the description of NMR phenomena as a quantum information processing technique.

The density operator ρ of a collection of identical, independent nuclei (an ensemble) is defined in such way that the macroscopic average of the expectation value of any observable A over the ensemble is given by [11]:

$$\langle A \rangle = \mathrm{Tr}\{\rho A\} \tag{2.5.1}$$

It should be emphasized that the left-hand side of this equation represents the statistical average over the entire ensemble and not the expectation value for a given particular system in the ensemble. Notice that one uses here for the ensemble average of the expectation values the same symbols used previously for the expectation values over a single system. This is common practice, and the context should make clear if one is referring to the whole ensemble or to an individual system.

In NMR experiments, the observables of interest are generally the components of the macroscopic nuclear magnetization, which are proportional to the ensemble average values of the components of the nuclear spin operator: $\langle I_x \rangle$, $\langle I_y \rangle$, and $\langle I_z \rangle$. The magnetization in the x-direction of an ensemble of nuclear spins, for example, is given by:

$$\langle M_x \rangle \propto \mathrm{Tr}\{\rho I_x\} \tag{2.5.2}$$

The time evolution of the density operator is given by the Liouville–von Neumann equation, which can be deduced from the Schrödinger equation [2,4]:

$$\frac{d\rho}{dt} = \frac{i}{\hbar}[\rho, \mathcal{H}] \tag{2.5.3}$$

2.5. Density matrix formalism: populations, coherences, and NMR observables

where $\mathcal{H}$ is the Hamiltonian of the system.

If the Hamiltonian $\mathcal{H}$ commutes with the density operator, then ρ is constant. If $\mathcal{H}$ is time-independent, then ρ will be given by:

$$\rho(t) = e^{-(i/\hbar)\mathcal{H}t} \rho(0) e^{(i/\hbar)\mathcal{H}t} \qquad (2.5.4)$$

The unitary operator $U = e^{-(i/\hbar)\mathcal{H}t}$ is called the *evolution operator*, or the *propagator* of the system [9]. In terms of U, Equation (2.5.4) reads:

$$\rho(t) = U\rho(0)U^\dagger \qquad (2.5.5)$$

In a more general case, when the Hamiltonian is not time-independent but can be split in a finite number of time-independent terms (not necessarily commuting between themselves) acting during finite intervals, the evolution of the density operator can be calculated by the expression [13,14]:

$$\rho(t) = e^{-(i/\hbar)\mathcal{H}_n t_n} \ldots e^{-(i/\hbar)\mathcal{H}_2 t_2} e^{-(i/\hbar)\mathcal{H}_1 t_1} \rho(0) e^{(i/\hbar)\mathcal{H}_1 t_1}$$
$$\times e^{(i/\hbar)\mathcal{H}_2 t_2} \ldots e^{(i/\hbar)\mathcal{H}_n t_n} \qquad (2.5.6)$$

This expression is very useful for understanding NMR experiments consisting of sequences of RF pulses and free evolution periods.

In thermal equilibrium, the density operator is simply related to the Hamiltonian of the system by [4,11]:

$$\rho_0 = \frac{e^{-\mathcal{H}/k_B T}}{\sum_m e^{-E_m/k_B T}} \qquad (2.5.7)$$

where the sum is extended over all Hamiltonian eigenstates and E_m represents the eigenvalues of $\mathcal{H}$. The sum in the denominator is called the *partition function* of the system: $Z = \sum_m = e^{-E_m/k_B T}$.

In a given orthonormal basis, the density operator has, as any other operator, a matrix representation, which is called the *density matrix*:

$$\rho = \begin{bmatrix} \rho_{11} & \rho_{12} & \cdots \\ \rho_{21} & \rho_{22} & \cdots \\ \vdots & \vdots & \ddots \end{bmatrix} \qquad (2.5.8)$$

The density operator of any statistical ensemble must satisfy some general requirements: (1) it is a Hermitian operator; (2) the diagonal elements of its matrix representation are greater than or equal to zero; and (3) the sum of these diagonal elements equals to unity [11,12].

The diagonal elements of the density matrix are called the *populations*, whereas the off-diagonal elements are called the *coherences*. The populations have a physical interpretation related to the probability of finding a member of the ensemble in a given state, when performing a macroscopic measurement. Specifically, the element ρ_{mm} gives the probability of finding a member of the ensemble in the state specified by the quantum number m. The previously mentioned condition that the sum of the diagonal elements is equal to unity is, therefore, equivalent to the usual normalization condition imposed on probability evaluation [2].

This physical meaning of the populations was already introduced in Equation (2.2.4) in the context of nuclear paramagnetism, where the populations were interpreted as the fractional number of spins in each of the $m = \pm 1/2$ states. Surely one cannot speak of the exact number of nuclei that lies in each particular state; most nuclei are in general in superposition states [5]. However, it is common practice to talk about populations as referring to the mean number of nuclei in each state, expressing actually the chance of finding a nucleus in that state when performing a macroscopic measurement.

Thus, in the context of NMR, the populations are related to the raising of longitudinal magnetization along a static magnetic field. On the other hand, the coherences are related to the existence of transverse magnetization, always arising after the application of some external RF excitation. This point will be made clear in the next section, with the detailed analysis of spin 1/2 systems.

In the basis of the Hamiltonian eigenstates, the thermal equilibrium density matrix constructed from Equation (2.5.7) is diagonal:

$$\rho_0 = \frac{1}{Z} \begin{pmatrix} e^{-E_1/k_B T} & 0 & \cdots \\ 0 & e^{-E_2/k_B T} & \cdots \\ \vdots & \vdots & \ddots \end{pmatrix} \quad (2.5.9)$$

This simple form for the density matrix implies that, in thermal equilibrium, the populations obey to the Boltzmann distribution, whereas the coherences are identically zero.

For an ensemble of identical nuclei with spin quantum number I placed in a static magnetic field $\mathbf{B}_0$, the Zeeman Hamiltonian is given by Equation (2.2.2): $\mathcal{H}_Z = -\hbar \omega_L I_z$. Using the basis formed by the eigenstates of the operator I_z (see Equation (2.1.2)), the populations in thermal equilibrium are given by:

$$[\rho_0]_{mm} = \frac{e^{m\hbar\omega_L/k_B T}}{\sum_{s=-I}^{I} e^{s\hbar\omega_L/k_B T}} \quad (2.5.10)$$

In the so-called *high-temperature limit*, where the thermal energy $k_B T$ is much greater than the Zeeman level spacing $\hbar \omega_L$ (which is always true for temperatures above c.a. 1 K under ordinary magnetic fields of a few teslas; see Problem P.2.3), it is surely legitimate to keep only the leading terms in the polynomial expansions of the exponentials above:

$$e^{m\hbar\omega_L/k_B T} \cong 1 + \frac{m\hbar\omega_L}{k_B T} \quad (2.5.11)$$

$$\sum_{s=-I}^{I} e^{s\hbar\omega_L/k_B T} \cong 2I + 1 \quad (2.5.12)$$

It is helpful to define the ratio $\Delta = \frac{\hbar\omega_L}{k_B T}$, which is a dimensionless parameter that measures the deviation of populations from a uniform value due the application of the static magnetic field. It is this parameter (typically around 10^{-5}) that determines the magnitude of the thermal equilibrium magnetization characteristic of nuclear paramagnetism.

In the high-temperature limit, the thermal equilibrium density operator for a system of nuclei under action of a static magnetic field is therefore given by:

$$\rho_0 = \left(\frac{1}{2I+1}\right)\mathbf{1} + \left(\frac{1}{2I+1}\Delta\right)I_z \qquad (2.5.13)$$

where $\mathbf{1}$ is the identity matrix. As one can see, the density operator in this situation is simply related to the operator I_z. The first term describes a uniform background, independent of the application of the magnetic field. As the identity operator commutes with all operators, Equation (2.5.3) shows that this term does not evolves in time and can be ignored for purposes of time evolution calculations. The second term is called the *deviation density matrix* and it is usually represented by $\Delta\rho$. It is this traceless operator that is acted upon by the evolution operator in any NMR experiment. In thermal equilibrium, the matrix $\Delta\rho$ is therefore directly proportional to the operator I_z, which is associated with the appearance of the longitudinal magnetization. The thermal equilibrium density matrix deviation can thus be written as:

$$\Delta\rho_0 = \alpha I_z \qquad (2.5.14)$$

where $\alpha = \frac{\Delta}{2I+1}$. This constitutes the starting point for all NMR pulse sequences, to be detailed in the next sections.

2.6 NMR OF NON-INTERACTING SPINS 1/2

The *matrix representations* of the nuclear spin operators in the $|I, m\rangle$ basis for the simple $I = 1/2$ case are given by:

$$I_x = \frac{1}{2}\begin{bmatrix} 0 & 1 \\ 1 & 0 \end{bmatrix} \quad I_y = \frac{1}{2}\begin{bmatrix} 0 & -i \\ i & 0 \end{bmatrix} \quad I_z = \frac{1}{2}\begin{bmatrix} 1 & 0 \\ 0 & -1 \end{bmatrix} \qquad (2.6.1)$$

These matrices are thus directly related to the famous Pauli matrices, largely used in problems involving the spin of the electron [2].

From Equation (2.5.14), the deviation density matrix in thermal equilibrium under a static magnetic field is:

$$\Delta\rho_0 = \frac{\alpha}{2}\begin{bmatrix} 1 & 0 \\ 0 & -1 \end{bmatrix} = \frac{\hbar\omega_L}{4k_BT}\begin{bmatrix} 1 & 0 \\ 0 & -1 \end{bmatrix} \qquad (2.6.2)$$

As described previously, the analysis of the effect of RF pulses is best conducted by using the concept of the rotating frame. The transformation of the density operator from the laboratory to the rotating frame (which rotates around the z-axis with frequency $\mathbf{\Omega} = -\Omega\mathbf{k}$) is accomplished with the use of the rotation operator $e^{i\Omega t I_z}$ [4]:

$$\rho^{Rot} = e^{-i\Omega t I_z} \rho^{Lab} e^{i\Omega t I_z} \qquad (2.6.3)$$

where ρ^{Rot} and ρ^{Lab} indicate the density matrices in the rotating and laboratory frames, respectively. Applying this relation to the thermal equilibrium density matrix deviation

$\Delta \rho_0$ leads to exactly the same matrix in both frames, since the operators $e^{i\Omega t I_z}$, I_z, and $e^{-i\Omega t I_z}$ commute. This is obviously an immediate consequence of the fact that rotations around the z-axis do not change the longitudinal magnetization. Therefore:

$$\Delta \rho_0^{Rot} = \frac{\hbar \omega_L}{4 k_B T} \begin{bmatrix} 1 & 0 \\ 0 & -1 \end{bmatrix} \quad (2.6.4)$$

From this point on, one will drop the superscript *Rot*, with the assumed convention that, unless otherwise stated, all density operators are given in the rotating-frame.

The effect of a RF pulse can be obtained by transforming the complete Hamiltonian $\mathcal{H}_Z + \mathcal{H}_{RF}$ from the laboratory to the rotating-frame [4], which is equivalent to expressing the effective Hamiltonian associated with the rotating-frame effective field given in Equation (2.3.6):

$$\mathcal{H}_{eff} = -\hbar(\omega_L - \Omega) I_z - \hbar \omega_1 I_x \quad (2.6.5)$$

This describes a RF pulse applied along the x-direction of the rotating-frame. If the pulse is applied along another direction (i.e., with another phase), the only change will be in the angular momentum component appearing in the second term of the right-hand member of such equation. The most important aspect of this effective Hamiltonian is that it is time-independent, so that the solution given by Equation (2.5.4) for the density operator can now be used. In other words, the rotating-frame transformation has removed the time dependence from the RF Hamiltonian, which corresponds to the classical view given in Section 2.3 of a stationary magnetic field B_1^- in the rotating-frame [4,5].

If the nutation frequency $\omega_1 = \gamma_n B_1$ is much larger than the resonance offset (given by $\omega_L - \Omega$), then the effective Hamiltonian is approximated by:

$$\mathcal{H}_{eff} \cong -\hbar \omega_1 I_x \quad (2.6.6)$$

This expression is exactly true on resonance, i.e., when $\Omega = \omega_L$. In the case of strong RF pulses and for small resonance offsets, the approximation is usually appropriate. Considering thus a strong RF pulse with duration t_p applied along the x-direction, its effect can be calculated by Equation (2.5.5) and the Hamiltonian (2.6.6). The evolution operator in this case is then:

$$U = e^{-(i/\hbar)\mathcal{H}_{eff} t_p} = e^{i \omega_1 t_p I_x} = R_x(-\theta_p) \quad (2.6.7)$$

The operator $R_x(\theta_p) = e^{-i \theta_p I_x}$ is a rotation operator [5], which produces a nutation of an angle $\theta_p = \omega_1 t_p$ around the x-axis (according to the right-hand rule) of the rotating-frame. The density operator following this pulse is therefore:

$$\rho(t_p) = R_x(-\theta_p) \rho_0 R_x(\theta_p), \quad (2.6.8)$$

where it was used the property $R_x^\dagger(\theta) = R_x(-\theta)$, which follows from the Hermitian nature of the angular momentum operators.

The explicit form of the matrix representation of the operator $R_x(\theta_p)$ can be obtained with some algebraic manipulation of the properties of exponential operators in the simple

2.6. NMR of non-interacting spins 1/2

case of $I = 1/2$ [5]. This matrix is given below, together with those corresponding to the rotation operators around other axes:

$$R_x(\theta_p) = \begin{bmatrix} \cos(\theta_p/2) & -i\sin(\theta_p/2) \\ -i\sin(\theta_p/2) & \cos(\theta_p/2) \end{bmatrix} \quad (2.6.9)$$

$$R_y(\theta_p) = \begin{bmatrix} \cos(\theta_p/2) & -\sin(\theta_p/2) \\ \sin(\theta_p/2) & \cos(\theta_p/2) \end{bmatrix} \quad (2.6.10)$$

$$R_{\phi_p}(\theta_p) = \begin{bmatrix} \cos(\theta_p/2) & -i\sin(\theta_p/2)e^{-i\phi_p} \\ -i\sin(\theta_p/2)e^{i\phi_p} & \cos(\theta_p/2) \end{bmatrix} \quad (2.6.11)$$

The last rotation operator corresponds to a RF pulse applied with phase ϕ_p, i.e., a pulse whose magnetic field vector is aligned in the rotating-frame with an axis making an angle ϕ_p with the x-axis.

We can now use these explicit forms to calculate the effect of RF pulses on the deviation matrix densities for spin 1/2 nuclei. In the case of a $\pi/2$ pulse with the magnetic field B_1^- aligned with the x-direction of the rotating frame, we have:

$$\Delta\rho(t_p) = R_x(-\pi/2)\Delta\rho_0 R_x(\pi/2)$$

$$= \frac{1}{\sqrt{2}}\begin{bmatrix} 1 & i \\ i & 1 \end{bmatrix} \times \frac{\hbar\omega_L}{4k_BT}\begin{bmatrix} 1 & 0 \\ 0 & -1 \end{bmatrix} \times \frac{1}{\sqrt{2}}\begin{bmatrix} 1 & -i \\ -i & 1 \end{bmatrix}$$

$$= \frac{\hbar\omega_L}{4k_BT}\begin{bmatrix} 0 & -i \\ i & 0 \end{bmatrix} = \frac{\hbar\omega_L}{2k_BT}(I_y) \quad (2.6.12)$$

Therefore, after the pulse the deviation density matrix is proportional to the operator I_y, which indicates a magnetization pointing along the y-direction. In fact, this is exactly what is achieved by a geometrical rotation of the magnetization initially along the z-direction through an angle of $\pi/2$ around the x-axis in a left-hand sense. This rotation can be viewed, in a semi-classical description, as result of the torque exerted by the field B_1^- on the longitudinal magnetization in the rotating frame. As this is as clockwise rotation around the x-axis, such pulse is known as a "$-x$ $\pi/2$ pulse" or, in a more convenient notation, a $(\pi/2)_{-x}$ pulse. (The symbol $(\theta_p)_{\phi_p}$ is a convenient way to characterize a RF pulse, yielding the nutation angle θ_p and the phase ϕ_p of the pulse.) Obviously, a $(\pi/2)_x$ pulse has the opposite effect, i.e., it drives a magnetization initially aligned with the z-axis to the $-y$-axis.

The same result could be obtained by using the following general property of rotation operators [5,9]:

$$e^{-i\theta I_1} I_2 e^{i\theta I_1} = I_2 \cos\theta + I_3 \sin\theta \quad (2.6.13)$$

where I_1, I_2, and I_3 indicate the components of angular momentum operators that commute cyclically, i.e., $[I_1, I_2] = iI_3$, $[I_2, I_3] = iI_1$, and $[I_3, I_1] = iI_2$. Writing this rule for the specific case of a RF pulse applied along the x-axis starting form thermal equilibrium, one obtains[2]:

$$e^{i\theta_p I_x} I_z e^{-i\theta_p I_x} = I_z \cos\theta_p + I_y \sin\theta_p \qquad (2.6.14)$$

Applying Equation (2.6.14) with $\theta_p = \pi/2$ leads to the same result as Equation (2.6.12), that is, the deviation density matrix has changed from I_z to I_y with the application of the pulse. Writing the complete density matrices before and after the pulse, one obtains:

$$\rho_0 = \begin{bmatrix} \frac{1}{2} + \frac{\hbar\omega_L}{4k_B T} & 0 \\ 0 & \frac{1}{2} - \frac{\hbar\omega_L}{4k_B T} \end{bmatrix} \xRightarrow{(\pi/2)_{-x}} \rho(t_p) = \begin{bmatrix} \frac{1}{2} & -i\frac{\hbar\omega_L}{4k_B T} \\ i\frac{\hbar\omega_L}{4k_B T} & \frac{1}{2} \end{bmatrix} \qquad (2.6.15)$$

The physical interpretation of the net effect of the $\pi/2$ pulse is now clear: the pulse promotes an *equalization of the populations* of the $m = \pm 1/2$ states and, at the same time, *creates coherence* between these states. This coherence is associated with the transverse magnetization appearing in the y-direction of the rotating-frame. The equalization of populations is easily understood using the transition rates between the $m = \pm 1/2$ states described by Equation (2.3.2). The system absorbs energy from the RF source until the end of the pulse, when a *saturation* situation occurs, with the same mean number of nuclei in each state. However, the production of coherence is difficult to describe using this simple picture. Saturation also occurs for example for thermal equilibrium in the absence of static magnetic fields ($\omega_L = 0$ in Equation (2.6.15)), but in this case there is obviously no coherence. Therefore, in order to understand the complete effect of the $\pi/2$ pulse one needs more than the simple picture of spins "going up or down". This simple example shows the power of the density operator approach to understand pulse sequences in NMR experiments.

The effect of other RF pulses can be readily determined by similar methods, either by using the explicit matrix forms given in Equation (2.6.11) or by applying Equation (2.6.14). Another quite common situation is that of a π pulse applied along, say, the x-direction. Putting $\theta_p = \pi$ in Equation (2.6.14), one obtains for the deviation density matrix the result $\Delta\rho_p \sim -I_z$. In terms of the complete density matrices[3]:

$$\rho_0 = \begin{bmatrix} \frac{1}{2} + \frac{\hbar\omega_L}{4k_B T} & 0 \\ 0 & \frac{1}{2} - \frac{\hbar\omega_L}{4k_B T} \end{bmatrix} \xRightarrow{(\pi)_{-x}} \rho(t_p) = \begin{bmatrix} \frac{1}{2} - \frac{\hbar\omega_L}{4k_B T} & 0 \\ 0 & \frac{1}{2} + \frac{\hbar\omega_L}{4k_B T} \end{bmatrix} \qquad (2.6.16)$$

The interpretation is now straightforward: the π pulse leads to the *inversion of the populations* of the $m = \pm 1/2$ states, with no creation of coherence. Therefore, after such pulse the magnetization is simply inverted, pointing now in the $-z$-direction, with no transverse magnetization appearing as consequence of the pulse.

---

[2] Note that there are two compensating changes of signal with respect to Equation (2.6.13): the first due to the inverse order of the rotation operators in the left-hand side of both equations, and the second because I_x and I_z are in the "wrong" order in the left-hand member, which corresponds to using Equation (2.6.13) with $I_1 = I_x$, $I_2 = I_z$, and $I_3 = -I_y$, since $[I_x, I_z] = -iI_y$.

[3] Note that the effects of the pulses $(\pi)_{-x}$ and $(\pi)_x$ are exactly the same.

2.6. NMR of non-interacting spins 1/2

The effects of pulses with phases other than the four transverse conventional ones ($x, y, -x, -y$) can be calculated by using the matrix given in (2.6.11). Alternatively, the rotation operator $R_{\phi_p}(\theta_p)$ can be written in the form [5]:

$$R_{\phi_p}(\theta_p) = R_z(\phi_p) R_x(\theta_p) R_z(-\phi_p), \qquad (2.6.17)$$

where the rotation operator $R_z(\phi_p)$ is:

$$R_z(\phi) = e^{-i\phi I_z} = \begin{bmatrix} e^{-i\phi/2} & 0 \\ 0 & e^{i\phi/2} \end{bmatrix} \qquad (2.6.18)$$

Then, the effect of $R_{\phi_p}(\theta_p)$ can be obtained by sequentially applying Equation (2.6.13) and the time evolution of the density operator:

$$\rho(t_p) = R_{\phi_p}(-\theta_p) \rho_0 R_{\phi_p}(\theta_p) \qquad (2.6.19)$$

The case of off-resonance pulses is a little bit more involved, since now the axis of nutation is given by the effective magnetic field in Equation (2.3.6) and, therefore, is not contained in the transverse plane. The extent to which this axis is tilted out of the transverse plane and the sense of the tilt is determined by the frequency offset ($\omega_L - \Omega$). Also the nutation frequency is no longer given by $\omega_1 = \gamma_n B_1$, but depends on the offset. It can be shown that the rotation operator describing such off-resonance pulse can be written as a product of five rotations about orthogonal axes [5]:

$$R_{\phi_p}^{off}(t_p) = R_z(\phi_p) R_y(\beta_p) R_z(\omega_{eff} t_p) R_y(-\beta_p) R_z(-\phi_p) \qquad (2.6.20)$$

where $\beta_p = \arctan[\omega_1/(\omega_L - \Omega)]$ is the angle that measures the deviation of the nutation axis from the transverse plane, $\omega_{eff} = \sqrt{\omega_1^2 + (\omega_L - \Omega)^2}$ is the effective nutation frequency, and ϕ_p is the phase of the pulse. Therefore, the presence of non-negligible offsets leads to an increase in the nutation frequency and a change in the rotation axis direction. It is easy to see that, if a given pulse duration has been carefully adjusted to give, say, a π pulse on resonance, the same pulse will not have the same performance for nuclei far from resonance.

In practical cases, the extent to which off-resonance effects are important depend on the dispersion of Larmor frequencies in the system. As we will see in the next section, due to internal interactions, magnetically distinct nuclei in diamagnetic substances have slightly distinct Larmor frequencies (an effect named *chemical shift*). Therefore, in any sample containing chemically distinct nuclei, there will be always some nuclei that are off resonance. If the chemical shifts are small compared to the nutation frequency, then approximation (2.6.6) is good and off-resonance effects can be neglected. For example, typical chemical shifts for spin 1/2 nuclei in diamagnetic substances fall in the range of Hz, whereas the nutation frequency $\omega_1/2\pi$ is typically of some kHz (for strong pulses). Therefore, $\omega_{eff} \cong \omega_1$ for all nuclei of interest in this case. The situation is completely different in the case of metallic and/or magnetic materials, where the dispersion of Larmor frequencies can be much larger [4]. Also, for nuclei with $I > 1/2$ experiencing a strong quadrupolar interaction (to be described later) the difference in effective nutation frequencies cannot be disregarded [10].

This feature can be a problem, but also offers a means of applying *selective pulses* to excite only some selected resonance lines in a given window defined by the frequency and bandwidth of the RF pulse. Therefore, it is quite common to use pulses with large intensity (named *hard pulses*) for non-selective irradiation (i.e., excitation of all nuclei irrespective of their particular resonant frequencies) and low-power pulses (named *soft pulses*) to selective excitation [5].

After the application of the RF pulse (or sequence of RF pulses), the time evolution of the system of non-interacting spins 1/2 is again governed solely by the Zeeman Hamiltonian, which, in the rotating frame is:

$$H_Z = -\hbar(\omega_L - \Omega)I_z \quad (2.6.21)$$

The time evolution of the density matrix in this case is straightforward. It corresponds to a free precession (ignoring for the time being all relaxation effects), with frequency equal to the frequency offset:

$$\rho(\tau) = e^{i(\omega_L - \Omega)\tau I_z} \rho(t_p) e^{-i(\omega_L - \Omega)\tau I_z} \quad (2.6.22)$$

where τ is the time from the end of application of the pulse. After a $(\pi/2)_{-x}$, for example, one has $\Delta\rho(t_p) = \frac{\hbar\omega_L}{2k_B T} I_y$. Then, $\Delta\rho(\tau)$ describes a transverse magnetization precessing around the z-axis in the rotating-frame with frequency equal to $(\omega_L - \Omega)$ in the negative sense:

$$\Delta\rho(\tau) = e^{i(\omega_L - \Omega)\tau I_z} \left(\frac{\hbar\omega_L}{2k_B T} I_y\right) e^{-i(\omega_L - \Omega)\tau I_z}$$

$$= \frac{\hbar\omega_L}{2k_B T}\left[I_y \cos(\omega_L - \Omega)\tau + I_x \sin(\omega_L - \Omega)\tau\right] \quad (2.6.23)$$

The motion of the magnetization in the laboratory frame can be easily obtained now from the inverse of Equation (2.6.3) and the result is obviously a transverse magnetization precessing around the z-axis with frequency ω_L (in the negative sense). This time-varying magnetization can be detected by a RF coil placed in the transverse plane and, considering also the effect of transverse relaxation, this gives rise to the decaying signal (FID) mentioned in Section 2.3.

We can summarize the results of this section saying that the effects of RF pulses can be described in the rotating frame in terms of the rotation operators (usually around the $x, y, -x,$ and $-y$ axes) applied to the deviation density matrix starting from thermal equilibrium. The evolution of the system after or between the RF pulses is described as a free precession around the z-axis, with a frequency that depends on the frequency offset and is therefore different for nuclei experiencing distinct local fields (due to chemical shifts, for example).

2.7 NUCLEAR SPIN INTERACTIONS

Up to now, we have described the nuclear spins as isolated entities, interacting only with the externally applied magnetic fields. If this were the whole story, NMR would not have

2.7. Nuclear spin interactions

many interesting applications, except perhaps as a means of directly determining the gyromagnetic ratio of a given nucleus. However, fortunately the nuclear spins are not at all isolated from each other and from the local environment where they are located. Otherwise, each nuclear spin experiences a number of electromagnetic fields originated from internal interactions present in the material. These interactions influence the exact value of the resonance frequency of each nucleus, and so the measurement of the spectrum of frequencies for a nucleus in a given material constitutes a way to achieve information on the internal interactions between the nucleus and its environment for that particular substance. This feature is in the core of all applications of NMR as a tool for structural and chemical characterization of materials. Furthermore, as it will be detailed later, these internal interactions provide the way through which logical gates can be implemented in NMR quantum computing schemes. In what follows we present the basics of the main interactions involving the nuclear spin in a given material, without going deeply into the physics of any of the interactions. More detailed descriptions can be found in the classic texts by Slichter [4] and Abragam [8].

Under conditions at which NMR experiments are usually performed, the interactions between the nucleus and the electromagnetic fields present in its environment (including the interactions with electrons, other nuclei, other ions, and so on) are well described using the concept of the *nuclear spin Hamiltonian* ($\mathcal{H}_{nuclear}$). This Hamiltonian contains only terms that depend on the orientation of the nuclear spin and, therefore, its matrix representation is usually given in the $|m\rangle$ basis, which corresponds to eigenstates of the Zeeman Hamiltonian ($\mathcal{H}_Z$). It is convenient to write the nuclear spin Hamiltonian in the form:

$$\mathcal{H}_{nuclear} = \mathcal{H}_{ext} + \mathcal{H}_{int} \tag{2.7.1}$$

where $\mathcal{H}_{ext}$ represents the interactions of the nucleus with applied electromagnetic fields (*external* interactions) and $\mathcal{H}_{int}$ corresponds to *internal* interactions with the local environment of the nucleus. The two contributions to $\mathcal{H}_{ext}$ are the Zeeman and the RF Hamiltonians, already discussed before:

$$\mathcal{H}_{ext} = \mathcal{H}_Z + \mathcal{H}_{RF} \tag{2.7.2}$$

Usually, these correspond to the dominant terms in the nuclear spin Hamiltonian, so that in many cases of interest (but not always), the internal interactions can be treated by perturbation methods. The effect of the RF pulse is better described in the rotating frame, where $\mathcal{H}_{RF}$ plays the dominant role. If the magnitude of $\mathcal{H}_{RF}$ is so large that all terms in $\mathcal{H}_{int}$ can be neglected during the pulse, then the result of the application of the RF pulse is straightforward, as described before for the case of isolated spins 1/2. This situation is not always true, especially when large quadrupolar couplings are present, so in these cases the combined effect of $\mathcal{H}_{RF}$ and the relevant terms in $\mathcal{H}_{int}$ must be taken into account even during the RF pulse.

There are several contributions to the internal Hamiltonian, depending on the physical characteristics of the analyzed material. In the case of diamagnetic insulating substances (which correspond the vast majority of current applications of NMR in liquids and in solids), the main interactions are usually classified according to:

$$\mathcal{H}_{int} = \mathcal{H}_{CS} + \mathcal{H}_D + \mathcal{H}_J + \mathcal{H}_Q \tag{2.7.3}$$

In such expression, $\mathcal{H}_{CS}$ is the chemical-shift interaction of the nucleus with the orbital motion of the surrounding electrons; $\mathcal{H}_D$ is the direct (through space) dipolar interaction between nuclei; $\mathcal{H}_J$ is the electron-mediated interaction between nuclei; and $\mathcal{H}_Q$ is the quadrupolar interaction between a nucleus with spin $> 1/2$ and the electric field gradient at the nuclear position. Each of these terms are briefly described below and simplified forms for the Hamiltonians are given.

For paramagnetic substances (or samples containing paramagnetic centres), there are other terms in $\mathcal{H}_{int}$ involving the interactions between the nucleus and the spin magnetic dipole moment of unpaired electrons, such as the paramagnetic shift and the Knight shift (for interaction with conduction electrons in metals). In the case of solids presenting any type of magnetic ordering (as in ferro or antiferromagnetic materials), there are strong *hyperfine* magnetic fields at the nucleus, which are in many instances much larger than the external magnetic field and this opens the possibility of realization of NMR without externally applied static magnetic fields (method known as "zero-field NMR"). The interested reader is referred to the books by Guimarães [7] and Turov & Petrov [15] for a detailed description of this subject.

2.7.1 Chemical shift

The magnetic field actually experienced at the nuclear position is not equal to the external magnetic field. Even disregarding bulk magnetic susceptibility effects, which are minor in diamagnetic substances [4], the disturbance of the orbital motion of nearby electrons gives rise to an induced magnetic field that adds to the externally applied magnetic field, leading to a *local* magnetic field given by:

$$\mathbf{B}_{loc} = (1 - \tilde{\sigma})\mathbf{B}_0 \qquad (2.7.4)$$

The quantity $\tilde{\sigma}$ is known as the *chemical shielding tensor* associated with that particular nuclear site. The tensorial character of $\tilde{\sigma}$ implies that $\mathbf{B}_{loc}$ is in general in a direction different from that of $\mathbf{B}_0$, which reflects the anisotropy of the molecular environment of the considered nucleus. As this is a purely magnetic interaction, analogous to the Zeeman one, the Hamiltonian $\mathcal{H}_{CS}$ is given by:

$$\mathcal{H}_{CS} = -\boldsymbol{\mu} \cdot (-\tilde{\sigma}\mathbf{B}_0) \cong \gamma_n \hbar \sigma_{zz} B_0 I_z \qquad (2.7.5)$$

The last step in the equation above is known as *secular approximation* and it is a generally appropriate simplification valid as consequence of the much larger magnitude of the Zeeman interaction with the external magnetic field as compared to the chemical shift one [5].

It is important to stress that the component σ_{zz} depends on the relative orientation of the electron cloud in the molecule with respect to the external magnetic field. In a monocrystalline solid there is only one value of the parameter σ_{zz} for each orientation of the specimen. For an isotropic liquid substance, the average of all possible molecular orientations leads to an average value for the chemical shift known as the *isotropical chemical shift* (σ_{iso}) [4]:

$$\mathcal{H}_{CS} \cong \gamma_n \hbar \sigma_{iso} B_0 I_z \qquad (2.7.6)$$

2.7. Nuclear spin interactions

The parameter σ_{iso} is related to the *trace* of the tensor $\tilde{\sigma}$, which is usually written in a molecular reference frame where this tensor is diagonal, known as the *principal axis system* (PAS) of the tensor $\tilde{\sigma}$:

$$\sigma_{iso} = (\sigma_{XX} + \sigma_{YY} + \sigma_{ZZ})/3 \tag{2.7.7}$$

where σ_{XX}, σ_{YY}, and σ_{ZZ} represent the Cartesian components of the tensor $\tilde{\sigma}$ in the PAS.

Therefore, the net effect of the chemical shift interaction in this case is the production of a small correction added to the magnetic field. Therefore, the practical consequence is a *shift* of the resonance frequency away from the Larmor frequency of a free isolated nucleus:

$$\omega = \omega_L(1 - \sigma_{iso}) \tag{2.7.8}$$

A similar expression (but involving σ_{zz} instead of σ_{iso}) also applies to the case of a monocrystalline material. For polycrystalline or powdered samples, on the other hand, the continuous distribution of orientations of the several crystallites with respect to the direction of $\mathbf{B}_0$ causes an anisotropic broadening of the resonance spectrum, known as broadening due to chemical shift anisotropy (CSA) [10].

It is the dependence of the resonance frequency on the specific molecular environment of each nucleus expressed by Equation (2.7.8) that makes the NMR technique so widespread and useful as a tool for identification and characterization of chemical groups in liquid substances (and also in solids if some special techniques are employed, as it will be detailed later). The resonance frequency is usually expressed in practice as a relative shift measured with reference to the resonance frequency (ω_{ref}) of a standard substance:

$$\delta = \frac{\omega - \omega_{ref}}{\omega_{ref}} \tag{2.7.9}$$

The δ values, simply called as the chemical shifts of the resonance lines, are usually expressed in parts per million (ppm), which indicates the order of magnitude of the typical shifts in diamagnetic substances.

2.7.2 Dipolar coupling

Any two magnetic dipole moments interact directly through the magnetic fields created by each one on the position of the other. The magnetic field created by a classical point dipole at a point located by the vector $\mathbf{r}$ with origin on the dipole is (in SI units) [16]:

$$\mathbf{B}_{dip} = \frac{\mu_0}{4\pi} \frac{3(\boldsymbol{\mu} \cdot \mathbf{e})\mathbf{e} - \boldsymbol{\mu}}{r^3} \tag{2.7.10}$$

where $\boldsymbol{\mu}$ is the magnetic dipole moment, r and $\mathbf{e}$ are respectively the magnitude of the vector $\mathbf{r}$ and the unit vector in its direction, and μ_0 is the magnetic permeability of free space.

The interaction of a dipole $\boldsymbol{\mu}_1$ with such field created by a dipole $\boldsymbol{\mu}_2$ is given by the classical Zeeman interaction energy: $-\boldsymbol{\mu}_1 \cdot \mathbf{B}_{dip}^{(2)}$. The Hamiltonian describing the dipolar interaction can thus be written in the form:

$$H_{dip} = \frac{\mu_0}{4\pi} \frac{\gamma_{n1}\gamma_{n2}\hbar^2}{r_{12}^3} \left[\mathbf{I}_1 \cdot \mathbf{I}_2 - 3(\mathbf{I}_1 \cdot \mathbf{e}_{12})(\mathbf{I}_2 \cdot \mathbf{e}_{12}) \right] \quad (2.7.11)$$

This expression can be rearranged by grouping the terms involving each combination of the Cartesian components of the operators $\mathbf{I}_1$ and $\mathbf{I}_2$, associated with the two spins. When this is done, only the dominant terms are retained in the *secular* approximation. For the *heteronuclear* case (i.e., the interaction between two unlike nuclei with $\gamma_{n1} \neq \gamma_{n2}$) this leads to the simple expression:

$$H_{dip} = -\frac{\mu_0}{4\pi} \frac{\gamma_{n1}\gamma_{n2}\hbar^2}{r_{12}^3} I_{1z} I_{2z} (3\cos^2\theta_{12} - 1) \quad (2.7.12)$$

where θ_{12} is the angle between the vector $\mathbf{e}_{12}$ and the external magnetic field $\mathbf{B}_0$. The main characteristics of this Hamiltonian are its dependence with the inverse cube of the distance between the spins (which restricts usually this interaction to spins located within short distances) and the orientational dependence. In the case of a single crystal, the resulting spectrum (for each nucleus) is composed of two resonance lines separated by an amount proportional to $\gamma_{n1}\gamma_{n2}/r_{12}^3$, known as Pake doublet [4]. In an isotropic liquid, the complete averaging of the term $3\cos^2\theta_{12} - 1$ leads to zero contribution due to the direct dipolar interaction for the NMR spectrum in such case (Problem P2.8). On the other hand, in a polycrystalline solid or powder, the superposition of the contributions of all crystallites gives rise to a broad powder spectrum, with singularities resembling the Pake doublet found in monocrystals [10]. The homonuclear dipolar interaction involves further terms [4], but the qualitative features described above remain the same.

2.7.3 J-coupling

The J-coupling (also called *indirect* or *scalar* coupling) is also an interaction between the nuclear magnetic dipole moments of neighbor nuclei, but in this case the interaction is not direct, being mediated by the electron cloud involved in the chemical bonds between the corresponding atoms. The main practical distinction between J- and direct couplings resides on the fact that the former possesses an isotropic part that survives to the random molecular motion in isotropic substances, being therefore easily observable in NMR spectra of liquids. As consequence, besides the chemical shift that allows the identification of each chemical environment, the spectrum shows a further splitting that allows the assessment of the details of the chemical bonds connecting neighbor atoms, usually inside the same molecule [5]. In solids, this interaction is in general overwhelmed by the more intense direct dipolar interaction, not being thus normally observed (some examples of J-coupling in solids can be found in [10]).

The Hamiltonian describing the J-coupling between two nuclear spins $\mathbf{I}_1$ and $\mathbf{I}_2$ is generally written in the form:

$$H_J = 2\pi\hbar \mathbf{I}_1 \cdot \tilde{J} \cdot \mathbf{I}_2 \quad (2.7.13)$$

2.7. Nuclear spin interactions

The tensor $\tilde{J}$ possesses non-vanishing trace, which gives rise to the isotropic contribution mentioned above in the NMR spectra of liquids. In terms of the PAS components of the tensor $\tilde{J}$ the following definition is used:

$$J = (J_{XX} + J_{YY} + J_{ZZ})/3 \tag{2.7.14}$$

The secular form of the Hamiltonian $\mathcal{H}_J$ in liquids for the simple heteronuclear case is then [5]:

$$H_J = 2\pi\hbar J I_{1z} I_{2z} \tag{2.7.15}$$

When this Hamiltonian is taken as perturbation to the Zeeman Hamiltonian (including isotropic chemical shift effects), it is found that each line is split in a multiplet that depends on the number of identical nuclei coupled by the same constant J as well as the spin of each of these nuclei. For two distinct spin 1/2 nuclei, for example, each line is split into two other lines separated by an amount (in frequency units) equal to the magnitude of J. The parameter J can be positive or negative, which means the coupling can favor either an antiparallel or parallel alignment of the nuclear spins, respectively. The physical meaning of the J-coupling is related to second-order effects in the interaction between the nucleus and the spin magnetic moment of the electrons, under the influence of the external magnetic field. These electron spin-dependent effects vanish to first order in diamagnetic materials, but through second-order perturbation theory it can be shown that there is a non-vanishing coupling manifested as an interaction between the nuclear spins [4].

2.7.4 Quadrupolar coupling

As mentioned previously, all nuclei with spin $I > 1/2$ possess non-spherical charge distribution and so they are subjected to electrostatic interaction with neighbor electrons and ions. Through its electric quadrupole moment, the nucleus interacts with the electric field gradient (EFG) at the nuclear position. Although such interaction is of electrostatic origin, it influences the spatial orientation of the nucleus and thus it must be related to the nuclear spin coordinates. A simple example of a quadrupolar electrostatic coupling is depicted in Figure 2.5 and an instructive classic calculation of the orientational dependence of the interaction energy in this case is proposed in Problem P2.9.

The EFG is described by a second rank tensor $\tilde{V}$ whose components correspond to the second derivative of the scalar electric potential V evaluated at the nucleus:

$$V_{\alpha\beta} = \left(\frac{\partial^2 V}{\partial x_\alpha \partial x_\beta}\right)_0 \tag{2.7.16}$$

where the subscript indicates a derivative calculated at the nuclear position, taken as the origin, and x_α and x_β correspond to Cartesian coordinates. By construction, the EFG tensor is symmetric and traceless [4]. In its PAS, therefore, there are only two independent parameters characterizing the tensor $\tilde{V}$. Usually these are chosen to be the parameters eq and η defined below:

$$eq = V_{ZZ} \tag{2.7.17}$$

$$\eta = \frac{V_{XX} - V_{YY}}{V_{ZZ}} \tag{2.7.18}$$

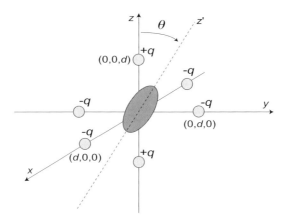

Figure 2.5 A non-spherically symmetric charge distribution (like a quadrupolar nucleus) interacting with an EFG produced by nearby point charges (like ions in a crystal lattice). The resulting coupling energy depends on the angle θ that specifies the orientation of the distribution with respect to the EFG.

where the upper case letters indicate PAS coordinates and the Cartesian coordinates in this system are ordered following the convention $|V_{ZZ}| \geqslant |V_{XX}| \geqslant |V_{YY}|$. The parameter eq, sometimes simply named the *electric field gradient*, is a measure of the maximum strength of the second derivative of the potential V. The parameter η is called the *asymmetry parameter* of the EFG tensor, giving thus information about the deviation of the EFG from axial symmetry about the Z-axis.

It can be generally shown that the quadrupolar coupling is described by the following Hamiltonian, written in the PAS associated with the EFG tensor [4]:

$$\mathcal{H}_Q = \frac{e^2qQ}{4I(2I-1)}\big[3I_Z^2 - \mathbf{I}^2 + \eta\big(I_X^2 - I_Y^2\big)\big] \tag{2.7.19}$$

For application of the perturbation theory, it is necessary to transform this expression to the laboratory frame, where the dominant Hamiltonian $\mathcal{H}_Z$ is proportional do the spin operator I_z. When such transformation is conducted, using Wigner matrices and Euler angles [10], the resulting form for $\mathcal{H}_Q$ in the laboratory frame, assuming axial symmetry for the EFG tensor ($\eta = 0$) and using the secular approximation to keep only terms that commute with $\mathcal{H}_Z$, is:

$$\mathcal{H}_Q = \frac{e^2qQ}{8I(2I-1)}(3\cos^2\theta - 1)\big(3I_z^2 - \mathbf{I}^2\big) \tag{2.7.20}$$

where θ is the angle between the Z-axis of the PAS of the EFG tensor and the external magnetic field $\mathbf{B}_0$ that defines the z-axis in the laboratory-frame.

This Hamiltonian is usually written in terms of a parameter with frequency dimensions, named ω_Q, defined as:

$$\omega_Q = \frac{e^2qQ}{8I(2I-1)\hbar}(3\cos^2\theta - 1) \tag{2.7.21}$$

2.7. Nuclear spin interactions

so that:

$$\mathcal{H}_Q = \hbar\omega_Q(3I_z^2 - \mathbf{I}^2) \qquad (2.7.22)$$

It is important to mention here that this simple form is appropriate only in situations where first-order perturbation theory can be safely applied. If the magnitude of the quadrupolar coupling is large, as it occurs commonly in solids with non-cubic symmetry, then it is necessary to use second order corrections. In such cases other terms in the expression for $\mathcal{H}_Q$ with different angular dependence must be considered and the complete form of the Hamiltonian is much more complicated than the expression given above.

In the cases where first-order calculation is appropriate, the effects of the Hamiltonian $\mathcal{H}_Q$ given in Equation (2.7.22) can be readily computed as both operators I_z and $\mathbf{I}^2$ commute with the main Hamiltonian H_Q. The result is that the energy levels are not equally spaced as they were in the case of the Zeeman interaction only. Otherwise, the energy depends on the quantum number m and the parameter ω_Q in the form:

$$E_m = -m\hbar\omega_L + \hbar\omega_Q[3m^2 - I(I+1)] \qquad (2.7.23)$$

The energy difference between adjacent levels gives the frequency of observable transitions, which means that there are $2I$ transitions with different frequencies. The situation is illustrated in Figure 2.6 for the case $I = 3/2$. It is important to observe that the energy values of symmetric levels ($\pm m$) are shifted by the same amount as consequence of the quadrupolar coupling. Therefore, the central transition ($1/2 \leftrightarrow -1/2$) frequency is not affected by the quadrupolar coupling to first-order, which means that second-order contributions are specially important for calculation of the frequency corresponding to the central transition in half-integer spin nuclei [10].

The frequency separation between the different peaks depends on the parameter ω_Q, which by its turn depends on the orientation of the EFG with respect to B_0. In a single crystal there is only one value of θ and therefore all peaks are distinctly observed. In a liquid crystal sample, there occurs a partial spatial averaging of the term $(3\cos^2\theta - 1)$, but again all peaks are usually observed. For an isotropic liquid sample, on the other hand, the complete averaging to zero of such term indicates that the quadrupolar coupling cannot be observed directly in the NMR spectrum (although it still has important effects on the nuclear spin relaxation [5]). For a polycrystalline material, the superposition of all orientations possible for the collection of crystallites leads to a typically broadened spectrum with sharp singularities, similar to that described for the chemical shift case. The study of the methods to either remove or, more interestingly, to take advantage of typically quadrupolar lineshapes in NMR spectra for attaining structural information on solid substances is one of the more rapidly growing research areas of solid state NMR [10].

2.7.5 General form of the internal Hamiltonians

All internal nuclear spin Hamiltonians described above present common features that can be explored for a theoretical description of their effects onto the appearance of a NMR

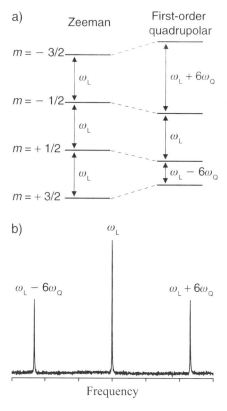

Figure 2.6 (a) Illustration of the energy-level splitting due to Zeeman plus first-order quadrupolar coupling in the case of a spin 3/2 nucleus. (b) A typical NMR spectrum corresponding to the energy levels given in part (a). Note that this type of spectrum with sharp peaks is observed only in oriented samples, such as single crystals or liquid crystal samples.

spectrum. By appropriate manipulations, it can be shown that all these Hamiltonians can be written in the form [10,17]:

$$\mathcal{H}_{int} = C \sum_{\alpha,\beta=1}^{3} I_\alpha R_{\alpha\beta} A_\beta \qquad (2.7.24)$$

In this expression, C is a constant specific for each interaction, depending on gyromagnetic ratios and nuclear electric quadrupole moments. I_α is a Cartesian component of the nuclear spin, with the sum extended to all three Cartesian coordinates. $R_{\alpha\beta}$ represents the components of a 3×3 Cartesian tensor of second rank that specifies the detailed nature of each interaction, some examples of which appeared in the expressions given previously ($\tilde{\sigma}$, $\hat{V}$, and $\tilde{J}$). Finally, A_β is a Cartesian component of a vector that can be the same nuclear spin vector (quadrupolar interaction), another nuclear spin vector (dipolar interaction or J-coupling), or the external magnetic field (chemical shift interaction). Some of the tensors represented by $R_{\alpha\beta}$ are traceless (cases of dipolar and quadrupolar interactions), whereas others (cases of chemical shift and J-coupling) posses non-vanishing trace,

2.8. NMR of two coupled spins 1/2

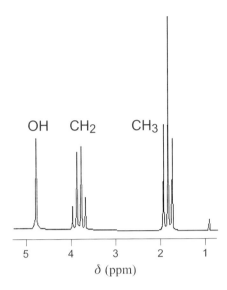

Figure 2.7 A typical ^{1}H NMR spectrum of ethanol (CH$_3$–CH$_2$–OH), showing the effects of chemical shift and J-coupling in liquid substances. The three well separated clusters of peaks refer to protons in each of the three different chemical groups in the molecule (CH$_3$, CH$_2$, and OH), evidencing the effect of chemical shift and its order of magnitude. The multiplet structures within the CH$_3$ and CH$_2$ groups of peaks are due to J-coupling between protons inside the same molecule.

which leads in the latter case to the appearance of isotropic shifts in NMR spectra in liquid substances.

With this general common form, the behavior of all internal Hamiltonians under rotations can be nicely described using irreducible spherical tensorial representations for the tensor and vector products given in Equation (2.7.24) [4,17]. This formalism allows the complete understanding of the behavior and evolution of all these Hamiltonians during complex experimental manipulations carried out in high-resolution NMR, all of which involves rotations performed either spatially or in the spin space.

To summarize the main contents of this section, we show in Figure 2.7 the typical appearance of NMR spectra in liquid substances, evidencing the role played by chemical shift and J-coupling interactions, which, as discussed above, are the only ones that survive to the random molecular motion in liquids. On the other hand, Figure 2.8 exhibits some typical powder patterns associated with the anisotropic nature of these interactions, as commonly found in solid-state NMR spectra of polycrystalline samples.

2.8 NMR OF TWO COUPLED SPINS 1/2

As described in Section 2.6, the dynamics of an ensemble of non-interacting nuclei with spin 1/2 is described with use of the density matrix approach in a two-fold matrix space, which allows the understanding of the behavior of such collection of nuclei when submitted to static magnetic field, RF pulses, free evolution times, and so on. This description is complete only if one has a material composed of identical units (molecules, ionic groups, etc.) with just one chemically distinct NMR-active nucleus in each unit and with all inter-

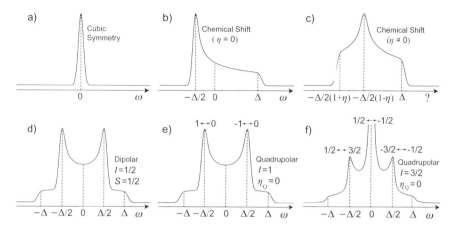

Figure 2.8 Typical powder-pattern spectra observed in polycrystalline solids: Chemical Shift under (a) cubic, (b) axial, and (c) non-axial symmetries; (d) Dipolar interaction between two spins 1/2 (I and S); Quadrupolar interaction for spins e) 1 and f) 3/2, considering an EFG with axial symmetry. The zero of the frequency scale in each case corresponds to the frequency associated with isotropic average (as occurring in liquids). The parameter Δ depends on the anisotropy of the tensors describing each of the interactions.

nal nuclear spin interactions vanishing for this nucleus (except for the isotropical chemical shift, which only leads to a shifted resonance frequency). This occurs in the case of chemically simple substances and/or when some kind of averaging process or symmetry requirement is effective in vanishing those interactions. Some specific examples are water (one ^{1}H site), tetra-methyl-silane (Si(CH$_3$)$_4$, with one ^{13}C, one ^{29}Si, and one ^{1}H site), cubic sodium chloride (NaCl, with one site for the quadrupolar nucleus ^{23}Na, but vanishing quadrupolar interaction because of the cubic symmetry), and others, which give so simple NMR spectra that they are usually employed as references for establishing chemical shift scales. Even in the liquid state, most substances present both chemical shifts and J-couplings, as introduced in Section 2.7, so that the nuclei feel intramolecular interactions with neighbours and cannot anymore be described as isolated entities. The next simplest case is then the study of two J-coupled chemically distinct nuclei in a liquid substance. This description could be applicable for example to molecules like as chloroform (CHCl$_3$), where there are only two chemically and magnetically distinct nuclei (^{13}C and ^{1}H) interacting through the J-coupling. A detailed description of the concepts of chemical and magnetic equivalence, the motional processes that can lead to vanishing of J-couplings, and other chemical aspects of the problem can be found in references [4,5]. Our aim in this section is to give the basic description of the dynamics of two coupled nuclei with spin 1/2, involving now a four-fold matrix space, and thus generalizing the description of Section 2.6. The extension to larger numbers of coupled spins can then be immediately guessed, although the algebraic aspects would be much more complicated.

Let us consider two J-coupled nuclear spins denoted as I and S. The simplest case to examine is that of heteronuclear interaction (^{13}C and ^{1}H nuclei in the chloroform molecule, for example), so that the magnitude of the J-coupling is much smaller than the difference between the resonant frequencies of the two nuclei. This is named the "AX system" in the

2.8. NMR of two coupled spins 1/2

Chemistry literature.[4] The secular Hamiltonian in the heteronuclear case is then given by (see Equation (2.7.15)):

$$\mathcal{H} = -\hbar\omega_I I_z - \hbar\omega_S S_z + 2\pi\hbar J I_z S_z \tag{2.8.1}$$

The resonant frequencies ω_I and ω_S include the effects of isotropical chemical shifts. The eigenstates and eigenvalues of this Hamiltonian are easily obtained from the matrix representations of each operator I_z and S_z. As these operators commute and act on different vector spaces [2], the eigenstates of the Hamiltonian above are simply given by the direct product of the eigenstates of the operators I_z and S_z. The eigenstates are then written as $|m_I, m_S\rangle$, where $m_I = \pm 1/2$, $m_S = \pm 1/2$ are the eigenvalues of I_z and S_z, respectively. The eigenvalues of the Hamiltonian, i.e., the energy levels associated with the Hamiltonian (2.8.1), are directly evaluated from the acting of the operators I_z and S_z on the $|m_I, m_S\rangle$ basis, leading to:

$$|+1/2, +1/2\rangle: \quad E_{+1/2,+1/2} = \hbar\left(-\frac{\omega_I}{2} - \frac{\omega_S}{2} + \frac{\pi J}{2}\right) \tag{2.8.2}$$

$$|+1/2, -1/2\rangle: \quad E_{+1/2,-1/2} = \hbar\left(-\frac{\omega_I}{2} + \frac{\omega_S}{2} - \frac{\pi J}{2}\right) \tag{2.8.3}$$

$$|-1/2, +1/2\rangle: \quad E_{-1/2,+1/2} = \hbar\left(\frac{\omega_I}{2} - \frac{\omega_S}{2} - \frac{\pi J}{2}\right) \tag{2.8.4}$$

$$|-1/2, -1/2\rangle: \quad E_{-1/2,-1/2} = \hbar\left(\frac{\omega_I}{2} + \frac{\omega_S}{2} + \frac{\pi J}{2}\right) \tag{2.8.5}$$

In the case of positive values for J, γ_I, and γ_S, and if $\omega_I > \omega_S$, then the levels above are arranged from the top to the bottom in order of increasing energy. Transitions between these levels are allowed (i.e., give rise to NMR signal) according to the selection rules $\Delta m_{I,S} = \pm 1$. Therefore, there are four peaks in the full NMR spectrum, at the frequencies $\omega_I \pm \pi J$ and $\omega_S \pm \pi J$. The separation within each doublet is, in angular frequency units, equal to $2\pi J$, or, in frequency units, equal to J.

It is clear then that one needs to describe all relevant operators (related to the spin components) using their matrix representations in the four-fold vector space generated by the state vectors $|m_I, m_S\rangle$. These matrices can be constructed by evaluating each matrix element or they can be built by the direct tensorial product of the corresponding 2×2 matrices that describe the dynamics of the separate spin 1/2 systems [5,12]. The direct tensorial product of two $n \times n$ matrices A and B lead to a $n^2 \times n^2$ matrix $C = A \otimes B$ whose elements are

---

[4] Two nuclei of the same chemical species, with a J-coupling comparable to the difference between their chemically shifted resonant frequencies are known as "AB system" and the treatment is a little bit more involved [5, 12].

related to the products between the elements of A and B by:

$$C_{1;1} = A_{1;1}B_{1;1}; \ldots C_{1;n} = A_{1;1}B_{1;n};$$
$$C_{1;n+1} = A_{1;2}B_{1;1}; \ldots C_{1;n^2} = A_{1n}B_{1n} \tag{2.8.6}$$
$$C_{2;1} = A_{1;1}B_{2;1}; \ldots C_{2;n} = A_{1;1}B_{2;n};$$
$$C_{2;n+1} = A_{1;2}B_{2;1}; \ldots C_{2;n^2} = A_{1n}B_{2n} \tag{2.8.7}$$
$$(\ldots) \tag{2.8.8}$$

It is easier then to think of C as being a "matrix of matrices". For instance, the matrix representation of the operator I_z is given by:

$$I_z = I_z^{(2\times 2)} \otimes \mathbf{1}^{(2\times 2)} = \frac{1}{2}\begin{bmatrix} 1 & 0 \\ 0 & -1 \end{bmatrix} \otimes \begin{bmatrix} 1 & 0 \\ 0 & 1 \end{bmatrix}$$

$$= \frac{1}{2}\begin{bmatrix} 1\begin{bmatrix} 1 & 0 \\ 0 & 1 \end{bmatrix} & 0\begin{bmatrix} 1 & 0 \\ 0 & 1 \end{bmatrix} \\ 0\begin{bmatrix} 1 & 0 \\ 0 & 1 \end{bmatrix} & -1\begin{bmatrix} 1 & 0 \\ 0 & 1 \end{bmatrix} \end{bmatrix} = \frac{1}{2}\begin{bmatrix} 1 & 0 & 0 & 0 \\ 0 & 1 & 0 & 0 \\ 0 & 0 & -1 & 0 \\ 0 & 0 & 0 & -1 \end{bmatrix} \tag{2.8.9}$$

where $I_z^{(2\times 2)}$ is the matrix given in Equation (2.6.1) (which represents the operator I_z in the 2×2 space of a single spin 1/2) and $\mathbf{1}^{(2\times 2)}$ is the 2×2 identity matrix.
On the other hand, the matrix representation for S_z is:

$$S_z = \mathbf{1}^{(2\times 2)} \otimes S_z^{(2\times 2)} = \begin{bmatrix} 1 & 0 \\ 0 & 1 \end{bmatrix} \otimes \frac{1}{2}\begin{bmatrix} 1 & 0 \\ 0 & -1 \end{bmatrix}$$

$$= \frac{1}{2}\begin{bmatrix} 1\begin{bmatrix} 1 & 0 \\ 0 & -1 \end{bmatrix} & 0\begin{bmatrix} 1 & 0 \\ 0 & -1 \end{bmatrix} \\ 0\begin{bmatrix} 1 & 0 \\ 0 & -1 \end{bmatrix} & 1\begin{bmatrix} 1 & 0 \\ 0 & -1 \end{bmatrix} \end{bmatrix} = \frac{1}{2}\begin{bmatrix} 1 & 0 & 0 & 0 \\ 0 & -1 & 0 & 0 \\ 0 & 0 & 1 & 0 \\ 0 & 0 & 0 & -1 \end{bmatrix} \tag{2.8.10}$$

The density operator of the system can be constructed using the results of Section 2.5 (see Equation (2.5.7)). At thermal equilibrium and in the high temperature limit, the density matrix obtained from the Hamiltonian given in Equation (2.8.1) is:

$$\rho_0 = \frac{1}{Z}e^{-\mathcal{H}/k_B T} \cong \frac{1}{4}\mathbf{1} + \frac{1}{4}\frac{\hbar\omega_I}{k_B T}I_z + \frac{1}{4}\frac{\hbar\omega_S}{k_B T}S_z \tag{2.8.11}$$

The small terms involving J were neglected in this expression. The deviation density matrix is:

2.8. NMR of two coupled spins 1/2

$$\Delta\rho_0 = \frac{1}{4}\frac{\hbar\omega_I}{k_B T}I_z + \frac{1}{4}\frac{\hbar\omega_S}{k_B T}S_z$$

$$= \frac{1}{8}\begin{bmatrix} \frac{\hbar\omega_I}{k_B T}+\frac{\hbar\omega_S}{k_B T} & 0 & 0 & 0 \\ 0 & \frac{\hbar\omega_I}{k_B T}-\frac{\hbar\omega_S}{k_B T} & 0 & 0 \\ 0 & 0 & -\frac{\hbar\omega_I}{k_B T}+\frac{\hbar\omega_S}{k_B T} & 0 \\ 0 & 0 & 0 & -\frac{\hbar\omega_I}{k_B T}-\frac{\hbar\omega_S}{k_B T} \end{bmatrix} \quad (2.8.12)$$

It is clear in this matrix how the differences in populations between the several (m_I, m_S) levels appear in the matrix density. The differences involving the term $\frac{\hbar\omega_I}{k_B T}$ between the pairs $(\Delta\rho_{011}, \Delta\rho_{022})$ and $(\Delta\rho_{033}, \Delta\rho_{044})$ reflect the population differences related to spin I, whereas the differences involving the term $\frac{\hbar\omega_S}{k_B T}$ between the pairs $(\Delta\rho_{011}, \Delta\rho_{033})$ and $(\Delta\rho_{022}, \Delta\rho_{044})$ reflect population differences related to spin S.

The concept of the rotating frame is used here again to describe the effects of RF pulses. But now we have actually *two* independent rotating frames, each one related to either frequency ω_I or ω_S. The use of two independent rotating frames in this theoretical description is related in practice to the use of *double-resonance* probes for performing actual NMR experiments, i.e., probes that can be tuned independently to two very different frequencies. This point will be further detailed in Section 2.12. Each rotating frame precesses about the common z-axis with frequency Ω_I or Ω_S. The effective Hamiltonian for a RF pulse applied along the x-direction to the spin I in the double rotating frame is therefore:

$$\mathcal{H}_{\text{eff}}^{(I_x)} = -\hbar(\omega_I - \Omega_I)I_z - \hbar(\omega_S - \Omega_S)S_z + 2\pi\hbar J I_z S_z - \gamma_I \hbar B_1^{(I)} I_x \quad (2.8.13)$$

The expression corresponding to a RF pulse applied to the spin S is analogous. Usually, for strong RF pulses applied close to resonance, the term involving the J-coupling as well as those corresponding to frequency offsets are small compared to the term involving the RF field, so that the effective Hamiltonian during the pulse can be approximated by $-\gamma_I \hbar B_1^{(I)} I_x$ or $-\gamma_S \hbar B_1^{(S)} S_x$. Therefore, the effects of RF pulses are simply described as rotations around the transverse axis defined by the phase of the pulse. Also, as the resonance frequencies in the heteronuclear case are far apart one from another, each pulse applied at the frequency of one spin has no effect on the other spin, and therefore it is completely disregarded when computing the time evolutions of the respective density matrices. For example, the operators below describe, respectively, the effects of $(\pi/2)_x$ pulses applied to the I and S spins:[5]

$$R_x^{(I)}(\pi/2) = R_x^{(2\times 2)}(\pi/2) \otimes \mathbf{1}^{(2\times 2)} = \frac{1}{\sqrt{2}}\begin{bmatrix} 1 & 0 & -i & 0 \\ 0 & 1 & 0 & -i \\ -i & 0 & 1 & 0 \\ 0 & -i & 0 & 1 \end{bmatrix} \quad (2.8.14)$$

$$R_x^{(S)}(\pi/2) = \mathbf{1}^{(2\times 2)} \otimes R_x^{(2\times 2)}(\pi/2) = \frac{1}{\sqrt{2}}\begin{bmatrix} 1 & -i & 0 & 0 \\ -i & 1 & 0 & 0 \\ 0 & 0 & 1 & -i \\ 0 & 0 & -i & 1 \end{bmatrix} \quad (2.8.15)$$

---

[5] It is important to note that in the case of strong coupling between identical nuclei, the RF pulses act simultaneously on both spins and the matrices describing such pulses are completely different from the above ones [5].

After such $(\pi/2)_x$ pulses, the deviation density matrix can be obtained using either the operators $R_x^{(I)}(\pi/2)$ or $R_x^{(S)}(\pi/2)$ and following the same reasoning used in Section 2.6 for non-interacting spins 1/2 (see Equation (2.6.12), for example). Starting from $\Delta\rho_0$ given in Equation (2.8.12), after a $(\pi/2)_x$ applied to spin I one has $\Delta\rho(t_p) \sim -I_y$, whereas for a $(\pi/2)_x$ applied to spin S, $\Delta\rho(t_p) \sim -S_y$.

Next, it is necessary to consider the evolution of the density matrix during the periods between the RF pulses. The time evolution occurs under action of the frequency offsets (related to chemical shifts of each spin), which is analogous to the case of non-interacting spin 1/2 systems, but now there is also evolution under the J-coupling interaction. Using the Hamiltonian given in (2.8.1) (properly transformed to the rotating frame) and the matrix representations for I_z and S_z given in Equations (2.8.9) and (2.8.10), respectively, one arrives at the evolution operator evaluated a time τ after the end of application of the RF pulse:

$$U = e^{-(i/\hbar)H\tau}$$

$$= \begin{bmatrix} e^{i(\omega_I - \Omega_I)\tau/2} \\ \times e^{i(\omega_S - \Omega_S)\tau/2} e^{-i\pi J\tau/2} & 0 & 0 & 0 \\ 0 & \begin{array}{c} e^{i(\omega_I - \Omega_I)\tau/2} \\ \times e^{-i(\omega_S - \Omega_S)\tau/2} e^{i\pi J\tau/2} \end{array} & 0 & 0 \\ 0 & 0 & \begin{array}{c} e^{-i(\omega_I - \Omega_I)\tau/2} \\ \times e^{i(\omega_S - \Omega_S)\tau/2} e^{i\pi J\tau/2} \end{array} & 0 \\ 0 & 0 & 0 & \begin{array}{c} e^{-i(\omega_I - \Omega_I)\tau/2} \\ \times e^{-i(\omega_S - \Omega_S)\tau/2} e^{-i\pi J\tau/2} \end{array} \end{bmatrix}$$

(2.8.16)

The first two exponentials in each element can be interpreted as rotations around the z-axis at the frequency offsets $\pm(\omega_I - \Omega_I)$ and $\pm(\omega_S - \Omega_S)$, where the plus or minus sign depends on the m_I and m_S values. As for the $e^{\pm i\pi J\tau/2}$ terms, they also describe precessions around the z-axis, but now with a frequency that depends, for each spin, on the state of the *other* spin in the coupled pair. For the elements 1; 1 and 4; 4 (which correspond to the $|+1/2, +1/2\rangle$ and $|-1/2, -1/2\rangle$ states, respectively) one has a parallel coupling between the spins I and S, which leads to a change in the precession frequency by the amount $-\pi J$ both for I and S spins. On the other hand, the elements 2; 2 and 3; 3 (corresponding to $|+1/2, -1/2\rangle$ and $|-1/2, +1/2\rangle$ states, respectively) involve an antiparallel coupling between the spins I and S, leading now to a change in precession frequency by the amount $+\pi J$ both for I and S spins. It is clear then that we have now a coupled evolution of the spins: the acting (by application of RF pulses and allowing some time evolution) on a given spin will produce a measurable effect on the other spin. Therefore, determining how a nuclear spin pair behaves under a given RF pulse sequence allows the assessment of the magnitudes and signs of the scalar coupling constants, and thus of the possible interconnectivity between the involved atoms. This is the basis of many multiple resonance experiments performed in modern NMR applications aiming structural elucidation of complex molecules.

Furthermore, the existence of this internal coupling mechanism opens the possibility of construction of logical gates using the spins as logical indicators. This feature is in the essence of NMR applications in Quantum Computing and will be detailed in later

2.8. NMR of two coupled spins 1/2

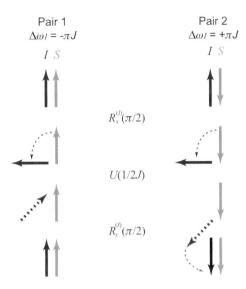

Figure 2.9 Conditional inversion of a spin in a weakly coupled spin 1/2 system, following the pulse sequence given in (2.8.17).

Chapters. Now we just present a simple example of this coupled evolution for a basic pulse sequence that allows a vector model interpretation of the evolution, without carrying out the detailed matrix calculations. This experiment is related to a method known in NMR Chemistry as INEPT (Insensitive Nuclei Enhanced via Population Transfer) [5,43,12]. The pulse sequence is as follows:

$$R_x^{(I)}(\pi/2) \quad U(1/2J) \quad R_y^{(I)}(\pi/2) \tag{2.8.17}$$

The $\pi/2$ pulses are applied only to the I spin, and the signal is also recorded for that spin. The $U(1/2J)$ operator indicates an evolution under the J-coupling for the interval $\tau = 1/2J$. Let us consider two different spin pairs, one with parallel (let us call it pair 1) and the other (pair 2) with antiparallel orientation of the I and S spins. In a real sample these would correspond to different molecules, with the nuclei coupled in a different relative orientation in each one. The vector model description for the time evolution of the pairs 1 and 2 is shown in Figure 2.9 in the left- and right-hand sides, respectively. Considering a rotating frame with a frequency exactly midway between the doublet peaks for spin I, which means that $\Omega_I = \omega_I$, then the precession frequency is $\Delta\omega_I = -\pi J$ for pair 1 and $\Delta\omega_I = +\pi J$ for pair 2. After the $(\pi/2)_x$ pulse, the I spins from both pairs 1 and 2 are driven to the transverse plane into the $-y$ direction. During the subsequent time they precess in opposite directions; after the interval $\tau = 1/2J$, the precession angle reaches the absolute value of $\pi/2$ and the I spin magnetization vectors point either to the $-x$ (for pair 1) or to the $+x$ (for pair 2) direction. Now the $(\pi/2)_y$ pulse is applied and drives the I spins belonging to each pair to opposite directions. For pair 1, the spin ends in the $+z$ direction, whereas it ends in the $-z$ direction for pair 2. Therefore, this simple pulse sequence leads to a conditional inversion of spin I and thus it is capable of distinguishing the signals arising from pairs containing nuclei coupled in different ways.

With the evolution and rotation operators, all pulse sequences describing experiments in weakly coupled AX systems can be properly followed. To determine the state of the system after a given sequence of events involving pulses and free evolution periods, it is usual to expand the density matrix in a series of operators constructed from the products involving the spin operators I_x, I_y, I_z, S_x, S_y, and S_z, as well as the identity matrix. These operators are known as *product operators*. In the case of the AX system, there are 16 product operators, some examples of which being I_x, S_z, $I_y S_y$, $I_z S_x$, and so on. Each particular product has a physical meaning related to the populations and coherences of the coupled spin system. Each coherence is classified according to the difference in the quantum number of the total angular momentum operator of the system ($I_z + S_z$) between the states connected by the coherence. In this way, one can have zero, single, or double quantum coherences in the AX system. For example, the operators I_x and I_y represent single quantum coherences for spin I, whereas S_x and S_y are related to single quantum coherences for spin S. The occurrence of such single quantum coherences points to observable transverse magnetization for the respective spin species. The operator $I_x S_z$ is known as antiphase x-magnetization for spin I, whereas $I_z S_z$ represents an antiphase z-magnetization for both spins. The operator $I_x S_x$ contains a combination of zero and double quantum coherences, which are not directly observable experimentally but play an important role in many NMR experiments because they evolve in time along the pulse sequence and can be converted into observable single quantum coherences by suitable RF pulses [4,5].

2.9 NMR OF QUADRUPOLAR NUCLEI

In this section we will give the basics of the dynamics of nuclei with $I > 1/2$ experiencing a non-vanishing quadrupolar interaction with an electric-field gradient (EFG). As mentioned previously, this situation is quite common in crystalline solids with non-cubic symmetry and also in liquid crystals. In isotropic liquids, on the other hand, the quadrupolar interaction is, as mentioned before, averaged out by the rapid motion of the molecules, leading to vanishing quadrupolar effects. Most examples will be given taking $I = 3/2$ as a specific case, since this is of great importance both in quantum computation schemes and also in many solid-state NMR investigations. The extension to other spin values is mostly straightforward, although more laborious.

For an ensemble of nuclei with spin $I > 1/2$ experiencing no quadrupolar interaction, such as in an isotropic liquid or a crystal with cubic symmetry, the equations describing the time evolution of the density matrix under action of static and RF magnetic fields are a natural extension of the $I = 1/2$ case. The Hamiltonian contains only the Zeeman and RF terms; the effects of RF pulses are described by rotations of the spin operators around the transverse axes in the rotating frame, whereas free evolution corresponds to rotations around the z-axis.

To be more specific, let us detail the $I = 3/2$ case. The matrix representations of the nuclear spin operators in the $|m\rangle$ basis are given by:

$$I_x = \frac{1}{2}\begin{bmatrix} 0 & \sqrt{3} & 0 & 0 \\ \sqrt{3} & 0 & 2 & 0 \\ 0 & 2 & 0 & \sqrt{3} \\ 0 & 0 & \sqrt{3} & 0 \end{bmatrix} \quad I_y = \frac{i}{2}\begin{bmatrix} 0 & -\sqrt{3} & 0 & 0 \\ \sqrt{3} & 0 & -2 & 0 \\ 0 & 2 & 0 & -\sqrt{3} \\ 0 & 0 & \sqrt{3} & 0 \end{bmatrix}$$

2.9. NMR of quadrupolar nuclei

$$I_z = \frac{1}{2}\begin{bmatrix} 3 & 0 & 0 & 0 \\ 0 & 1 & 0 & 0 \\ 0 & 0 & -1 & 0 \\ 0 & 0 & 0 & -3 \end{bmatrix} \tag{2.9.1}$$

The thermal equilibrium density matrix is:

$$\rho_0 = \frac{1}{4}\mathbf{1} + \frac{1}{4}\frac{\hbar\omega_L}{k_B T}I_z = \begin{bmatrix} \frac{1}{4}+\frac{3\hbar\omega_L}{8k_B T} & 0 & 0 & 0 \\ 0 & \frac{1}{4}+\frac{\hbar\omega_L}{8k_B T} & 0 & 0 \\ 0 & 0 & \frac{1}{4}-\frac{\hbar\omega_L}{8k_B T} & 0 \\ 0 & 0 & 0 & \frac{1}{4}-\frac{3\hbar\omega_L}{8k_B T} \end{bmatrix} \tag{2.9.2}$$

And the deviation density matrix corresponding to an ensemble of identical nuclei at thermal equilibrium is:

$$\Delta\rho_0 = \frac{1}{4}\frac{\hbar\omega_L}{k_B T}I_z = \frac{1}{4}\begin{bmatrix} \frac{3\hbar\omega_L}{2k_B T} & 0 & 0 & 0 \\ 0 & \frac{\hbar\omega_L}{2k_B T} & 0 & 0 \\ 0 & 0 & -\frac{\hbar\omega_L}{2k_B T} & 0 \\ 0 & 0 & 0 & -\frac{3\hbar\omega_L}{2k_B T} \end{bmatrix} \tag{2.9.3}$$

The effects of RF pulses are analysed in the rotating frame, precessing around the z-axis with frequency Ω. The effective Hamiltonian for a RF pulse applied along the x-direction in this case is given simply by:

$$\mathcal{H}_{eff} = -\hbar(\omega_L - \Omega)I_z - \gamma_n\hbar B_1 I_x \tag{2.9.4}$$

On resonance, i.e., when $\Omega = \omega_L$, we have $\mathcal{H}_{eff} = -\hbar\omega_1 I_x$ and the time-evolution is accomplished by the operator $U = e^{-(i/\hbar)\mathcal{H}_{eff} t_p} = e^{i\omega_1 t_p I_x} = R_x(-\theta_p)$, associated with left-hand rotations around the x-axis with nutation frequency $\omega_1 = \gamma_n B_1$. The expressions for the rotation operators can be obtained from the evaluation of the exponential series expansion:

$$R_x(\theta_p) = e^{-i\theta_p I_x} = \sum_{k=0}^{\infty}(-i)^k \frac{\theta_p^k}{k!}I_x^k \tag{2.9.5}$$

This series can be evaluated either by numerical methods or by collecting the partial series of common spin operators (such as I_x, I_x^2, and so on) [5,12]. For the present case, $I = 3/2$, we have for a $\pi/2$ pulse, for example:

$$R_x(\pi/2) = \frac{1}{2\sqrt{2}}\begin{bmatrix} 1 & -i\sqrt{3} & -\sqrt{3} & i \\ -i\sqrt{3} & -1 & -i & -\sqrt{3} \\ -\sqrt{3} & -i & -1 & -i\sqrt{3} \\ i & -\sqrt{3} & -i\sqrt{3} & 1 \end{bmatrix} \tag{2.9.6}$$

It can be easily shown from direct matrix products that $R_x(-\pi/2)I_z R_x(\pi/2) = I_y$. This result shows thus that, starting from thermal equilibrium, where the deviation density matrix is given by Equation (2.9.3), one can generate single-quantum coherences in the density operator by applying a RF pulse with proper phase and duration. Therefore, a transverse magnetization can be excited and detected in the same way as the spin 1/2 case. These considerations show then that, in the absence of quadrupolar interaction, the action of RF pulses in the spin 3/2 case is completely analogous to the case of spin 1/2.

When there is a non-vanishing quadrupolar coupling, the situation is much more complicated. Now the Hamiltonian in the laboratory frame, before the application of the RF pulse, is given by:

$$\mathcal{H} = -\hbar\omega_L I_z + \hbar\omega_Q (3I_z^2 - \mathbf{I}^2) \tag{2.9.7}$$

where, as before, ω_L is the Larmor frequency (including the corresponding chemical shifts) and ω_Q is the effective quadrupolar coupling constant, which includes an orientational dependence as shown in Equation (2.7.21). For a RF pulse applied in the x-direction, the effective Hamiltonian in the rotating-frame is:

$$\mathcal{H}_{eff} = -\hbar(\omega_L - \Omega)I_z + \hbar\omega_Q(3I_z^2 - \mathbf{I}^2) - \hbar\omega_1 I_x \tag{2.9.8}$$

When we are close to resonance ($\omega_L \cong \Omega$) and the RF amplitude is sufficiently large ($\omega_1 \gg \omega_Q$), the effective Hamiltonian in the rotating-frame is once more given simply by $\mathcal{H}_{eff} \cong -\hbar\omega_1 I_x$. Therefore, in this case we have again the result that RF pulses cause rotations around the x-axis, leading to coherences that can be readily detected as transverse magnetization. For $I = 3/2$, the population differences between the adjacent levels $(3/2, 1/2)$, $(1/2, -1/2)$, and $(-1/2; -3/2)$ are converted into single quantum coherences and therefore all contribute to the transverse magnetization. The resultant NMR spectrum in this case is composed by three peaks, centered at the frequencies ω_L, $\omega_L - 6\omega_Q$, and $\omega_L + 6\omega_Q$. This can readily be computed by applying first-order perturbation theory to the Hamiltonian (2.9.7), considering the quadrupolar term as a perturbation to the main Zeeman Hamiltonian [4].

We have therefore established the important result that, in principle, even in the presence of the quadrupolar interaction, RF pulses with sufficiently high power can excite all allowed transitions and the whole NMR spectrum can then be obtained. These RF pulses are termed *hard* or *non-selective* pulses. It is important to stress that this can only be achieved if the condition $\omega_1 \gg \omega_Q$ is satisfied (with both ω_1 and ω_Q being much smaller than ω_L).

The opposite extreme regime, when $\omega_1 \ll \omega_Q$, involves the so-called *soft* or *selective* RF pulses. In this case, pulses of long duration and low power can be used to excite just one of the transitions, which is achieved by a suitable choice of the resonance offset (Ω) and also of the pulse length. The main aspect to be considered here is that each transition has associated with it a specific frequency and a different effective nutation frequency, which depends on the values of I and m. For selective excitation of a transition between the levels $m + 1$ and m, it is found that the ideal soft $\pi/2$ pulse must be shorter than the corresponding hard $\pi/2$ pulse (which excites all transitions) by a factor of $\sqrt{I(I+1) - m(m+1)}$. For excitation of just the central transition $(1/2 \longleftrightarrow -1/2)$ for half-integer spin nuclei, the length of an ideal soft $\pi/2$ pulse is reduced by a factor of $I + 1/2$ in comparison to a hard pulse. Thus, for $I = 3/2$, the ideal pulse for the selective excitation of the central transition

2.9. NMR of quadrupolar nuclei

is half the length of a $\pi/2$ pulse used for non-selective excitation, whereas the excitation of the $3/2 \longleftrightarrow 1/2$ and $-1/2 \longleftrightarrow -3/2$ transitions requires a pulse shorter by a factor of $\sqrt{3}$ in comparison to the hard pulse [18,19].

The theoretical basis for such features can be obtained by analysis of the time evolution in the rotating frame under action of the Hamiltonian (2.9.8). This requires the diagonalization of such Hamiltonian and the solution of the Liouville–von Neumann equation (Equation (2.5.3)). Usually, the results are properly described using the fictitious spin-1/2 formalism or the single-transition operator approach [18,20,21]. As example, we give below the matrix representation of the rotation operator corresponding to the selective excitation of the central transition in the case $I = 3/2$ for a $\pi/2$ pulse [22]:

$$R_x^{1/2 \longleftrightarrow -1/2}(-\pi/2) = \begin{bmatrix} e^{-i\omega_Q t_p} & 0 & 0 & 0 \\ 0 & \frac{1}{\sqrt{2}} e^{i\omega_Q t_p} & \frac{i}{\sqrt{2}} e^{i\omega_Q t_p} & 0 \\ 0 & \frac{i}{\sqrt{2}} e^{i\omega_Q t_p} & \frac{1}{\sqrt{2}} e^{i\omega_Q t_p} & 0 \\ 0 & 0 & 0 & e^{-i\omega_Q t_p} \end{bmatrix} \quad (2.9.9)$$

where t_p is the pulse length. The terms involving $e^{i\omega_Q t_p}$ are related to the evolution occurring during the pulse interval as consequence of the quadrupolar interaction. If the pulse length satisfies the criterion $\omega_Q t_p \ll 1$, then this evolution can be disregarded, and the matrix above assumes the form:

$$R_x^{1/2 \longleftrightarrow -1/2}(-\pi/2) = \begin{bmatrix} 1 & 0 & 0 & 0 \\ 0 & \frac{1}{\sqrt{2}} & \frac{i}{\sqrt{2}} & 0 \\ 0 & \frac{i}{\sqrt{2}} & \frac{1}{\sqrt{2}} & 0 \\ 0 & 0 & 0 & 1 \end{bmatrix} \quad (2.9.10)$$

This matrix can be written as

$$R_x^{1/2 \longleftrightarrow -1/2}(-\pi/2) = e^{i(\pi/2) I_x^{1/2 \longleftrightarrow -1/2}} \quad (2.9.11)$$

where the fictitious spin-1/2 operator $I_x^{1/2 \longleftrightarrow -1/2}$ is defined below:

$$I_x^{1/2 \longleftrightarrow -1/2} = \frac{1}{2} \begin{bmatrix} 0 & 0 & 0 & 0 \\ 0 & 0 & 1 & 0 \\ 0 & 1 & 0 & 0 \\ 0 & 0 & 0 & 0 \end{bmatrix} \quad (2.9.12)$$

It is clearly seen that the central fragments of this matrix correspond to the matrix representation of the operator I_x for the spin 1/2 case, as shown in Equation (2.6.1). Thus, the equations above show that the selective excitation actually changes only the populations of the $m = \pm 1/2$ levels, generating coherences only between these levels and therefore giving rise to a spectrum containing only the corresponding resonance line.

When $\omega_1 \sim \omega_Q$ (the so-called intermediate regime) the evolution of the density matrix in the rotating-frame exhibits a much more complex behavior as function of the pulse length and this is the basis of the method known as nutation NMR spectroscopy [10,19]. The distinction between selective ($\omega_1 \ll \omega_Q$), non-selective ($\omega_1 \gg \omega_Q$), and intermediate regimes

($\omega_1 \sim \omega_Q$) can bring some difficulties regarding the quantitativeness of NMR spectra in materials with more than one value of ω_Q. In a solid polycrystalline sample, the spread of ω_Q values due to the orientation distribution of the crystallites with respect to the $\mathbf{B}_0$ direction makes the selectivity condition difficult to be equally fulfilled for all crystallites in a given sample. The same obviously occurs if the material has different crystalline sites with widely differing quadrupolar parameters. The problem is that the intensities of the observed resonance lines will be largely dependent on the particular relation between ω_1 and ω_Q for each specific nuclear site in a solid material. The same pulse that is selective for, say, the central transition for a nucleus in a given site can be non-selective for a nucleus in another site. The relation between the observed intensities will not be therefore proportional to the relation between the respective amounts of nuclei in each site. In such cases, for uniform non-selective excitation, it is preferable to work with short high-power pulses, with flip angles significantly below $\pi/2$ (typically in the order of $\pi/12$ or $\pi/20$). It can be shown that for small flip angles the pulse response depends linearly on the pulse length and thus in this linear regime the relation between the intensities of the resonance lines is not greatly affected by differences in the ratio ω_Q/ω_1. The use of such small flip-angle pulses aims to guarantee the quantitativeness of the response of the nuclei in the material to the RF excitation, since the intensity of the detected signal depends in a complex way on the magnitude of both ω_1 and ω_Q [10,19,23].

The time evolution of the density operator after the end of the RF pulse (or sequence of RF pulses) can be easily calculated by the same methods applied to spin 1/2 systems. Considering the on-resonance case, the Hamiltonian in the rotating-frame is now:

$$\mathcal{H}_Q^{Rot} = \hbar\omega_Q(3I_z^2 - \mathbf{I}^2) \tag{2.9.13}$$

The corresponding evolution operator is:

$$U_H = e^{-i\omega_Q t(3I_z^2 - \mathbf{I}^2)} \tag{2.9.14}$$

Disregarding relaxation effects, the deviation density operator evaluated a time τ from the end of the pulse is:

$$\Delta\rho(\tau) = e^{-3i\omega_Q\tau I_z^2}\Delta\rho(t_p)e^{3i\omega_Q\tau I_z^2} \tag{2.9.15}$$

where the contribution of the term $\mathbf{I}^2$ was neglected, because it is proportional to the identity operator, $\mathbf{I}^2 = I(I+1)\mathbf{1}$, and therefore has no net effect on the time evolution of the density operator. After a non-selective $(\pi/2)_{-x}$, for example, one has $\Delta\rho(t_p) \sim I_y$. Then, the time evolution described by Equation (2.9.15) shows a precession in the rotating-frame associated with the quadrupolar interaction. The resulting density matrix allows the calculation of the expectation value of the transverse magnetization and, consequently, of the FID. For $I = 3/2$, it is straightforward to show that:

$$\langle I_x + iI_y\rangle = \text{Tr}\big[\rho(\tau)(I_x + iI_y)\big] = 3\cos(6\omega_Q\tau) + 2\cos(0.\tau) \tag{2.9.16}$$

Thus, the expectation value of the magnetization evolves in the rotating frame with the frequencies 0 and $\pm 6\omega_Q$. When viewed from the laboratory frame, these precessing magnetizations give rise to alternate electric signals detected with frequencies $\omega_L - 6\omega_Q$, ω_L, and $\omega_L + 6\omega_Q$ and with intensities respectively in the proportion 3 : 4 : 3. The Fourier transform of the associated FID thus contains three lines centered at these frequencies, whose separation is proportional to the quadrupolar coupling parameter ω_Q.

On the other hand, after a selective pulse applied to a given transition connecting the levels specified by m and $m-1$, the deviation density matrix contains coherences only between those levels. For $I = 3/2$ and considering a selective $(\pi/2)_{-x}$ pulse applied to the central transition $(1/2 \longleftrightarrow -1/2)$, the deviation matrix is proportional to a fictitious spin-1/2 operator $I_y^{1/2 \longleftrightarrow -1/2}$, whose form is analogous to the one given in Equation (2.9.12). The time evolution of such operator can be calculated from Equation (2.9.15), and the resulting FID from the calculation of the expectation value $\langle I_x + iI_y \rangle$. In this case, it is found that the FID involves only one oscillating signal with frequency ω_L. Therefore, the NMR spectrum now contains only one line, centered at frequency ω_L, but with intensity reduced as compared to the corresponding line detected in the non-selective case [18,23].

Also multiple-quantum (MQ) coherences (i.e., coherences corresponding to density matrix elements of the type ρ_{mn}, with $|m-n| > 1$) can be generated following the application of RF pulses in the case of nuclei experiencing strong quadrupolar coupling. (The same is true for systems of nuclear spins coupled by dipolar or J-couplings.) These coherences cannot be directly observed in an NMR experiment, but their production and evolution has many important consequences both in solid-state NMR applications [10,14] and in quantum computing processes involving quadrupolar nuclei (to be detailed later). In practice, MQ coherences are indirectly detected by applying RF pulses that convert them into single quantum coherences. The RF pulses for generation and for reconversion of MQ coherences need to be carefully designed with respect to amplitude, phase, and offset frequency to properly provide the desired coherence transfer pathways [13,20].

2.10 DENSITY MATRIX APPROACH TO NUCLEAR SPIN RELAXATION

The relaxation processes can be also described using the density operator formalism. To account for the longitudinal and transverse relaxations for a simple spin 1/2 system in a phenomenological way [24], the relaxation effects on the time evolution of the density matrix, $\rho(t)$, can be estimated by:

$$\frac{d\rho}{dt} = \frac{i}{\hbar}[\rho, H_0] - \frac{1}{\hbar}R \tag{2.10.1}$$

where $\mathcal{H}_0$ is the static nuclear spin Hamiltonian which defines the NMR spectrum, and R is a phenomenological relaxation matrix whose components are given by:

$$R_{ij} = \hbar \frac{[\rho_{ij}(t) - \rho_{ij}(0)]}{T_{ij}} \tag{2.10.2}$$

For single spin 1/2 systems, T_{ij} is related to the relaxation times by:

$$T_{ij} = T_1 \delta_{ij} + T_2(1 - \delta_{ij}) \tag{2.10.3}$$

However, in a more general situation, T_1 and T_2 can assume different values for the relaxation of each population and coherence.

Using a more general formalism to describe the relaxation for any spin system [4,8,12, 25], the time evolution of the density matrix under the influence of the relaxation Hamiltonian can be determined by:

$$\frac{d\rho(t)}{dt} = \frac{i}{\hbar}[\rho(t), \mathcal{H}_T(t)] \quad (2.10.4)$$

where

$$\mathcal{H}_T(t) = \mathcal{H}_0 + \mathcal{H}_R(t) \quad (2.10.5)$$

and $\mathcal{H}_R(t)$ is a perturbative time-dependent Hamiltonian. This last term contains all the interactions providing the pathways which allow the nuclear spin non-equilibrium density matrix going back to the equilibrium.

To simplify the solution of Equation (2.10.4), let's use the interaction representation [9] to remove time-independent terms of $\mathcal{H}_T(t)$ and write the effective relaxation Hamiltonian $\mathcal{H}_R^*(t)$ as:

$$\mathcal{H}_R^*(t) = e^{(i/\hbar)\mathcal{H}_0 t} H_R(t) e^{-(i/\hbar)\mathcal{H}_0 t} \quad (2.10.6)$$

and the new density matrix as relaxation density matrix $\rho^R(t)$, defined by:

$$\rho^R(t) = e^{(i/\hbar)\mathcal{H}_0 t} \rho(t) e^{-(i/\hbar)\mathcal{H}_0 t} \quad (2.10.7)$$

Using the interaction representation, $\rho^R(t)$ will evolve under the new Liouville equation:

$$\frac{d\rho^R(t)}{dt} = \frac{i}{\hbar}[\rho^R(t), \mathcal{H}_R^*(t)] \quad (2.10.8)$$

While the diagonal elements $\rho_{ii}^R(t)$ account for the time dependence of the populations of the eigenstates associated with the longitudinal relaxation, the off-diagonal elements $\rho_{ij}^R(t)$ describe the time dependence of the coherences between the different eigenstates associated with the transverse relaxation [26].

The derivative of the diagonal components, $\frac{d\rho_{ii}^R(t)}{dt}$, gives the rate at which the population of one eigenstate changes. Considering that the total population of the spin system is constant, the following constraint should be respected: $\frac{d}{dt}\sum_i \rho_{ii}^R(t) = 0$. The derivative of the off-diagonal components, $\frac{d\rho_{ij}^R(t)}{dt}$, with $i \neq j$, accounts for the rate at which each coherence decays.

Normally, the experimentally observed longitudinal and transverse relaxations involve zero- and single-quantum processes ($\Delta m = i - j = 0$ or ± 1). The other time-dependent density matrix elements $\rho_{ij}^R(t)$ with $|\Delta m| > 1$ are classified as *multiple-quantum coherences*. Although these coherences are not directly observable, it is possible to measure every element of the relaxation density matrix by employing specially designed NMR methods. These coherences are very important for multiple-quantum experiments [13] and quantum computation NMR applications, as it will be described later.

The random molecular motional processes, which make the nuclear spin Hamiltonian $\mathcal{H}_R^*(t)$ time dependent, are responsible for the nuclear spin relaxation. In order to understand the effects of the relaxation Hamiltonian $\mathcal{H}_R^*(t)$, which represents a weak and statistically random time-dependent perturbation on the spin system, it is necessary to use the density matrix treatment known as the Redfield theory [27,28]. The Redfield theory is a semi-classical treatment in which the spin system and the lattice are considered as quantum and classical entities, respectively.

The Redfield theory results in the *master equation* [26]:

$$\frac{d\rho_{mn}^R(t)}{dt} = \sum_{k,l} e^{i(\omega_{mn} - \omega_{kl})t} R_{mnkl} \rho_{kl}^R(t) \qquad (2.10.9)$$

where $\omega_{mn} = (E_m - E_n)/\hbar$ is the resonance frequency connecting the m and n states associated with the dominant static Hamiltonian H_0. The elements R_{mnkl} are the components of the relaxation matrix defined by:

$$R_{mnkl} = J_{mknl}(\omega_{mk}) + J_{mknl}(\omega_{nl}) - \delta_{mk} \sum_p J_{pnpl}(\omega_{pl}) - \delta_{nl} \sum_p J_{pmpk}(\omega_{pk})$$

$$(2.10.10)$$

where the components $J_{mnkl}(\omega)$ are the spectral density functions:

$$J_{mnkl}(\omega) = \int_0^\infty \overline{\langle m|\mathcal{H}_R^*(0)|n\rangle \langle k|\mathcal{H}_R^*(\tau)|l\rangle} e^{-i\omega\tau} d\tau = \int_0^\infty P_{mnkl}(\tau) e^{-i\omega\tau} d\tau$$

$$(2.10.11)$$

where $P_{mnkl}(\tau)$ is the correlation function and τ is the correlation time.

In order to implement the relaxation calculations some additional definitions should be provided: i) the nuclear spin interactions acting as relaxation mechanisms and ii) a molecular model motion, including iii) the distribution of correlation times at which the motion is occurring. These features will depend on aspects such as the temperature, the physical state of the sample, and the magnitude of the external applied magnetic field, among others. A more detailed analysis of such relaxation mechanisms can be found in References [4,8,25,26].

2.11 SOLID-STATE NMR

In liquid isotropic samples the molecules typically execute fast and random motions so that the anisotropic components of the spin interactions are averaged out. Then, the only remaining contribution for the NMR spectra comes from the isotropic terms in the chemical shift and J-coupling interactions, resulting in spectra composed of very well defined resonance lines. Given that these motions are restricted in solid samples, the anisotropic components are not averaged or are only partially averaged. Since the resonance of each nucleus depends on the local field at its site, and the intensity of these local fields depends on the orientation of the neighboring nuclei, of the electron clouds, and/or of the electric gradient fields, there will be a considerable spread in the resonance frequencies resulting in broad spectra, where it is difficult to distinguish the different isotropic terms (usually the isotropic chemical shift terms). Another problem for obtaining spectra from solid samples for rare nuclei (low natural abundance) with small gyromagnetic ratio, such as ^{13}C, is associated with the low sensitivity related to two problems: small NMR signal and long spin-lattice relaxation times [29]. In order to circumvent these problems three methods are routinely applied in solid-state NMR experiments: Dipolar (Homo and Heteronuclear) Decoupling and Magic-Angle Spinning (MAS) for improving the resolution of the spectra, and Cross-Polarization (CP) for increasing the sensitivity.

2.11.1 Dipolar decoupling

The dipolar decoupling technique is used to suppress the magnetic dipolar interaction between nuclear spins. When observing rare ^{13}C nuclei, for example, the homonuclear dipolar interaction (^{13}C–^{13}C) is negligible due to their small natural abundance and the only important dipolar interaction comes from the abundant ^{1}H nuclei. As mentioned above, this interaction gives rise to large broadening in solid samples, due to its anisotropic character. To suppress the ^{13}C–^{1}H dipolar interaction, the method used is called *heteronuclear decoupling* and was proposed by Sarles and Cotts in 1958 [30]. It consists in the continuous wave (CW) irradiation with RF at the ^{1}H resonance frequency while observing the ^{13}C signal, being therefore an example of a *double-resonance* experiment. In order to effectively suppress the ^{13}C–^{1}H dipolar interaction, it is necessary to apply RF $\mathbf{B}_1$ fields satisfying the condition $\gamma_H B_1 > \Delta \nu_{dip}$, where γ_H is the gyromagnetic ratio of the abundant nuclei (^{1}H in this case) and $\Delta \nu_{dip}$ is the linewidth due to the heteronuclear dipolar interaction. In the cases of residual linewidths arising from insufficient proton decoupling power, a simple two-pulse phase modulation (TPPM) scheme greatly improves the quality of the spectra [31]. In the case of homonuclear dipolar interaction, normally ^{1}H–^{1}H dipolar interaction, the method employed is denominated *homonuclear decoupling*. There are several ways to suppress this interaction and the most common ones are those based on multiple pulse sequences, such as WaHuHa [32], MREV-8 [33,34], and BR-24 [35], or on the Lee–Goldburg method [36].

2.11.2 Magic-angle spinning (MAS)

In 1959 Andrew et al. [37] and Lowe [38] proposed, independently, this method to suppress the magnetic dipolar interaction in solids. In 1962, Andrew and Eades [39] showed that MAS could also be applied to eliminate other anisotropic interactions (to first order). To introduce this method, let us take as example again the magnetic dipolar interaction between the nuclei ^{13}C and ^{1}H. The z-component of the dipolar magnetic field produced by the ^{1}H nucleus on the ^{13}C site is given by:

$$B_{dip} = \frac{\mu_0}{4\pi} \frac{\mu_H}{r_{CH}^3} (3\cos^2\theta - 1) \qquad (2.11.1)$$

where μ_H is the magnetic dipole moment of the ^{1}H nucleus, r_{CH} is the modulus of the distance vector connecting both nuclei, and θ is the angle between $\mathbf{r}_{CH}$ and $\mathbf{B}_0$. The term $(3\cos^2\theta - 1)$ describes the anisotropy of the dipolar interaction. If this term were equal to zero, the dipolar interaction would be eliminated. One way to do that is choosing

2.11. Solid-state NMR

$\theta = \theta_m \cong 54.74°$, where θ_m is called the *magic-angle*. Of course, it is impossible to put all the spin pairs aligned along this specific orientation, except in very special cases of oriented monocrystalline samples.

The method proposed to reach this condition consists in fast spinning the whole sample around an axis making and angle equal to θ_m with the external magnetic field $\mathbf{B}_0$, with a spinning speed $\nu_r > \Delta\nu_{dip}$. In this situation, all the internuclear vectors $\mathbf{r}_{CH}$ will be, on average, along the magic-angle orientation, resulting in an average dipolar field:

$$\langle B_{dip} \rangle = \frac{\mu_0}{4\pi} \frac{\mu_H}{r_{CH}^3} \langle 3\cos^2\theta - 1 \rangle = 0 \qquad (2.11.2)$$

A similar reasoning can be applied to the other nuclear spin interactions, as all these interactions have a common tensorial structure and the correspondent tensors behave in an analogous way under rotations (see Section 2.7). The relevant angle θ in each case will be the angle between one of the axes (usually a symmetry axis) of the principal axis system (PAS) of the interaction tensor and the external magnetic field $\mathbf{B}_0$. As a general rule, the MAS procedure will be effective in removing the anisotropic line broadening if the spinning speed is larger than the correspondent linewidth introduced by the specific interaction when the sample is static [40].

In the case of ^{13}C–^{1}H dipolar interaction, $\Delta\nu_{dip}$ is in the range 1–100 kHz. For ^{13}C chemical shift anisotropy (CSA) with $B_0 \sim 10$ T (as the magnitude of this interaction depends on the magnetic field strength), it is typically found that $\Delta\nu_{CSA}$ is in the range 1–10 kHz. For usual MAS systems, the maximum spinning speed is around 25 kHz. In this situation, it would be relatively easy to suppress the broadening due to chemical shift anisotropy from the ^{13}C spectra, but it would be difficult to suppress ^{13}C–^{1}H dipolar interaction. For suppressing the ^{13}C–^{1}H dipolar interaction, it is necessary to apply both heteronuclear decoupling and MAS. In cases where it is impossible to spin the sample with $\nu_r > \Delta\nu$, spinning sidebands will be observed in the spectrum [40].

For nuclei experiencing strong quadrupolar interactions, the broadening can be much larger, reaching the range of MHz. Therefore, MAS is not capable of removing completely the broadening in such cases. Moreover, as second-order effects involve other angular dependences different from the simple $(3\cos^2\theta - 1)$ term, MAS is not effective for removing these effects either. Therefore, other techniques are necessary for high-resolution NMR of quadrupolar nuclei in solid samples, such as double rotation (DOR) or two dimensional experiments such as dynamic angle spinning (DAS), multiple-quantum magic-angle spinning (MQ-MAS), and others [10].

2.11.3 Cross-polarization (CP)

As already discussed, it is experimentally difficult to obtain spectra from solid samples for rare nuclei (low natural abundance) or with small gyromagnetic ratios, such as ^{13}C. The low sensitivity is associated with two problems: small NMR signal and long spin-lattice relaxation times [29]. In 1973, Pines at al. proposed the method called Cross-Polarization in order to circumvent these problems [41]. This method involved once more a *double-resonance* experiment, based on the transference of polarization from the abundant ^{1}H spins, with short spin-lattice relaxation times T_1, to the rare ^{13}C nuclei. After this polarization transfer, the rare nucleus signal intensity is increased by a factor equal to the ratio between the gyromagnetic ratios of both nuclei ($\gamma_{abundant}/\gamma_{rare}$), when compared with the excitation with a single $\pi/2$ pulse. In the case of ^{13}C and ^{1}H this factor is about 4. Despite the fact that the NMR experiment is performed for the ^{13}C nuclei, the repetition rate for signal averaging is determined by the short relaxation time T_1 of ^{1}H, because the ^{13}C magnetization is now determined by the ^{1}H nuclei. In order to establish an efficient contact between ^{1}H and ^{13}C nuclei for the polarization transfer, it is necessary to satisfy the Hartmann–Hahn condition: $\gamma_H B_{1H} = \gamma_C B_{1C}$, where γ_n and B_1 represent respectively the gyromagnetic ratio and RF amplitude corresponding to each nucleus. This means that the RF fields have to be applied simultaneously in such a way that the nutation frequencies of both nuclei around the respective resonant RF fields are the same. As a result, a resonance exchange of energy between ^{13}C and ^{1}H nuclei can readily occur through a mutual spin flip mechanism [29].

2.11.4 The CP-MAS experiment

The combination of Heteronuclear Decoupling, MAS, and CP techniques in only one experiment was proposed in 1976 by Schaefer and Stejskal [42] and marked the birth of high-resolution solid-state NMR spectroscopy for rare nuclei. Figure 2.10 shows schematically the procedure to perform this clever experiment, using once more ^{1}H and ^{13}C nuclei as examples of abundant and rare nuclei, respectively.

To summarize the experiment, one applies a $\pi/2$ pulse to ^{1}H, which is followed by a change of 90° in the phase of the RF field of ^{1}H relative to the first pulse. This leads to the *spin-locking* of the abundant spin system. Still under the spin-locking condition, which was used to lower the ^{1}H spin temperature [4] much below the lattice temperature, the Hartmann–Hahn condition is established, and this situation is kept until the polarization transfer from ^{1}H to ^{13}C spins is complete. After that, the observation of the ^{13}C FID is performed under heteronuclear decoupling. In order to increase the signal-to-noise ratio of ^{13}C FID, the CP-MAS experiment is repeated as many times as necessary, just waiting the time necessary for the total ^{1}H spin-lattice relaxation between acquisitions. MAS is continuously applied during the experiment.

Based on methods which rely on the manipulation of the anisotropic terms of the nuclear spin interactions and on the combination of different basic NMR techniques, such as MAS and decoupling, an enormous number of solid-state NMR pulse sequences were proposed in the last 20 years. Solid-state NMR provides powerful techniques for elucidating details of segmental dynamics and local conformation in solid materials. NMR methods allow the study of dynamics occurring in a wide frequency range (from the order of 1 Hz to

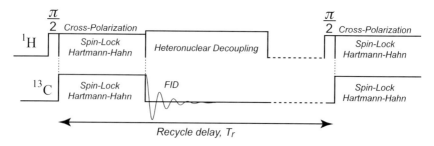

Figure 2.10 CP-MAS pulse sequence for ^{13}C–^{1}H pairs. The sample is under MAS during all the experiment.

100 MHz). Besides, NMR also provides a series of methods capable of obtaining reliable measurements of torsion angles between localized sites, short and intermediate range structure, as well as interatomic distances. A remarkable characteristic of new NMR methods is the possibility of getting important information about molecular structure and dynamics of materials in an almost model free fashion, making NMR a unique tool for the characterization of complex materials. Besides, despite its 50 years of history, the development of new NMR methods is still and opened area, being one of the main challenges for NMR researchers.

2.12 THE EXPERIMENTAL SETUP

To perform NMR experiments involving spectroscopy, relaxation studies, or quantum computing, it is necessary to have an equipment (generally named *NMR spectrometer*) composed basically of: (i) A magnet, to generate the $\mathbf{B}_0$ field. This is not necessary only in the case of zero-field NMR performed with magnetically ordered samples or with materials that give pure nuclear quadrupole resonance (NQR) signals [1]. (ii) A probe, containing the coil where the sample to be analyzed is placed. And (iii) an RF system, composed of a transmitter and a receiver, which allows manipulating the RF phase, frequency, and amplitude for exciting the nuclei and detecting the induced signals. In modern spectrometers, the whole experiment is controlled and monitored by computer interfaces that allow automatic recording and processing of the NMR signal.

A basic electronic block diagram of the transmitter and receiver sections of a NMR spectrometer is shown in Figure 2.11. For transmission (Figure 2.11a), the synthesizer generates the RF and enables the control, with high precision and fast switching of both, of the frequency (0.1 Hz resolution) and phase (in steps 0.1°) by the use of remote control interfaces. The high-quality RF is generated with these features at the frequency Ω, which is used to excite the nuclei and also as reference for the receiver. After the synthesizer, the signal reaches the RF modulator. The RF is amplitude modulated with rectangular or special functions, in order to produce RF pulses. At this same point of the circuit, the RF can also be phase modulated only in 4 steps of 0°, 90°, 180° and 270°. This phase modulation is extensively used for the phase cycling necessary for removing artifacts or implementing many special pulse sequences. Afterward, the RF pulses are amplified and sent to the probe. Depending on the function used for the amplitude modulation, the time length and amplification level, the pulses can be hard (non-selective) or soft (selective).

To avoid sending high power RF pulses to the high sensitivity pre-amplifiers of the receiver, duplexers are used to direct the intense RF pulses to the probes and the weak NMR signal to the receiver. In order to have the maximum power transfer from the transmitter to the probe, the former should be tuned at the Larmor frequency (ω_L) and its impedance should match the output impedance of the high power amplifier (50Ω). It is desirable that the RF circuit used to tune the probe (tank circuit) have a band pass wide enough to permit the homogeneous RF excitation of all the NMR frequencies of interest in a given experiment (that is, to have not a too high quality factor Q) [44].

As the NMR experiment can be performed with a single or several nuclei of different species simultaneously, the probes should be simultaneously tunable at several different frequencies. Usually, the experiments are performed in single-, double-, or triple-resonance, but typically only one species of nucleus is observed at the corresponding Lar-

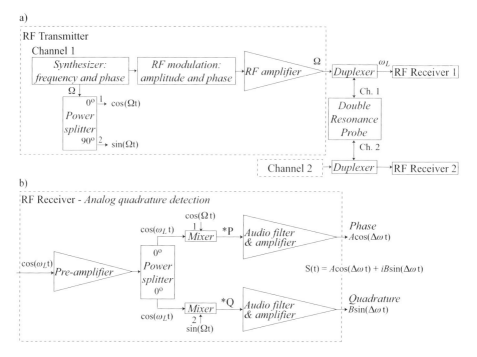

Figure 2.11 Block diagram of a NMR spectrometer.

mor frequency. For some experiments, two nuclei have to be excited and observed and, consequently, two receivers should be used to detect the signals induced at two different Larmor frequencies. This is exemplified in Figure 2.11, with the indications of two transmitter channels and two receivers.

The receivers have some special purposes. Their pre-amplifiers should be able to provide a high NMR signal amplification without introducing too much noise. The remaining RF circuit is used to generate the phase and quadrature signals, Figure 2.11b. The power-splitter in the receiver divides the NMR signal detected by the probe with the frequency ω_L into two components with the same phases and amplitudes (0°): $\cos(\omega_L t)$ and $\cos(\omega_L t)$. These are mixed with the reference RF signals coming from the output of a second power-splitter (shown in Figure 2.11a) directly connected to the synthesizer, which separates the RF into two equal amplitude components in quadrature (0°, 90°), i.e, $\cos(\Omega t)$ and $\sin(\Omega t)$, respectively. The main aim of this mixing stage is to bring the observed NMR signal down to the audio range (Hz–kHz) and to allow phase-sensitive detection. These signals are mixed in the points 1 and 2 of the RF receiver, Figure 2.11b, and at the points indicated by $^*P$ and $^*Q$, the following most significant RF signals are present:

At point $^*P$, in phase signals:

$$\cos\big[(\omega_L - \Omega)t\big] + \cos\big[(\omega_L + \Omega)t\big] + \text{high order harmonics} \qquad (2.12.1)$$

At point $^*Q$, in quadrature signals:

$$\sin\big[(\omega_L - \Omega)t\big] + \sin\big[(\omega_L + \Omega)t\big] + \text{high order harmonics} \qquad (2.12.2)$$

2.12. The experimental setup

To avoid the high frequency signals (i.e., the signals originated from the term involving the sum $\omega_L + \Omega$ as well as other high order harmonics), an audio filter/amplifier is used to detect only the difference frequency signals in phase and quadrature, $\cos[(\omega_L - \Omega)t] = \cos\Delta\omega t$ and $\sin[(\omega_L - \Omega)t] = \sin\Delta\omega t$, respectively. These signals in quadrature are sent to the analog-to-digital converter (ADC) and digital signal processor (DSP), which could be a computer or a dedicated digital system, in the form of a complex function:

$$S(t) = A\cos\Delta\omega t + iB\sin\Delta\omega t \qquad (2.12.3)$$

where A and B are constants.

The real and imaginary components of the signal $S(t)$ can be interchanged and their signs inverted in the memory locations to allow phase cycling of the received NMR signal, a procedure called signal routing. In this way, the phase of the complex NMR signal can be changed in steps of 90°, covering the main values 0°, 90°, 180° and 270°. Together with the phase cycling of the RF pulses, the complete phase cycling for implementing the pulse sequences is now possible.

The reason for implementing the quadrature detection is related to the Fourier transformation of the NMR signal. If the NMR signal were detected in the single mode (only with $S(t) = A\cos\Delta\omega t$), the Fourier transform would lead to two peaks at $\pm\Delta\omega$, thus duplicating the number of lines in the spectrum [45]. To avoid the superposition or proximity of this pair of peaks, which could make impossible to interpret a spectrum composed by several lines, one should make the detection far from the resonance condition ($\Delta\omega = 0$), making necessary a detection with a bandwidth of, at least, two times the maximum frequency observed, $2\Delta\omega_{max}$. Given that the random noise affecting the NMR signal is proportional to $\sqrt{2\Delta\omega_{max}}$, the signal-to-noise ratio would decrease by a factor $\sqrt{2}$. This is an important issue from the point of view of optimizing sensitivity, considering that the NMR signal is normally very weak and it would be necessary to double the number of scans to increase the signal-to-noise by this same factor [44]. A simple way to solve this problem is to make quadrature detection ($S(t) = A\cos\Delta\omega t + iB\sin\Delta\omega t$, with $A = B$), because in this case the Fourier transform of the complex signal would generate only one peak for each NMR line. This allows then the detection to be effected close to resonance, with a two-fold reduction in the bandwidth to just $\Delta\omega_{max}$.

Normally, the audio filter/amplifier gains are slightly different, the receiver electronics introduces offsets to each channel, and the quadrature detection is not perfect, making the NMR signal in the practice looks like:

$$S(t) = \left[A\cos(\Delta\omega t) + a\right] + i\left[B\sin(\Delta\omega t + \delta) + b\right] \qquad (2.12.4)$$

where a and b are the offsets observed for each channel, and δ is the quadrature mismatch. These technical problems produce the following artifacts in the NMR spectrum:

- $A \neq B$ or $\delta \neq 0$: the lines show up in duplicate, like in the single mode detection, but one of them is much smaller than the other one;
  a and/or $b \neq 0$: an intense sharp feature appears at the centre of the spectrum (i.e., at zero frequency with respect to the carrier frequency Ω).

These artifacts can be eliminated by a phase cycling scheme called CYCLOPS [46]. To make the discussion of this phase cycling easier, let's consider that δ is usually sufficiently

Exp. number	Transm. phase	Received signal phase quad.		Signal routing		True signal, including offsets (a and b), and amplitude mismatching ($A \neq B$)
		M_x	M_y	C1	C2	
1	x	$A \sin \omega t + a$	$B \cos \omega t - B$	M_x	M_y	$[A \sin \omega t - a] + i[B \cos \omega t + b]$
2	y	$-A \cos \omega t + a$	$B \sin \omega t + B$	M_y	$-M_x$	$[B \sin \omega t + b] + i[A \cos \omega t - a]$
3	$-x$	$-A \sin \omega t - a$	$-B \cos \omega t - B$	$-M_x$	$-M_y$	$[A \sin \omega t - a] + i[B \cos \omega t - b]$
4	$-y$	$A \cos \omega t + a$	$-B \sin \omega t + B$	$-M_y$	M_x	$[B \sin \omega t - b] + i[A \cos \omega t + a]$

Adding the four signals: $2[(A + B) \sin \omega t] + i2[(A + B) \cos \omega t]$

Figure 2.12 Acquisition steps for performing phase cycling with CYCLOPS.

small (when operating the spectrometer within the correct RF frequency ranges) and disregard it. CYCLOPS involves the consecutive acquisition of the four signals summarized in Figure 2.12, where, additionally to the RF phase cycling, the detected phase and quadrature signals can be addressed or routed selectively to the memory locations C1 and C2 (signal routing), according to the designed full phase cycling (RF phase and memory locations).

Since the NMR acquisition always involves signal averaging, the phase cycling normally does not extend the experiment time excessively, only imposing the restriction that the total number of acquired transients must be an entire multiple of the number of steps in the phase cycle. In the case of CYCLOPS, after every four scans along the averaging process, one gets the perfect quadrature detection.

Phase cycling is a fundamental procedure in most NMR experiments and is used not only for removing instrument artifacts, but also for selecting or suppressing signals, specially for achievement of specific coherence transfer pathways [5,13]. In NMR experiments, one must be aware of the importance of phase cycling, which sometimes is more difficult to understand than the basic aspects of the pulse sequences.

Nowadays, in modern NMR spectrometers, the analog quadrature detection is being discontinued. Instead, a digital detection of the NMR signals is being implemented, as shown schematically in Figure 2.13. For that, the NMR signal is pre-amplified, and its frequency is shifted, by the use of a mixer, to an intermediate fixed RF frequency ω_{IF}. Next, the signal is directly sent to the ADC and DSP. The digital signal processor performs all the computational processing, including, if necessary, the digital quadrature transformation. In this case, the quadrature artifacts do not exist and there are various ways to perform the digital filtering and processing of the NMR signal.

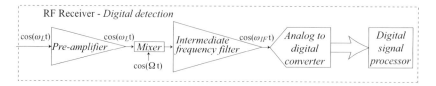

Figure 2.13 Schematic structure of a RF receiver designed for digital detection at the intermediate frequency ω_{IF}.

2.13 APPLICATIONS OF NMR IN SCIENCE AND TECHNOLOGY

The NMR phenomenon not only took significant part of the development of the modern physics from the beginning to the middle of the 20th century, when the Stern–Gerlach ($\approx$ 1922), Rabi–Cohen ($\approx$ 1934), Bloch ($\approx$ 1945) and Purcell ($\approx$ 1945) experiments were performed, but also is giving fundamental contributions nowadays in several fundamental scientific areas (physics, chemistry, biology, medicine, etc.). By the use of complex multiple pulse sequences involving or not several nuclei at the same time, which include cross-talks among the nuclei through dipolar interactions, selective suppression of spin interactions (e.g. homo- or hetero-nuclear decoupling, magic angle spinning, spin echo techniques, etc.), and signal acquisition in a multidimensional way, dynamical, structural and morphological information can be obtained in several time (from ns to s) and distance (from Å to mm) scales for samples in the solid, liquid, or gaseous phase. Due to these reasons, there are several last generation spectrometers in the market, which permit the use of NMR, not only in Quantum Computing, but also in:

1. Materials science (polymers and derivatives, proteins, molecular sieves, etc.);
2. Soil science;
3. Petroleum science (including well-logging);
4. Analytical Chemistry;
5. Medicine (imaging, functional imaging, in vivo spectroscopy);
6. Microscopy and Atomic-Force-Like microscopy;
7. And several other applications, making NMR one of the most important methods for fundamental and applied research.

PROBLEMS WITH SOLUTIONS

P2.1 - Obtain the result (2.2.1), starting from the equation of motion of a rigid body submitted to an external torque.

Solution
Consider a body with angular momentum $\mathbf{L}$ and magnetic dipole moment $\boldsymbol{\mu} = \gamma \mathbf{L}$ in the presence of a magnetic field $\mathbf{B}_0$. The torque on the magnetic dipole moment is:

$$\boldsymbol{\tau} = \boldsymbol{\mu} \times \mathbf{B}_0 = \gamma \mathbf{L} \times \mathbf{B}_0$$

The dynamics is governed by the equation of motion $\boldsymbol{\tau} = \frac{d\mathbf{L}}{dt}$. Therefore we have:

$$\frac{d\mathbf{L}}{dt} = \gamma \mathbf{L} \times \mathbf{B}_0$$

Taking $\mathbf{B}_0 = B_0 \mathbf{k}$ and writing the equations for the vector components of $\mathbf{L}$:

$$\frac{dL_x}{dt} = \gamma L_y B_0$$

$$\frac{dL_y}{dt} = -\gamma L_x B_0$$

$$\frac{dL_z}{dt} = 0$$

The last equation above shows that L_z is time independent, i.e., it is a constant of motion. The first two equations are coupled, and are similar in form to the differential equation of a two-dimensional harmonic oscillator. Taking another derivative of these similar equations and substituting the results, we obtain:

$$\frac{d^2 L_x}{dt^2} = \gamma \frac{dL_y}{dt} B_0 = -(\gamma B_0)^2 L_x$$

$$\frac{d^2 L_y}{dt^2} = -\gamma \frac{dL_x}{dt} B_0 = -(\gamma B_0)^2 L_y$$

Now we have two uncoupled differential equations identical to the differential equation of an one-dimensional harmonic oscillator. Therefore, the solutions are:

$$L_x(t) = A \sin(\omega_L t + \delta)$$
$$L_y(t) = A \cos(\omega_L t + \delta)$$

with the parameter ω_L given by $\omega_L = \gamma B_0$ (assuming $\gamma > 0$). The phases were chosen so as to satisfy the equations involving the first-derivatives of L_x and L_y and, for the same reason, the constants A and δ need to be the same in both equations above.

Choosing an appropriate time origin and changing the names of the constants, we arrive at:

$$L_x(t) = L_\perp \sin(\omega_L t)$$
$$L_y(t) = L_\perp \cos(\omega_L t)$$
$$L_z(t) = L_\parallel$$

where $L_\parallel$ and $L_\perp$ indicate the parallel and transverse components of the angular momentum with respect to the field. The vector $\mathbf{L}$ thus precesses around the z axis, keeping constant its magnitude ($= \sqrt{L_\parallel^2 + L_\perp^2}$) and z component ($L_\parallel$). The sense of precession is negative, i.e., left-handed with respect to the z axis. To see this, observe that at $t = 0$, $L_x = 0$ and $L_y = L_\perp$ is at its maximum value. After some time, L_y decreases and L_x increases, indicating the sense of precession. Therefore, we can write the vector equation for the precession frequency:

$$\boldsymbol{\omega}_L = -\gamma \mathbf{B}_0,$$

which is the Larmor equation we wanted to obtain.

P2.2 - Starting from the evolution operator associated with the Zeeman Hamiltonian (2.2.2), obtain the time evolution of the expectation values $\langle I_x \rangle$, $\langle I_y \rangle$, and $\langle I_z \rangle$. Interpret the result to show the Larmor precession.

Solution

The Schrödinger equation allows the assessment of the time evolution of the vector state of the system:

$$|\psi_t\rangle = U(t, 0)|\psi_0\rangle$$

where the subscripts indicate the time instants corresponding to each state. The evolution operator associated with the Zeeman Hamiltonian is:

$$U(t, 0) = e^{-i(H/\hbar)t} = e^{i\omega_L t I_z}$$

Therefore:

$$|\psi_t\rangle = e^{i\omega_L t I_z}|\psi_0\rangle$$

The time evolution of the expectation value of any operator can be calculated by:

$$\langle A \rangle_t = \langle \psi_t | A | \psi_t \rangle = \langle \psi_0 | e^{-i\omega_L t I_z} A e^{i\omega_L t I_z} | \psi_0 \rangle$$

For the I_z operator, we have:

$$\langle I_z \rangle_t = \langle \psi_t | I_z | \psi_t \rangle = \langle \psi_0 | e^{-i\omega_L t I_z} I_z e^{i\omega_L t I_z} | \psi_0 \rangle = \langle \psi_0 | I_z | \psi_0 \rangle = \langle I_z \rangle_0$$

This means therefore that the expectation value of the operator I_z remains unaltered. As for the I_x component, we have:

$$\langle I_x \rangle_t = \langle \psi_t | I_x | \psi_t \rangle = \langle \psi_0 | e^{-i\omega_L t I_z} I_x e^{i\omega_L t I_z} | \psi_0 \rangle$$

The term $e^{-i\omega_L t I_z} I_x e^{i\omega_L t I_z}$ can be calculated from the "sandwich theorem", Equation (2.6.13):

$$e^{-i\omega_L t I_z} I_x e^{i\omega_L t I_z} = I_x \cos \omega_L t + I_y \sin \omega_L t$$

Therefore:

$$\langle I_x \rangle_t = \langle \psi_0 | I_x \cos \omega_L t + I_y \sin \omega_L t | \psi_0 \rangle = \langle \psi_0 | I_x | \psi_0 \rangle \cos \omega_L t + \langle \psi_0 | I_y | \psi_0 \rangle \sin \omega_L t$$

Or:

$$\langle I_x \rangle_t = \langle I_x \rangle_0 \cos \omega_L t + \langle I_y \rangle_0 \sin \omega_L t$$

A similar calculation leads to:

$$\langle I_y \rangle_t = -\langle I_x \rangle_0 \sin \omega_L t + \langle I_y \rangle_0 \cos \omega_L t$$

If we take, for example, $\langle I_y \rangle_0 = 0$ as an initial condition, then we arrive at:

$$\langle I_x \rangle_t = \langle I_x \rangle_0 \cos \omega_L t$$
$$\langle I_y \rangle_t = -\langle I_x \rangle_0 \sin \omega_L t$$
$$\langle I_z \rangle_t = \langle I_z \rangle_0$$

These equations clearly show a precession motion of the vector $\langle \mathbf{I} \rangle$ (i.e., of the expectation value of the nuclear spin) about the z axis in the left-hand direction with angular frequency of magnitude ω_L, which is in accordance with the classic Larmor precession previously derived.

P2.3 - Verify the numeric value of the Boltzmann factor (Equation (2.2.4)) for protons in a 5 T magnetic field at room temperature. (The gyromagnetic ratio for the proton is 26.7522×10^7 rad s^{-1} T^{-1}.)

Solution
The calculation is straightforward and we will assume that the temperature and magnetic field values are known with infinite accuracy (to avoid rounding problems):

$$\omega_L = \gamma B_0 = (26.7522 \times 10^7 \text{ rad s}^{-1} \text{ T}^{-1}) \times 5\text{T} = 1.33761 \times 10^9 \text{ rad s}^{-1}$$

This corresponds to a frequency of about 213 MHz. Therefore, at room temperature (300 K), we have:

$$e^{-\hbar \omega_L / kT} = e^{-(1.05457 \times 10^{-34})(1.33761 \times 10^9)/(1.38065 \times 10^{-23} \times 300)} = 0.99997$$

This result very close to the unity shows how small is the difference in populations between the Zeeman levels at usual conditions of temperature and magnetic field.

P2.4 - Obtain Equation (2.2.5), starting from the result (2.2.4).

Solution
Let us take the populations n_- and n_+ related by the expressions below:

$$\frac{n_-}{n_+} = e^{-\hbar\omega_L/k_B T}$$

$$n_- + n_+ = n_0$$

The solutions to these equations are:

$$n_- = \frac{n_0}{1+e^{\hbar\omega_L/k_B T}} \cong \frac{n_0}{2+\frac{\hbar\omega_L}{k_B T}} = \frac{n_0}{2}\frac{1}{1+\frac{\hbar\omega_L}{2k_B T}} \cong \frac{n_0}{2}\left(1-\frac{\hbar\omega_L}{2k_B T}\right)$$

$$n_+ = \frac{n_0}{1+e^{-\hbar\omega_L/k_B T}} \cong \frac{n_0}{2-\frac{\hbar\omega_L}{k_B T}} = \frac{n_0}{2}\frac{1}{1-\frac{\hbar\omega_L}{2k_B T}} \cong \frac{n_0}{2}\left(1+\frac{\hbar\omega_L}{2k_B T}\right)$$

The approximations above are valid in the high-temperature limit.

The magnetization can be calculated by adding the contributions from each sub-ensemble of nuclei corresponding to the $m=-1/2$ and $m=+1/2$ levels. The nuclei with $m=+1/2$ contribute with the magnetic dipole moment component $\mu_z^{(+)}=+\gamma\hbar/2$, whereas those with $m=-1/2$ contribute with $\mu_z^{(-)}=-\gamma\hbar/2$:

$$M_0 = n_+\mu_z^{(+)} + n_-\mu_z^{(-)} = (n_+ - n_-)\frac{\gamma\hbar}{2} \cong \frac{n_0}{2}\frac{\gamma\hbar}{2}\frac{\hbar\omega_L}{k_B T} = \frac{n_0\gamma^2\hbar^2 B_0}{4k_B T}$$

This is the Curie expression for the magnetization of an ensemble of spin 1/2 nuclei.

P2.5 - Evaluate from (2.2.5) the numerical value of the magnetic susceptibility (defined as $\chi = \mu_0 M_0/B_0$, where $\mu_0 = 4\pi \times 10^{-7}$ T.m.A^{-1} is the permeability of free space) of protons in water at room temperature. Compare with the corresponding value for free unpaired electrons with the same concentration and under the same conditions. Compare also with typical values of magnetic susceptibility in diamagnetic materials (see, for example, [1,7]).

Solution
The expression for χ is:

$$\chi = \frac{\mu_0 n_0 \gamma^2 \hbar^2}{4k_B T}$$

The volume concentration of protons (n_0) is obtained from the density of pure water and its molecular weight, which gives:

$$n_0 = (1.0 \text{ g/cm}^3)(18 \text{ g/mol})^{-1}(6.02 \times 10^{23} \text{ mol}^{-1})(2) = 6.7 \times 10^{28} \text{ m}^{-3}$$

The factor of 2 is to take into account the existence of two hydrogen atoms in each molecule. Therefore, we have in SI units:

$$\chi = 4.0 \times 10^{-9} \quad \text{(for protons in water)}$$

For comparison, the magnetic susceptibility absolute values of typical diamagnetic substances fall into the range 10^{-5}–10^{-6}, which means a value much larger than the static nuclear magnetic susceptibility.

For free electrons, the magnitude of the magnetic dipole moment is obtained from the Bohr magneton ($\mu_B = \frac{e\hbar}{2m_e} = 9.2740 \times 10^{-24}$ J T^{-1}) and the g factor (2.0023). The proton gyromagnetic ratio γ should then

Problems with solutions

be replaced by $\frac{g\mu_B}{\hbar}$ in the expression for χ. A similar calculation of χ for free electrons using the same concentration n_0 above then leads to:

$$\chi = 1.7 \times 10^{-3} \quad \text{(for free electrons)}$$

P2.6 - Verify the numeric values of the Larmor frequencies for protons and electrons, both in a 5 T static magnetic field.

Solution
For protons, using the γ value given in Problem 2.3, we have:

$$\omega_L^{(p)} = \gamma B_0 = (26.7522 \times 10^7 \text{ rad s}^{-1} \text{ T}^{-1}) \times 5\text{T} = 1.34 \times 10^9 \text{ rad s}^{-1}$$

Therefore, the frequency is about 213 MHz (radiofrequency).
For free electrons, the magnitude of the magnetic dipole moment is obtained from the Bohr magneton ($\mu_B = \frac{e\hbar}{2m_e} = 9.2740 \times 10^{-24} \text{ J T}^{-1}$) and the g factor (2.0023). Therefore:

$$\omega_L^{(e)} = \frac{g\mu_B}{\hbar} B_0 = (17.608 \times 10^{10} \text{ rad s}^{-1} \text{ T}^{-1}) \times 5\text{T} = 8.80 \times 10^{11} \text{ rad s}^{-1}$$

Therefore, the frequency is about 140 GHz (microwave).

P2.7 - Do again *Problem P2.2* now employing the density matrix formalism. That is, use Equation (2.5.1) to obtain again the equations for the time evolution of the expectation values $\langle I_x \rangle$, $\langle I_y \rangle$, and $\langle I_z \rangle$ and interpret the result to show the Larmor precession.

Solution
As described in the text, the time evolution of any ensemble operator can be determined using the deviation density matrix. Under action of the Zeeman Hamiltonian, the deviation density operator evolves in time as:

$$\Delta\rho_t = e^{-(i/\hbar)\mathcal{H}t} \Delta\rho_0 e^{(i/\hbar)\mathcal{H}t} = e^{i\omega_L t I_z} \Delta\rho_0 e^{-i\omega_L t I_z}$$

In thermal equilibrium, for example, we have $\Delta\rho_0 = \alpha I_z$ (Equation (2.5.14)) and thus

$$\Delta\rho_t = e^{i\omega_L t I_z} (\alpha I_z) e^{-i\omega_L t I_z} = \alpha I_z.$$

The deviation density matrix in this case is therefore time-independent, as expected since we are dealing with thermal equilibrium.
Let us consider a more general situation with arbitrary initial deviation density matrix $\Delta\rho_0$ evolving in time under the Zeeman Hamiltonian. The calculation of the ensemble averages of the expectation values of the spin operator I_z is given below:

$$\langle I_z \rangle_t = \text{Tr}\{\Delta\rho_t I_z\} = \text{Tr}\{e^{i\omega_L t I_z} \Delta\rho_0 e^{-i\omega_L t I_z} I_z\}$$

$$= \text{Tr}\{e^{-i\omega_L t I_z} I_z e^{i\omega_L t I_z} \Delta\rho_0\} = \text{Tr}\{e^{-i\omega_L t I_z} e^{i\omega_L t I_z} I_z \Delta\rho_0\}$$

$$= \text{Tr}\{I_z \Delta\rho_0\} = \text{Tr}\{\Delta\rho_0 I_z\} = \langle I_z \rangle_0$$

We used here the general property of the trace operation: $\text{Tr}\{AB\} = \text{Tr}\{BA\}$. The result above shows once more that $\langle I_z \rangle$ is time independent, whatever the initial state of the ensemble.
The calculation of $\langle I_x \rangle$ is similar:

$$\langle I_x \rangle_t = \text{Tr}\{\Delta\rho_t I_x\} = \text{Tr}\{e^{i\omega_L t I_z} \Delta\rho_0 e^{-i\omega_L t I_z} I_x\} = \text{Tr}\{e^{-i\omega_L t I_z} I_x e^{i\omega_L t I_z} \Delta\rho_0\}$$

The term $e^{-i\omega_L t I_z} I_x e^{i\omega_L t I_z}$ can be calculated from the "sandwich theorem", Equation (2.6.13):

$$e^{-i\omega_L t I_z} I_x e^{i\omega_L t I_z} = I_x \cos\omega_L t + I_y \sin\omega_L t$$

Therefore:

$$\langle I_x \rangle_t = \text{Tr}\{(I_x \cos\omega_L t + I_y \sin\omega_L t)\Delta\rho_0\} = \text{Tr}\{I_x \Delta\rho_0\} \cos\omega_L t + \text{Tr}\{I_y \Delta\rho_0\} \sin\omega_L t$$

$$= \langle I_x \rangle_0 \cos\omega_L t + \langle I_y \rangle_0 \sin\omega_L t$$

The result for $\langle I_y \rangle_t$ is similar:

$$\langle I_y \rangle_t = -\langle I_x \rangle_0 \sin\omega_L t + \langle I_y \rangle_0 \cos\omega_L t$$

As in Problem P2.2, if we take $\langle I_y \rangle_0 = 0$ as an initial condition, then we arrive at:

$$\langle I_x \rangle_t = \langle I_x \rangle_0 \cos\omega_L t$$
$$\langle I_y \rangle_t = -\langle I_x \rangle_0 \sin\omega_L t$$
$$\langle I_z \rangle_t = \langle I_z \rangle_0$$

These are finally the equations of the precession motion of the vector $\langle \mathbf{I} \rangle$ (i.e., of the ensemble average of the expectation value of the nuclear spin) about the z axis in the left-hand direction with angular frequency of magnitude ω_L.

P2.8 - Show that in an isotropic liquid the term $3\cos^2\theta_{12} - 1$ in Equation (2.7.12) is averaged out due to random molecular motions and interpret the consequences of this result.

Solution

The average can be calculated by integrating over the surface of a sphere centered on one nucleus and considering that the internuclear vector can point to any direction on the surface of this sphere with equal chance. Let us use θ, ϕ, and Ω as the polar, azimuthal, and solid angles associated with the internuclear vector. Therefore:

$$\langle 3\cos^2\theta - 1 \rangle = \frac{\int d\Omega (3\cos^2\theta - 1)}{\int d\Omega} = \frac{1}{4\pi} \int_0^{2\pi} d\phi \int_0^{\pi} d\theta \sin\theta (3\cos^2\theta - 1)$$

$$= \frac{1}{2} \int_{-1}^{1} (3u^2 - 1)\, du = 0$$

The usual substitution $u = -\cos\theta$ was used in the last step. This result shows then that the isotropic molecular motion leads to a complete vanishing of the dipolar interaction. The same is true for all other interactions with similar angular dependencies (anisotropic part of the chemical shift, anisotropic part of the J-coupling, first-order quadrupolar interaction), as long as the motion is fast and isotropic as it occurs in liquids. If the motion is restricted, then the averaging process can be only partial (as in liquid crystals) or inexistent (as in rigid solids).

P2.9 - Obtain the electrostatic interaction energy for the charge distribution in Figure 2.5 and discuss its orientational dependence.

Solution

The electrostatic interaction energy between the charge distribution and the electric point charges of magnitude q_0 is calculated by the expression $W = \int \rho V\, d\Lambda$, where ρ is the charge density of the distribution, V is the electric potential generated by the point charges, and the integral in $d\Lambda$ is over the volume of the distribution. The power series expansion of the electric potential about the origin leads to:

$$W = \int \rho V\, d\Lambda = \left(\int \rho\, d\Lambda\right) V(0) + \sum_\alpha \left(\int \rho x_\alpha\, d\Lambda\right)\left(\frac{\partial V}{\partial x_\alpha}\right)_0$$

$$+ \frac{1}{2}\sum_{\alpha,\beta} \left(\int \rho x_\alpha x_\beta\, d\Lambda\right)\left(\frac{\partial^2 V}{\partial x_\alpha \partial x_\beta}\right)_0 + \cdots$$

The subscript indicates that the derivatives are all calculated at the centre of symmetry of the charge distribution, taken as the origin. The first term, involving the total charge of the distribution, is a constant, not related to the orientation of the distribution and it can be taken as zero with no loss of generality, as this is just a matter of choice of the reference for the electric potential. The second term involves the electric dipole moment of the distribution (first-order in x_α, which gives the length scale of the distribution). Due to symmetry requirements, all atomic nuclei have no electric dipole moments. Even if the charge distribution possesses a non-zero electric dipole moment, the first-order term gives no contribution because the first-derivatives of the potential $\left(\frac{\partial V}{\partial x_\alpha}\right)_0$ are equal (with a sign change) to the electric field components. But the resultant electric field is zero at the origin, as can be easily seen by adding the contributions from charges of opposite sign located on the Cartesian axes. Therefore, the first non-zero term in the expression of W is the second-order term, which is associated with the electric quadrupole moment of the distribution and the EFG produced by the point charges. By using the classic expression for the Coulomb electric potential V generated by point charges, it is easy to calculate the second-derivatives of V and to arrange them as components of a second-rank matrix $V_{\alpha\beta} = \left(\frac{\partial^2 V}{\partial x_\alpha \partial x_\beta}\right)_0$. The result is:

$$\mathbf{V} = \frac{1}{4\pi\epsilon_0} \frac{4q_0}{d^3} \begin{bmatrix} -1 & 0 & 0 \\ 0 & -1 & 0 \\ 0 & 0 & 2 \end{bmatrix}$$

The system of Cartesian axes chosen is therefore the principal axis system (PAS) of the tensor $\mathbf{V}$. Using the definitions given in the text, we have therefore:

$$eq = \frac{1}{4\pi\epsilon_0} \frac{8q_0}{d^3} \quad \text{and} \quad \eta = 0$$

The Laplace equation satisfied by the electric potential leads to $\sum_\alpha V_{\alpha\alpha} = 0$, which shows the traceless character of the tensor $\mathbf{V}$. It is then usual to write the second-order term of W in the form:

$$W^{(2)} = \frac{1}{2} \sum_{\alpha,\beta} \left(\int \rho x_\alpha x_\beta \, d\Lambda\right) V_{\alpha\beta} = \frac{1}{6} \sum_{\alpha,\beta} Q_{\alpha\beta} V_{\alpha\beta}$$

The components of the electric quadrupole moment of the distribution are defined by:

$$Q_{\alpha\beta} = \int (3x_\alpha x_\beta - r^2 \delta_{\alpha\beta}) \rho \, d\Lambda$$

where $r^2 = \sum_\alpha x_\alpha^2$ and $\delta_{\alpha\beta}$ is the Kronecker delta. It is easy then to see that the tensor $\mathbf{Q}$ is also traceless, i.e., $\sum_\alpha Q_{\alpha\alpha} = 0$.
Using the expression given above for $V_{\alpha\beta}$, we arrive at:

$$W^{(2)} = \frac{1}{6} \sum_{\alpha,\beta} Q_{\alpha\beta} V_{\alpha\beta} = \frac{1}{6} eq \left(\frac{-Q_{xx} - Q_{yy}}{2} + Q_{zz}\right) = \frac{1}{4} eq Q_{zz}$$

The term Q_{zz} is a component of the electric quadrupole tensor written in the PAS of the tensor $\mathbf{V}$. The PAS of the tensor $\mathbf{Q}$ is characterized by the symmetry axis of the distribution (axis z'), which makes an angle θ with z. In this system, we have:

$$Q_{z'z'} = \int (3z'^2 - r'^2) \rho \, d\Lambda = eQ$$

The parameter Q is characteristic of the distribution. The relation between Q_{zz} and $Q_{z'z'}$ can be obtained using the Wigner rotation matrices that describe the change from one coordinate system to another [4,10]. The result is:

$$Q_{zz} = \frac{3\cos^2\theta - 1}{2} Q_{z'z'}$$

Finally we get:

$$W^{(2)} = \frac{e^2 q Q}{8} (3\cos^2\theta - 1)$$

This expression clearly shows the angular dependency of the electrostatic energy of interaction between the charge distribution and the EFG. If the charge distribution rotates fast as a rigid body about its symmetry axis z', then this angular dependency refers to the orientation of the angular momentum itself, which is in close analogy with the electric quadrupolar nuclear spin interaction described in the text. Assuming $eQ > 0$, which means a distribution of positive charge density ($\rho > 0$) elongated along the z' axis, then the most stable situation (minimum of W) corresponds to $\cos\theta = 0$, i.e., when the distribution is as close as possible to the negative point charges in the xy plane.

REFERENCES

[1] C. Kittel, *Introduction to Solid State Physics*, 8th edition (John Wiley & Sons, New York, 2004).
[2] C. Cohen-Tannoudji, B. Diu, F. Laloe, *Quantum Mechanics* (John Wiley & Sons, New York, 1977).
[3] K.S. Krane, *Introductory Nuclear Physics* (John Wiley & Sons, New York, 1988).
[4] C.P. Slichter, *Principles of Magnetic Resonance*, 3rd edition (Springer, Berlin, 1990).
[5] M.H. Levitt, *Spin Dynamics* (John Wiley & Sons, Chichester, 2001).
[6] F. Reif, *Fundamentals of Statistical and Thermal Physics* (McGraw-Hill, USA, 1965).
[7] A.P. Guimaraes, *Magnetism and Magnetic Resonance in Solids* (John Wiley & Sons, New York, 1998).
[8] A. Abragam, *Principles of Nuclear Magnetism*, The International Series of Monographs on Physics (Oxford University Press, 1986).
[9] J.J. Sakurai, *Modern Quantum Mechanics* (Addison-Wesley, USA, 1994).
[10] K.J.D. MacKenzie, M.E. Smith, *Multinuclear Solid-State NMR of Inorganic Materials*, R.W. Cahn (Ed.), Pergamon Materials Series (Amsterdam, 2002).
[11] R. Balian, D.T. Haar, J.F. Gregg, *From Microphysics to Macrophysics: Methods and Applications of Statistical Physics* (Springer, Berlin, 1992).
[12] T.C. Farrar, J.E. Harriman, *Density Matrix Theory and its Application in NMR Spectroscopy* (The Farragut Press, Madison, 1992).
[13] R.R. Ernst, G. Bodenhausen, A. Wokaun, *Principles of Nuclear Magnetic Resonance in One and Two Dimensions* (Clarendon Press, Oxford, 1987).
[14] K. Schmidt-Rohr, H.W. Spiess, *Multidimensional Solid-State NMR and Polymers* (Academic Press, London, 1994).
[15] E.A. Turov, M.P. Petrov, *Nuclear Magnetic Resonance in Ferro and Antiferromagnets* (Halsted Press, New York, 1972).
[16] J.D. Jackson, *Classical Electrodynamics*, 3rd edition (John Wiley & Sons, New York, 1999).
[17] U. Haeberlen, *High Resolution NMR in Solids* (Academic Press, 1976).
[18] V.H. Schmidt, Pulse response in the presence of quadrupolar splitting, in: *Proc. Ampère International Summer School II*, 1971, pp. 75–83.
[19] M.E. Smith, E.R.H. van Eck, Recent advances in experimental solid state NMR methodology for half-integer spin quadrupolar nuclei, *Prog. Nucl. Magn. Reson. Spectr.* **34** (1999) 159.
[20] A. Wokaun, R.R. Ernst, Selective excitation and detection in multilevel spin systems: Application of single transition operators, *J. Chem. Phys.* **67** (1977) 1752.
[21] S. Vega, A. Pines, Operator formalism for double quantum NMR, *J. Chem. Phys.* **66** (1977) 5624.
[22] F.A. Bonk, E.R. de Azevedo, R.S. Sarthour, J.D. Bulnes, J.C.C. Freitas, A.P. Guimarães, I.S. Oliveira, T.J. Bonagamba, Quantum logical operations for spin 3/2 quadrupolar nuclei monitored by quantum state tomography, *J. Magn. Reson.* **175** (2005) 226.
[23] P.P. Man, J. Klinowski, A. Trokiner, H. Zanni, P. Papon, Selective and non-selective NMR excitation of quadrupolar nuclei in the solid-state, *Chem. Phys. Lett.* **151** (1988) 143.
[24] H.O. Wijewardane, C.A. Ullrich, Coherent control of intersubband optical bistability in quantum wells, *Appl. Phys. Lett.* **84** (2004) 3984.
[25] S.A. Smith, W.E. Palke, J.T. Gerig, The Hamiltonians of NMR: Part IV: NMR relaxation, *Concepts in Magnetic Resonance* **6** (1994) 137.
[26] T.C. Stringfellow, *Density Matrix Theory and Nuclear Spin Relaxation* (University of Wisconsin–Madison School of Pharmacy, Madison, 2003).
[27] A.G. Redfield, On the theory of relaxation processes, *IBM J. Res. Develop.* **1** (1957) 19.

References

[28] A.G. Redfield, *Advances in Magnetic Resonance* (Academic Press Inc., New York, 1965), vol. 1, pp. 1–32.

[29] E.O. Stejskal, J.D. Memory, *High Resolution NMR in the Solid State: Fundamentals of CP/MAS* (Oxford University Press, 1994).

[30] L.R. Sarles, R.M. Cotts, Double nuclear magnetic resonance and the dipole interaction in solids, *Phys. Rev.* **111** (1958) 853.

[31] A.E. Bennett, C.M. Rienstra, M. Auger, K.V. Lakshmi, R.G.J. Griffin, Heteronuclear decoupling in rotating solids, *Chem. Phys.* **103** (1995) 6951.

[32] J.S. Waugh, L.M. Huber, U. Haeberlen, Approach to high-resolution NMR in solids, *Phys. Rev. Lett.* **20** (1968) 180.

[33] P. Mansfield, Symmetrized pulse sequences in high-resolution NMR in solids, *J. Phys. C Solid State Phys.* **4** (1971) 1444.

[34] W.K. Rhim, D.D. Elleman, R.W. Vaughan, Analysis of multiple pulse NMR in solids, *J. Chem. Phys.* **59** (1973) 3740–3749.

[35] D.P. Burum, W.K. Rhim, Analysis of multiple pulse NMR in solids 3, *J. Chem. Phys.* **71** (1979) 944.

[36] M. Lee, W.I. Goldburg, Nuclear magnetic resonance line narrowing by a rotating RF field, *Phys. Rev.* **140** (1965) 1261.

[37] E.R. Andrew, A. Bradbury, R.G. Eades, Removal of dipolar broadening of nuclear magnetic resonance spectra of solids by specimen rotation, *Nature* **183** (1959) 1802.

[38] I.J. Lowe, Free induction decays of rotating solids, *Phys. Rev. Lett.* **2** (1959) 285–287.

[39] E.R. Andrew, R.G. Eades, Nuclear magnetic resonance in diamagnetic materials – possibilities for high-resolution nuclear magnetic resonance spectra of crystals, *Discuss. Faraday Soc.* (1962) 38.

[40] M.M. Maricq, J.S. Waugh, NMR in rotating solids, *J. Chem. Phys.* **70** (1979) 3300.

[41] A. Pines, M.G. Gibby, J.S. Waugh, Proton-enhanced NMR of dilute spins in solids, *J. Chem. Phys.* **59** (1973) 569.

[42] J. Schaefer, E.O. Stejskal, C-13 Nuclear magnetic resonance of polymers spinning at magic angle, *J. Am. Chem. Soc.* **98** (1976) 1031.

[43] L.M.K. Vandersypen, C.S. Yannoni, I.L. Chuang, Liquid state NMR quantum computing, in: D.M. Grant, R.K. Harris (Eds.), *Encyclopedia of Nuclear Magnetic Resonance* (John Wiley & Sons, Chichester, 2002), pp. 687–397.

[44] E. Fukushima, S.B.W. Roeder, *Experimental Pulse NMR: a Nuts and Bolts Approach* (Addison-Wesley, 1981).

[45] E.O. Brigham, *The Fast Fourier Transform: An Introduction to Its Theory and Application* (Prentice Hall, 1973).

[46] D. Reichert, G. Hempel, Receiver imperfections and CYCLOPS: An alternative description, *Concepts in Magnetic Resonance* **14** (2002) 130.

– 3 –

Fundamentals of Quantum Computation and Quantum Information

The fact that a body can act on to another, completely separated from each other, through the vacuum, with no intermediate interaction, sounds for me a great absurd, so that no man, with the philosophical capability of thinking can accept. – Isaac Newton

Those who do not get shocked with Quantum Mechanics did not understand it. – Niels Bohr

Whoever claims to have understood Quantum Mechanics is lying. – Richard Feynman

Quantum Mechanics: Calculus with black magic – Albert Einstein

In this chapter, a revision of some basic principles of quantum mechanics is presented, emphasizing those which are particularly interesting for Quantum Computation and Quantum Information. These concepts and results will be developed in the context of NMR QIP in the next chapters. The quantum logic gates of one and two qubits, along with their circuit notation are introduced. Two important applications of entanglement are discussed: superdense coding and teleport. The basic principles of the main quantum algorithms are presented, through their circuit notation. The realization of quantum computation in phase space is also discussed, and the use of the scattering circuit in order to obtain the discrete Wigner function of quantum systems, as well as the principles of quantum simulation. In the last section, a quantum algorithm for determining eigenvalues and eigenvectors is discussed. Most of this chapter is based in the excellent book of Michael Nielsen and Isaac Chuang [1], which must be consulted for those interested in a more detailed discussion. We keep the notation used in that book, which has become current in the literature. Other recommended textbooks about the subject can be found in References [2,3] and [4].

3.1 HISTORICAL DEVELOPMENT

Quantum Mechanics is a set of mathematical rules upon which physical theories are constructed. Applying the rules of quantum mechanics, it is possible to calculate the observables of an isolated physical system, at any instant in time, once the Hamiltonian is known [5]. However, there is no precise prescription for finding a Hamiltonian of a specific system.

It is correct to state that Quantum Mechanics is the most well succeeded theory in physics. Since its creation up to nowadays it has been applied to various branches, varying from particle physics to condensed matter, passing through nuclear and atomic physics, astrophysics, etc. The success of quantum mechanics in condensed matter physics has been particularly astonishing: it goes from material structures, transport properties of metals, insulators and semiconductors, magnetic ordering, superconductivity and optical properties of matter.

However, until the beginning of the seventies, the experiments performed in order to test the models and theories built using quantum mechanics dealt with systems that contained a large number of constituents. Therefore, these tests were always made in an indirect form. For instance, the phenomena of magnetism and superconductivity are explained by quantum mechanics, but the experiments carried out on magnetic samples and superconductors always involve a large number of particles: spins in the case of magnetic systems and Cooper pairs in the case of superconductors. This means that predictions of the observables in theses systems – magnetization, specific heat, electrical current, etc. – must be done using statistical averages, missing the information about the fundamental quantum correlations between the individual particles. However, since the seventies, developments in several areas allowed the experiments to be performed only with a few particles, making visible important quantum effects.

Quantum Computing (QC) may possibly be the most remarkable proposal of a practical application of quantum mechanics. For didactic purposes, we consider Quantum Information as the area of investigation in which the identification and the study of the quantum resources that can be used for information processing, whereas Quantum Computation as the application of such resources for building logical gates and algorithms. The elements of quantum information and quantum computation will be discussed in this chapter. The historical development of the QC and QI is summarized in Table 3.1, since its beginning

Table 3.1.

Year	Fact
1973	– The possibility of reversible classical computation was demonstrated by Charles Bennett.
1982	– First proposal of quantum computer presented by Paul Benioff based on Charles Bennett's work (1973).
1984	– Charles Bennett and Gilles Brassard create the quantum cryptography protocol BB84.
1985	– David Deutsch creates the first quantum algorithm.
1993	– Peter Shor creates the factorizing algorithm. In this same year quantum teleportation is created by Charles Bennett and collaborators.
1994	– Lov Grover discovers the quantum search algorithm.
1996	– A group of scientists working for IBM implements experimentally the protocol BB84 using photons in optical fibres, used in telecommunications.
1997	– Neil Gershenfeld and Isaac Chuang propose a way to prepare pseudo-pure states breaking out the path for CQ through Nuclear Magnetic Resonance (NMR).
1998	– This was an important year for NMR Quantum Computation. Several logical gates were successfully implemented using NMR. The quantum search algorithm and teleportation were also experimentally tested.
2001	– Shor factorizing algorithm is implemented using NMR.
2003	– Demonstration of entanglement between the spins of a nucleus and the electron, in the same molecule, combing the techniques of NMR and Electron Paramagnetic Resonance (EPR).
2004	– Single spin detection by magnetic resonance force microscopy. First step towards the quantum states determination of single qubits.
2005	– NMR on a chip. Multiple coherences of nuclear spin states created and detected electrically on a nanoscale gallium arsenide structure.

to nowadays, emphasizing the contributions of NMR to this area, due to the convenience of this particular technique for QIP [6].

3.2 THE POSTULATES OF QUANTUM MECHANICS

The theory of quantum mechanics is based upon four postulates, which are stated below [5]:

- Postulate I – All physical systems are associated with a complex vector space. This is known as Hilbert space, and its elements are complex vectors, $|\psi\rangle$, called kets, which represent the quantum state of the system. The conjugate of a ket is represented by a another vector called a bra, $\langle\psi|$.
- Postulate II – The time evolution of a quantum system, which does not interact with its neighborhood, is processed through unitary transformations as described in Equation (3.2.1), where $U^\dagger U = \mathbf{1}$, being $\mathbf{1}$ the identity matrix.

$$|\psi(t-t_0)\rangle = U(t-t_0)|\psi(t_0)\rangle \tag{3.2.1}$$

In the case that the Hamiltonian does not depend on t, the unitary evolution, U, is given by:

$$U(t-t_0) = \exp\left[-\frac{i}{\hbar}\mathcal{H}(t-t_0)\right] \tag{3.2.2}$$

Physically, unitary transformations represent processes that are reversible in time. In fact, the application of the conjugate operator $U^\dagger$ on both sides of Equation (3.2.1) will make the system return to its original quantum state:

$$|\psi(t_0)\rangle = U(t-t_0)^\dagger|\psi(t-t_0)\rangle \tag{3.2.3}$$

Another important property of the unitary transformations is the conservation of the scalar product:

$$\langle\psi(t_0)|U^\dagger U|\psi(t_0)\rangle = \langle\psi(t_0)|\psi(t_0)\rangle = \langle\psi(t-t_0)|\psi(t-t_0)\rangle \tag{3.2.4}$$

- Postulate III – Measurements, in quantum mechanics, are represented by sets of operators, called measurement operators $\{M_m\}$, where the index m reefers to one of the possible results. The probability, $p(m)$, for a particular value to be found in a measurement is the expected value of the corresponding measurement operator, which can be calculated using the systems ket, $|\psi\rangle$:

$$p(m) = \langle\psi|M_m^\dagger M_m|\psi\rangle \tag{3.2.5}$$

After the measurement, the quantum state of the system becomes:

$$|\psi_m\rangle = \frac{M_m}{\sqrt{p(m)}}|\psi\rangle \tag{3.2.6}$$

The normalization of the probabilities, $\sum_m p(m) = 1$, plus the hypothesis $\langle \psi | \psi \rangle = 1$ and Equation (3.2.5) imply to the completeness relation:

$$\sum_m M_m^\dagger M_m = 1 \qquad (3.2.7)$$

- Postulate IV – The Hilbert space elements of a quantum system composed of two subsystems $A - B$ is formed by the tensor product between the kets of the individual systems:

$$|\psi_{A-B}\rangle = |\psi_A\rangle \otimes |\psi_B\rangle \qquad (3.2.8)$$

This rule can be extended for a systems with N subsystems:

$$|\psi_{A-B-\cdots-N}\rangle = |\psi_A\rangle \otimes |\psi_B\rangle \otimes \cdots \otimes |\psi_N\rangle \qquad (3.2.9)$$

In this book we will deal only with discrete and finite Hilbert spaces.

3.3 QUANTUM BITS

The classical information unit is the *binary digit*, or bit. One bit can assume the logical values "0" or "1". In the computers, bits are physically represented by the presence or absence of electrical currents, travelling through the electronic components inside the chips. The presence of the current indicates that the bit is in the logical state "1" and its absence indicates that the bit is the logical state "0". Obviously, a bit cannot be at two logical states at the same time [7].

Analogously, the unit of information in Quantum Information and Quantum Computation is the quantum bit, or *qubit*, for short. A qubit can assume the logical values 0 or 1. However, it can also be in a logical state containing any linear combination of them, thanks to laws of quantum mechanics [8]. Physically, qubits can be represented by any quantum object with two well defined and distinct eigenstates. Examples of qubits are the photon polarization states, electrons in two-level atoms (as an approximation) and nuclear spins under the influence of a magnetic field.

The eigenstates of a qubit are represented by $|0\rangle$ and $|1\rangle$, defined by the two vectors:

$$|0\rangle = \begin{bmatrix} 1 \\ 0 \end{bmatrix}; \qquad |1\rangle = \begin{bmatrix} 0 \\ 1 \end{bmatrix} \qquad (3.3.1)$$

The set $\{|0\rangle, |1\rangle\}$ forms a two dimensional basis in Hilbert's space of one qubit, and is called *computational basis*. For the case of spin 1/2 particle, the logical state 0 can be represented by the spin up state ($|0\rangle \equiv |\uparrow\rangle$), whereas the logical state 1 can be represented by the spin down state ($|1\rangle \equiv |\downarrow\rangle$). Other orthonormal basis can be built from the computational basis, such as $|+\rangle$ and $|-\rangle$:

$$|+\rangle = \frac{1}{\sqrt{2}} \begin{bmatrix} 1 \\ 1 \end{bmatrix}; \qquad |-\rangle = \frac{1}{\sqrt{2}} \begin{bmatrix} 1 \\ -1 \end{bmatrix} \qquad (3.3.2)$$

3.4. Quantum logic gates

The generic state of a qubit is represented by a linear combination of the two eigenkets:

$$|\psi\rangle = \alpha|0\rangle + \beta|1\rangle \tag{3.3.3}$$

where the coefficients are complex numbers related to each other by, $|\alpha|^2 + |\beta|^2 = 1$. This state can be parametrized by angles θ and ϕ, such as $\alpha \equiv \cos\theta/2$ and $\beta \equiv \exp(i\phi)\sin\theta/2$:

$$|\psi\rangle = \cos\frac{\theta}{2}|0\rangle + e^{i\phi}\sin\frac{\theta}{2}|1\rangle \tag{3.3.4}$$

This representation allows a geometric visualization of the qubit quantum state as a point on the surface of a unit radius sphere, called *Bloch sphere*. The most important points on Bloch sphere are shown on the table below, adapted from Ref. [1].

θ	ϕ	$|\psi\rangle$	Observation
0	—	$|0\rangle$	North pole of the Bloch's sphere
π	—	$|1\rangle$	South pole of the Bloch's sphere
$\pi/2$	0 or π	$(|0\rangle \pm |1\rangle)/\sqrt{2}$	Equator line right on the x axis
$\pi/2$	$\pi/2$ or $-\pi/2$	$(|0\rangle \pm i|1\rangle)/\sqrt{2}$	Equator line right on the y axis

The power of quantum computation comes from the existence of superposition of states of qubits, particularly entanglement, and the ability to manipulate them through unitary transformations, as will be seen in the next sections.

3.4 QUANTUM LOGIC GATES

The Hilbert space for one qubit has only two dimensions. Quantum information processing requires unitary transformations operating on states of one and two qubits, called logic gates. Some important examples of unitary transformations of one qubit are the Pauli matrices:[1]

$$X = \begin{bmatrix} 0 & 1 \\ 1 & 0 \end{bmatrix}; \quad Y = \begin{bmatrix} 0 & -i \\ i & 0 \end{bmatrix}; \quad Z = \begin{bmatrix} 1 & 0 \\ 0 & -1 \end{bmatrix} \tag{3.4.1}$$

These matrices, plus the 2×2 identity matrix, form a basis in the 2×2 matrix space, so that any operation of one qubit can be decomposed as a linear combination of the four matrices. Notice that $X = X^\dagger$, $Y = Y^\dagger$ and $Z = Z^\dagger$, and also that $XX^\dagger = \mathbf{1}$, $YY^\dagger = \mathbf{1}$ and $ZZ^\dagger = \mathbf{1}$. The action of each one of these operations on a generic quantum state are written below:

$$X|\psi\rangle = \beta|0\rangle + \alpha|1\rangle$$
$$Y|\psi\rangle = -i\beta|0\rangle + i\alpha|1\rangle \tag{3.4.2}$$
$$Z|\psi\rangle = \alpha|0\rangle - \beta|1\rangle$$

---
[1] Here, as in the book of Michael Nielsen and Isaac Chuang, we will use two notations for the Pauli matrices: X, Y and Z whenever they represent quantum logic gates, and the usual σ_x, σ_y and σ_z in a more physical context.

There are also other three important one-qubit gates, which are the phase gate (S), the $\pi/8$ (T) gate and the Hadamard gate. The operators S and T are defined by the matrices:

$$S = \begin{bmatrix} 1 & 0 \\ 0 & i \end{bmatrix}; \quad T = \begin{bmatrix} 1 & 0 \\ 0 & e^{i\pi/4} \end{bmatrix} = e^{i\pi/8} \begin{bmatrix} e^{-i\pi/8} & 0 \\ 0 & e^{i\pi/8} \end{bmatrix} \quad (3.4.3)$$

and their actions on a qubit state are:

$$S|\psi\rangle = \alpha|0\rangle + i\beta|1\rangle$$
$$T|\psi\rangle = \alpha|0\rangle + e^{i\pi/4}\beta|1\rangle \quad (3.4.4)$$

The Hadamard gate can be decomposed into a linear combination of the X and Z operators:

$$H = \frac{1}{\sqrt{2}} \begin{bmatrix} 1 & 1 \\ 1 & -1 \end{bmatrix} = \frac{X+Z}{\sqrt{2}} \quad (3.4.5)$$

The application of Hadamard in one of the states of the computational basis creates a superposition state:

$$H|0\rangle = \frac{|0\rangle + |1\rangle}{\sqrt{2}} \quad H|1\rangle = \frac{|0\rangle - |1\rangle}{\sqrt{2}} \quad (3.4.6)$$

One important property of H is its self-reversibility: $H^2 = \mathbf{1}$.

3.4.1 Some examples of application of the postulates

A set of measurement operators for one qubit state is:

$$M_0 \equiv |0\rangle\langle 0|; \quad M_1 \equiv |1\rangle\langle 1| \quad (3.4.7)$$

These operators are projectors: they project a quantum state onto the computational basis. Notice that they are Hermitians, but not unitary. Therefore, they represent measurement processes which are not reversible. Using the postulate III, it is possible to calculate the probability of finding a qubit, initially in a superposition of states, in either state $|0\rangle$ and $|1\rangle$, as shown on Equations (3.4.8).

$$p(0) = \langle\psi|M_0^\dagger M_0|\psi\rangle = |\alpha|^2; \quad p(1) = \langle\psi|M_1^\dagger M_1|\psi\rangle = |\beta|^2 \quad (3.4.8)$$

After the measurement, the quantum state of the system will collapse to one of the following states:

$$|\psi_0\rangle = \frac{\alpha}{|\alpha|}|0\rangle \quad \text{or} \quad |\psi_1\rangle = \frac{\beta}{|\beta|}|1\rangle \quad (3.4.9)$$

The coefficients $\alpha/|\alpha|$ and $\beta/|\beta|$ are global phases and can be neglected.

3.4. Quantum logic gates

Using the postulate IV, it is possible to construct the Hilbert space for systems containing two or more qubits. For a two-qubit system, the dimension of the Hilbert space is 4×4, since it is composed by vectors (kets) and matrices (operators), calculated using the tensor product of each vector and matrix for the individual qubit, as may be seen on Equations (3.4.10) and (3.4.11), where both representations, kets and vectors, are shown:

$$\{|0\rangle, |1\rangle\} \otimes \{|0\rangle, |1\rangle\} = \{|00\rangle, |01\rangle, |10\rangle, |11\rangle\} \qquad (3.4.10)$$

$$|00\rangle \equiv \begin{bmatrix} 1 \\ 0 \\ 0 \\ 0 \end{bmatrix}; \quad |01\rangle \equiv \begin{bmatrix} 0 \\ 1 \\ 0 \\ 0 \end{bmatrix}; \quad |10\rangle \equiv \begin{bmatrix} 0 \\ 0 \\ 1 \\ 0 \end{bmatrix}; \quad |11\rangle \equiv \begin{bmatrix} 0 \\ 0 \\ 0 \\ 1 \end{bmatrix} \qquad (3.4.11)$$

The matrix representation for operators that act in only one qubit of a system containing two qubits can be constructed by calculating the tensorial product between one qubit operator and the 2×2 identity matrix:

$$O_a = O \otimes \mathbf{1}; \qquad O_b = \mathbf{1} \otimes O \qquad (3.4.12)$$

where O represents any qubit unitary operator ($O = X, Y, Z$, etc.). Here, the convention $|ab\rangle$ for the quantum state of composite system was used, so that O_a indicates that the operator acts on the first (a) qubit, while O_b acts on the second (b) one. These expressions can be easily generalized for an arbitrary number of qubits.

3.4.2 The controlled NOT – CNOT – gate

The controlled-not (CNOT) logic gate is essential for performing QIP. In fact, it can be proved that all quantum operations necessary for quantum computing can be achieved using only the CNOT and a set of one-qubit gates [1]. CNOT acts on a qubit of the system (called *target*), and changes its state, if the other qubit (called *control*) is in the state $|1\rangle$. If the control is in $|0\rangle$, nothing happens to the target. The matrix that represents the CNOT gates for a system containing two qubits are shown in Equations (3.4.13). For example, the CNOT$_a$ will change the state of the second (b) qubit if the first one is in the state $|1\rangle$.

$$\text{CNOT}_a = \begin{bmatrix} 1 & 0 & 0 & 0 \\ 0 & 1 & 0 & 0 \\ 0 & 0 & 0 & 1 \\ 0 & 0 & 1 & 0 \end{bmatrix} \quad \text{and} \quad \text{CNOT}_b = \begin{bmatrix} 1 & 0 & 0 & 0 \\ 0 & 0 & 0 & 1 \\ 0 & 0 & 1 & 0 \\ 0 & 1 & 0 & 0 \end{bmatrix} \qquad (3.4.13)$$

CNOT can be viewed also as the binary addition of two qubits, i.e., CNOT$_a|a,b\rangle = |a, a \oplus b\rangle$ and CNOT$_b|a,b\rangle = |a \oplus b, b\rangle$, where the symbol $\oplus$ represents the addition modulo 2, for which $0 \oplus 0 = 0$, $0 \oplus 1 = 1$ and $1 \oplus 1 = 0$.

$$\text{CNOT}_a|00\rangle = |00\rangle; \quad \text{CNOT}_b|00\rangle = |00\rangle$$

$$\text{CNOT}_a|01\rangle = |01\rangle; \quad \text{CNOT}_b|01\rangle = |11\rangle$$
$$\text{CNOT}_a|10\rangle = |11\rangle; \quad \text{CNOT}_b|10\rangle = |10\rangle \qquad (3.4.14)$$
$$\text{CNOT}_a|11\rangle = |10\rangle; \quad \text{CNOT}_b|11\rangle = |01\rangle$$

3.5 GRAPHICAL REPRESENTATION OF GATES AND QUANTUM CIRCUITS

Quantum circuits are diagrams that illustrate the operations necessary to implement a protocol, their time sequence and also the number of qubits present in the system [9]. They are composed of lines, one for each qubit, and symbols, which represent the quantum logic gates actions in one or more qubits. On Figure 3.1 it is shown the symbols used for one and two-qubit gates.

An example of a quantum circuit is illustrated in Figure 3.2. The upper line represents the qubit $|a\rangle$ and lower one the qubit $|b\rangle$. The operations appearing in the figure mean that the S gate is applied to the first qubit, whereas the gate T is applied to the second one. These operations are followed by the application of a two qubit operation U, and finally by the application of a Hadamard gate to the first qubit only. The whole process can be translated in mathematical language as $[H \otimes \mathbf{1}] \cdot U \cdot [S \otimes T]|ab\rangle$.

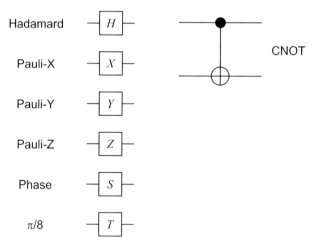

Figure 3.1 Quantum logic gates symbols for one and two-qubit operations. Adapted with permission from [1].

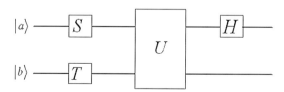

Figure 3.2 Generic graphical representation of a quantum circuit. Adapted with permission from [1].

3.5. Graphical representation of gates and quantum circuits

Figure 3.3 SWAP gate built from three CNOT gates. Adapted with permission from [1].

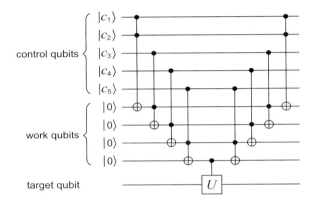

Figure 3.4 Various controlled operations applied to a target qubit. Adapted with permission from [1].

3.5.1 The SWAP logic gate

The logic gate called SWAP is built from a circuit containing only CNOT gates, as shown on Figure 3.3. The first CNOT is the controlled by the first qubit (CNOT$_a$), and second one is controlled by the second qubit (CNOT$_b$).

The action of a SWAP logic gate is defined by:

$$\text{SWAP}|ab\rangle = |ba\rangle \tag{3.5.1}$$

Having in mind that $a \oplus a \oplus b = b$ and $a \oplus b \oplus b = a$, one can demonstrate the action of the SWAP circuit:

$$\begin{aligned}
\text{CNOT}_a |a, b\rangle &= |a, a \oplus b\rangle \\
\text{CNOT}_b |a, a \oplus b\rangle &= |a \oplus a \oplus b, a \oplus b\rangle = |b, a \oplus b\rangle \\
\text{CNOT}_a |b, a \oplus b\rangle &= |b, a \oplus b \oplus b\rangle = |b, a\rangle
\end{aligned} \tag{3.5.2}$$

In general, many different kinds of controlled logic operations can be constructed, where the number of control qubits can vary, as well as the number of controlled (or target) qubits, as shown on Figure 3.4.

The operation illustrated on Figure 3.4 can be described as the application of the operator U on the target qubits, depending on the value of the product of the control qubits. This is mathematically represented by:

$$C_{xy}|x_1 \ldots x_n, y_1 \ldots y_m\rangle = |x_1 \ldots x_n\rangle U^{x_1 \ldots x_n} |y_1 \ldots y_m\rangle \tag{3.5.3}$$

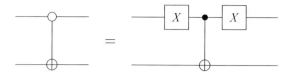

Figure 3.5 Controlled-NOT gate subject to the control qubit to be in the state $|0\rangle$. Adapted with permission from [1].

It is possible to construct controlled gates conditionally to the control qubit to be in the state $|0\rangle$. This is done simply by inverting the control qubit, using a X gate before and after the controlled operation, as shown in Figure 3.5.

There is an important relation between the two CNOT gates, of a two-qubit system, which is: $\text{CNOT}_a = H \otimes H \cdot \text{CNOT}_b \cdot H \otimes H$ (see [1]).

In classical computation, any logical operation can be done from combinations of the logic gate NAND (NOT-AND). The similar is true in quantum computing: any quantum operation can be implemented using a set of universal logic gates. Such a set is composed of the Hadamard (H), controlled NOT (CNOT), phase (S) and $\pi/8$ (T).

3.5.2 The Quantum Fourier Transform – QFT

The Fourier Transform operation is very useful, with a wide range of applications in physics, engineering, mathematics, etc. The discrete Fourier transform takes a vector of complex numbers to another vector, whose components are associated to the input vector, through the definition:

$$y_j \equiv \frac{1}{\sqrt{N}} \sum_{k=0}^{N-1} x_k e^{2\pi i j k / N} \qquad (3.5.4)$$

In Quantum Computing, the Quantum Fourier Transform (QFT) is behind the exponential gain in the speed of algorithms [10] such as Shor's factoring algorithm [11,12]. The operator QFT can be implemented using only $O(n^2)$ operations, whereas its classical analogue, the Fast Fourier Transform (FFT) requires about $O(n2^n)$ operations. Therefore, QFT is implemented exponentially faster than the FFT.

The QFT is an unitary transformation (see problems), which takes each eigenstate of the system to a superposition, as described by Equation (3.5.5), where n is the number of qubits in the system:

$$\text{QFT}|j\rangle = \frac{1}{\sqrt{2^n}} \sum_{k=0}^{2^n-1} e^{2\pi i j k / 2^n} |k\rangle \qquad (3.5.5)$$

An alternative definition of the QFT illustrates better the quantum operations needed for its implementation:

$$\text{QFT}|j_1 j_2 \cdots j_n\rangle = \frac{1}{\sqrt{2^n}} \left(|0\rangle + e^{2\pi i 0. j_n}|1\rangle\right)\left(|0\rangle + e^{2\pi i 0. j_{n-1} j_n}|1\rangle\right)$$
$$\cdots \left(|0\rangle + e^{2\pi i 0. j_1 j_2 \cdots j_n}|1\rangle\right) \qquad (3.5.6)$$

3.5. Graphical representation of gates and quantum circuits

where $0.j_l j_{l+1} \cdots j_m$ represents the binary fraction $\frac{j_l}{2} + \frac{j_{l+1}}{4} + \cdots + \frac{j_m}{2^{m-l+1}}$. This comes from the fact that it is always possible to describe a state $|j\rangle$ through its binary representation $|j_1 j_2 \cdots j_n\rangle$, reminding that $j = j_1 2^{n-1} + j_2 2^{n-2} + \cdots + j_n 2^0$. This demonstration is left as an exercise.

The relation above, Equation (3.5.6), is very useful, because it indicates the necessary operations to be performed on each qubit for the implementation of the QFT. For instance, starting from the first qubit, the first step is to apply a Hadamard gate, followed by a relative phase change, controlled by the other qubits, of the system. One can see from Equation (3.5.6) why a relative phase change is needed. This operation can be performed by some applications of the logic gate R_k:

$$R_k \equiv \begin{bmatrix} 1 & 0 \\ 0 & e^{2\pi i / 2^k} \end{bmatrix} \tag{3.5.7}$$

However, this operation must be controlled, i.e. the R_k gate is applied if the k-th qubit is in the $|1\rangle$ state (if the k-th qubit is in the state $|0\rangle$, no operation is applied). The sequence of operations for the k-th qubit is:

$$R_{n-k+1} \cdots R_3 R_2 H_k |j_1 \cdots j_n\rangle$$
$$= \frac{1}{\sqrt{2}} |j_1 \cdots j_{k-1}\rangle \left[|0\rangle + e^{2\pi i 0. j_k j_{k+1} \cdots j_n} |1\rangle \right] |j_{k+1} \cdots j_n\rangle \tag{3.5.8}$$

Let us exemplify the application of this sequence on the first qubit. First we apply a Hadamard, and consequently the state of the system becomes:

$$H_1 |j_1 j_2 \cdots j_n\rangle = \frac{1}{\sqrt{2}} \left[|0\rangle + e^{2\pi i 0. j_1} |1\rangle \right] |j_2 \cdots j_n\rangle \tag{3.5.9}$$

since $e^{2\pi i 0. j_1} = 1$ if $j_1 = 0$ and $e^{2\pi i 0. j_1} = -1$ if $j_1 = 1$. Next, we apply the controlled gates R_k to the same qubit, starting from R_2, i.e. controlled by the second qubit, and finishing with R_n, controlled to the last one:

$$R_2 H_1 |j_1 j_2 \cdots j_n\rangle = \frac{1}{\sqrt{2}} \left[|0\rangle + e^{2\pi i 0. j_1 j_2} |1\rangle \right] |j_2 \cdots j_n|\rangle$$

$$R_n \cdots R_2 H_1 |j_1 j_2 \cdots j_n\rangle = \frac{1}{\sqrt{2}} \left[|0|\rangle + e^{2\pi i 0. j_1 j_2 \cdots j_n} |1\rangle \right] |j_2 \cdots j_n|\rangle \tag{3.5.10}$$

After the application of these sequences the system will be in the state described by Equation (3.5.11). Therefore, a SWAP logic gate must be applied, in order to exchange the sates of individual qubits, accomplishing the QFT. The quantum circuit describing the QFT, may be seen on Figure 3.6.

$$|\psi\rangle = \frac{1}{\sqrt{2^n}} \left(|0\rangle + e^{2\pi i 0. j_1 j_2 \cdots j_n} |1\rangle \right)$$
$$\cdots \left(|0\rangle + e^{2\pi i 0. j_{n-1} j_n} |1\rangle \right) \left(|0\rangle + e^{2\pi i 0. j_n} |1\rangle \right) \tag{3.5.11}$$

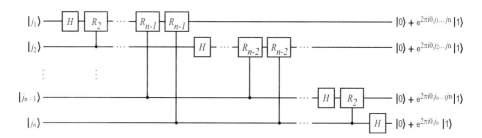

Figure 3.6 Quantum circuit to implement the QFT (the SWAP operation at the end is not shown). Adapted with permission from [1].

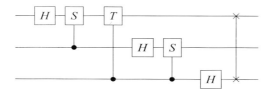

Figure 3.7 Quantum circuit to implement the QFT for 3 qubits. Adapted with permission from [1].

For a three-qubit system the QFT operator can be easily implemented using the basic known logic gates H, S and T, apart from the SWAP. The quantum circuit which illustrates this implementation is shown on Figure 3.7.

The most important quantum algorithm, the Shor algorithm [11], uses the QFT for finding the order of a number, which increases the speed of the factorization process. These are basically implemented by the same quantum circuit and are the main reasons for the exponential gain of speed in comparison with the classical factorizing algorithm.

3.6 QUANTUM STATE TOMOGRAPHY

Quantum state tomography is a technique which allows the determination of all the matrix elements of the density operator of a system. Such a procedure is very important for QIP, since at the end of an algorithm or protocol, one is usually interested in knowing the quantum state of the system. In Chapter 2, density matrix was introduced in the context of NMR, and in Chapter 4 the quantum state tomography will also be discussed in the context of NMR QIP. In this chapter, these concepts are presented within the QIP formalism.

3.6.1 The density matrix

In QIP one frequently has to deal with situations where the state vector of the system is not known, but only a set of possible states $\{|\psi_i\rangle\}$, each of which with a probability $\{p_i\}$ to occur. The set $\{p_i, |\psi_i\rangle\}$ is said to be a statistical ensemble. The appropriate mathematical tool to deal with such cases is the density matrix, ρ, defined as:

$$\rho \equiv \sum_i p_i |\psi_i\rangle\langle\psi_i| \tag{3.6.1}$$

where $p_i > 0$ and $\sum_i p_i = 1$. This operator has some important properties:

3.6. Quantum state tomography

1. The density matrix is a positive operator, that is, its eigenvalues are real and non-negative. Indeed, for any state $|\varphi\rangle$,

$$\langle\varphi|\rho|\varphi\rangle = \sum_i p_i \langle\varphi|\psi_i\rangle\langle\psi_i|\varphi\rangle = \sum_i p_i |\langle\varphi|\psi_i\rangle|^2 \geq 0 \qquad (3.6.2)$$

2. Because of the conservation of probabilities, the trace of ρ is equal to 1:

$$\text{Tr}(\rho) = \sum_i p_i \text{Tr}(|\psi_i\rangle\langle\psi_i|) = \sum_i p_i = 1 \qquad (3.6.3)$$

3. A quantum state is said to be pure if and only if $\text{Tr}(\rho^2) = 1$. This can be proved using directly the definition of ρ:

$$\rho^2 = \sum_i \sum_j p_i p_j |\psi_i\rangle\langle\psi_i|\psi_j\rangle\langle\psi_j| = \cdots$$

$$= \sum_i \sum_j p_i p_j \delta_{i,j} |\psi_i\rangle\langle\psi_j| = \sum_i p_i^2 |\psi_i\rangle\langle\psi_i| \qquad (3.6.4)$$

Therefore,

$$\text{Tr}(\rho^2) = \sum_i p_i^2 \text{Tr}(|\psi_i\rangle\langle\psi_i|) = \sum_i p_i^2 \leq 1 \qquad (3.6.5)$$

The equality in (3.6.5) is only satisfied if $p_i = 0$ to every i, except for a particular state j such as $p_j = 1$. Only an operator with the properties listed above can be considered a density matrix.

When dealing with composite systems, the density operator of the subsystems can be obtained through the partial trace operation over the density operator of the whole system. The partial trace operation is a sum over all the possible states of one subsystem. For instance, if ρ^{ab} is the density operator of a composite system $|ab\rangle$, the density operator of each subsystem is given by:

$$\rho^a \equiv \text{Tr}_b(\rho^{ab}); \qquad \rho^b \equiv \text{Tr}_a(\rho^{ab}) \qquad (3.6.6)$$

where Tr_a and Tr_b means the trace operation made only over the states of a or b, respectively.

3.6.2 Determining ρ

In order to experimentally determine all the elements of a density operator, it is necessary to perform several measurements. Therefore, if the quantum system has only one copy, it turns out to be impossible to measure ρ. For a simple quantum system, containing only one qubit, the set $\mathbf{1}$, σ_x, σ_y and σ_z form a orthonormal set of matrices upon which any operator ρ can be expanded as:

$$\rho = \frac{\text{tr}(\rho)\mathbf{1} + \text{tr}(\rho\sigma_x)\sigma_x + \text{tr}(\rho\sigma_y)\sigma_y + \text{tr}(\rho\sigma_z)\sigma_z}{2} \qquad (3.6.7)$$

Remembering that $\text{Tr}(\rho\sigma_z) = \langle\sigma_z\rangle$ is the expected value of the operator σ_z, it can be seen that many measurements are necessary for obtaining the expected value, since one single measurement will always result in $+1$ or -1. For a large sampling, the average over all the measured results will be equal to $\text{tr}(\rho Z)$, with a standard deviation less than $1/\sqrt{m}$, where m is the number of measurements. Consequently, the value of $\langle Z \rangle$ can be measured with an arbitrarily good precision. Obviously, this is also true for any other observable of the system.

This procedure can be generalized for an arbitrary number of qubits, and the density matrix can be written as:

$$\rho = \sum_{\vec{v}} \frac{\text{tr}(\sigma_{v_1} \otimes \sigma_{v_2} \otimes \cdots \otimes \sigma_{v_n} \cdot \rho)\sigma_{v_1} \otimes \sigma_{v_2} \otimes \cdots \otimes \sigma_{v_n}}{2^n} \qquad (3.6.8)$$

where the sum runs over the vectors ($\vec{v} = v_1, \ldots, v_n$), whose components are chosen from the set $\{\mathbf{1}, \sigma_x, \sigma_y, \sigma_z\}$, formed by the Pauli matrices plus the identity. The set $\{\sigma_{v_1} \otimes \sigma_{v_2} \otimes \cdots \otimes \sigma_{v_n}\}$ forms a basis in which any square matrix can be expanded upon.

Therefore, performing measurements of observables which are the products of the Pauli matrices, it is possible to determine all the elements of the density matrix operator ρ, with an arbitrary precision. This process is referred to as *Quantum State Tomography*, and is a procedure for measuring the quantum state of a system.

3.7 ENTANGLEMENT

The fourth postulate and the superposition principle allows the consideration of quantum states with form described in Equation (3.7.1). These states have interesting properties, and constitute an entirely new computational resource, of exclusively quantum nature [13].

$$|\psi^+\rangle = \frac{|00\rangle + |11\rangle}{\sqrt{2}} \qquad (3.7.1)$$

First, one can notice that there are no individual sates, of a two-qubit system, $|a\rangle$ and $|b\rangle$ such as $|\psi^+\rangle = |a\rangle \otimes |b\rangle$. Indeed, if there were such states, one could expand them in the computational basis $\{|0\rangle, |1\rangle\}$:

$$|a\rangle = \alpha|0\rangle + \beta|1\rangle$$
$$|b\rangle = \alpha'|0\rangle + \beta'|1\rangle$$

Therefore,

$$|ab\rangle = \alpha\alpha'|00\rangle + \beta\beta'|11\rangle + \alpha\beta'|01\rangle + \beta\alpha'|10\rangle$$

that implies

$$\alpha\alpha' = \beta\beta' = \frac{1}{\sqrt{2}} \quad \text{and} \quad \alpha\beta' = \beta\alpha' = 0$$

which is an inconsistency.

3.7. Entanglement

States like those of Equation (3.7.1) can not be factorized; they are called *entangled states*. As an example, let us calculate its density matrix:

$$|\psi^+\rangle\langle\psi^+| = \frac{|00\rangle\langle 00| + |00\rangle\langle 11| + |11\rangle\langle 00| + |11\rangle\langle 11|}{2}$$

$$= \frac{1}{2}\begin{pmatrix} 1 & 0 & 0 & 1 \\ 0 & 0 & 0 & 0 \\ 0 & 0 & 0 & 0 \\ 1 & 0 & 0 & 1 \end{pmatrix} \quad (3.7.2)$$

and calculate the Von Neumann's entropy of each qubit. This entropy is defined as:

$$S(\rho) = -\operatorname{Tr}(\rho \log \rho) = -\sum_j \lambda_j \log \lambda_j \quad (3.7.3)$$

Like any other entropy, $S(\rho)$ is a measurement of the knowledge we have about the system. The base of the logarithm on Equation (3.7.3) is 2, and λ_j are the eigenvalues of ρ. Since the state of Equation (3.7.2) is pure, $S(\rho) = 0$. However, if we calculate the entropy of each qubit using the partial trace concept, one finds that:

$$\rho_a = \rho_b = \frac{1}{2} = \frac{1}{2}|0\rangle\langle 0| + \frac{1}{2}|1\rangle\langle 1| \quad (3.7.4)$$

from which we obtain: $S(\rho_a) = S(\rho_b) = 1$. In other words, an entangled state is such that we have the maximum knowledge about the composite state, but no knowledge at all about its constituents!

There are other possible entangled states of a two qubit system:

$$|\psi^-\rangle = \frac{|00\rangle - |11\rangle}{\sqrt{2}}$$

$$|\varphi^+\rangle = \frac{|01\rangle + |10\rangle}{\sqrt{2}} \quad (3.7.5)$$

$$|\varphi^-\rangle = \frac{|01\rangle - |10\rangle}{\sqrt{2}}$$

The set of quantum states $\{|\psi^\pm\rangle, |\varphi^\pm\rangle\}$ forms a basis for a two-qubit system, called Bell's basis [14].

Entangled states such as the cat state have a perfect correlation between the observables of the individual qubits of the system. For instance, the expected values of X_a and X_b for the cat state, $|\psi^+\rangle$, are zero ($\langle X_a \rangle = 0$ and $\langle X_b \rangle = 0$):

$$\langle X_a \rangle = \langle \psi^+|X_a|\psi^+\rangle = \frac{\langle 00| + \langle 11|}{\sqrt{2}} \cdot \frac{|10\rangle + |01\rangle}{\sqrt{2}} = 0 \quad (3.7.6)$$

However, $\langle X_a X_b \rangle = 1$:

$$\langle X_a X_b \rangle = \langle \psi^+|X_a X_b|\psi^+\rangle = \frac{\langle 00| + \langle 11|}{\sqrt{2}} \cdot \frac{|11\rangle + |00\rangle}{\sqrt{2}} = 1 \quad (3.7.7)$$

This tells us that, in spite of the uncertainty about the sates of the individual qubits, they are perfectly correlated! This correlation is responsible for the non-local action of entangled states. For instance, suppose a measurement is made on the first qubit, a, represented by the operators $M_0^a = |0\rangle\langle 0| \otimes \mathbf{1}$ and $M_1^a = |1\rangle\langle 1| \otimes \mathbf{1}$. The probability of finding 0 is the same of finding 1:

$$p(0) = \langle\psi^+|M_0^{A\dagger}M_0^A|\psi^+\rangle = \frac{1}{2} = \langle\psi^+|M_1^{A\dagger}M_1^A|\psi^+\rangle = p(1) \qquad (3.7.8)$$

After the measurement, supposing that the result 0 was found, the system state becomes:

$$|\psi_0\rangle = \frac{M_0^A|\psi^+\rangle}{\sqrt{1/2}} = |00\rangle \qquad (3.7.9)$$

Therefore, a measurement of a qubit in an entangled state, *defines the state of the other*, upon which no measurement was performed! If the measurement is made in a different basis, for instance, $M_+^a \equiv |+\rangle\langle+|$ and $M_-^b \equiv |-\rangle\langle-|$, and the state $|+\rangle$ was found for qubit a, after the measurement, the quantum state of the system would be $|++\rangle$, implying that the qubit b is also in the same $|+\rangle$ state. For non-correlated systems, for instance

$$\bigl(|00\rangle + |01\rangle\bigr)/\sqrt{2} = |0\rangle \otimes \bigl(|0\rangle + |1\rangle\bigr)/\sqrt{2},$$

one can easily verify that the measurement of a does not affect b.

It is important to notice that the definition of entanglement for mixed states is more complicated than for pure states. Whereas product states are always non-entangled pure states, the same is not true for mixed states [15]. For a two-partite system, an *non-entangled* mixed state ρ is characterized by the existence of a set of probabilities $\{p_i\}$ and one-qubit density matrices $\{\rho_1^i, \rho_2^i\}$ such that one can write:

$$\rho = \sum_i p_i \rho_1^i \otimes \rho_2^i \qquad (3.7.10)$$

In opposite, an entangled mixed state is a state for which no such a decomposition exists.

Non-locality and Bell's inequality

The influence of the measurement result of a qubit affecting the state of another, as happens in an entangled state, is called *non-locality*. This strange property was pointed out for the first time in a very influential paper, published in 1935, by Albert Einstein, Boris Podolsky and Nathan Rose [13]. The paper aimed to demonstrate that Quantum Mechanics was an incomplete theory. According to the authors, a theory to be considered complete should contain what they defined *reality elements*. A reality element would be, still according to the authors, any physical quantity whose value could be predicted before performing a measurement on the system. For example, when a measurement of the observable σ_y is performed on a qubit, in an entangled cat state, the result determines the state of the other qubit, which could then be predicted before a measurement. Hence the observable σ_y is a reality element. However, before the measurement is performed on the *first* qubit,

3.7. Entanglement

no prediction can be made at all on either qubit, but only the probabilities of possible outcomes.

In the year of 1964 John Bell discovered a remarkable result [14], which sets the rules for deciding experimentally if the non-locality is indeed a fact of entangled systems. The result is expressed in the form of an inequality, which establishes an upper limit for the *local* correlations in a two-partite system. Since then, it is known as the Bell's inequality. For two qubits, Bell's inequality says that a determined quantity S, essentially a correlation function between observables, should not overcome the value 2. However, quantum mechanics is *non-local*, and predicts the value $S = 2\sqrt{2} \approx 2.83$, for an entangled state, therefore violating the Bell's inequality.

In the year of 1982 the Bell's inequality was tested in a famous experiment [16]. A French group led by Alain Aspect used entangled photons, produced by the decay of electrons from a excited state of ^{40}Ca, and demonstrated the non-local quantum correlations between the polarization of the two photons. They determined the value of the quantity S experimentally, and found $S = 2.70 \pm 0.05$, which is very close to the ideal result predicted by quantum mechanics.

3.7.1 Some applications of entanglement

Entangled states constitute a powerful natural resource for QIP. In this section, two of the most interesting applications are illustrated: superdense coding and the teleport.

Superdense coding

The superdense coding is a process, in which two bits of classical information are transmitted using only one quantum bit. Here the example of exchange of information between two parties Alice and Bob, is described. Suppose that initially Alice and Bob share qubits in an entangled cat state:

$$|\psi^+\rangle = \frac{|00\rangle + |11\rangle}{\sqrt{2}} \tag{3.7.11}$$

On the other hand, two classical bits have four possible sequences: 00, 01, 10 or 11, each of them representing a possible "message", which can be sent through a communication channel. Suppose that Alice wishes to send Bob the sequence 01. All she has to do is to apply the operator X on her qubit, of the entangled pair (say, the qubit a of the state $|ab\rangle$), transforming the state to:

$$|\varphi^+\rangle = \frac{|10\rangle + |01\rangle}{\sqrt{2}} \tag{3.7.12}$$

After the operation, she sends her qubit to Bob, who applies the operations CNOT_a (note that the control is in the first qubit) to the pair, and then H_a also on the first qubit:

$$\text{H}_a \cdot \text{CNOT}_a |\varphi^+\rangle = \text{H}_a \cdot \frac{|11\rangle + |01\rangle}{\sqrt{2}} = |01\rangle \tag{3.7.13}$$

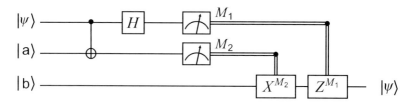

Figure 3.8 Quantum circuit to implement teleport. The double line represent classical bits, after the measurement is performed. Adapted with permission from [1].

After this, he performs a measurement in the computational basis, obtaining the sequence 01. In this way, Alice has sent two bits of classical information sending only one qubit. Any other sequence can be sent by just varying the first operation, i.e. the application of the X gate.

Teleport

Teleport is a process through which the state of a qubit is transferred to another, using the non-local properties of entangled states [17]. Differently from superdense coding, no qubit is transferred in teleport, but only a quantum state.

In the simplest case of teleport, three qubits are involved, two with Alice (let's label them $|\psi a\rangle$) and one with Bob (labelled $|b\rangle$). As in the superdense coding process, initially Alice and Bob qubits, $|a\rangle$ and $|b\rangle$, are in a cat state. Alice wishes to transmit to Bob the unknown state of a third qubit, $|\psi\rangle = \alpha|0\rangle + \beta|1\rangle$. Of course, she cannot measure $|\psi\rangle$, for she would only get 0 or 1, with the probabilities $|\alpha|^2$ and $|\beta|^2$, respectively. The quantum circuit that describes the teleport process is illustrated in Figure 3.8, where the top line represents the qubit Alice wants to teleport to Bob ($|\psi\rangle$), and the second and third lines, represent the entangled qubit pair, the second one with Alice and the third one with Bob.

At the beginning of the process, the quantum state of the three-qubit system is then given by:

$$|\Phi_0\rangle = |\psi\rangle|ab\rangle = |\psi\rangle\left[\frac{|00\rangle + |11\rangle}{\sqrt{2}}\right]$$

$$= \frac{1}{\sqrt{2}}[\alpha|0\rangle(|00\rangle + |11\rangle) + \beta|1\rangle(|00\rangle + |11\rangle)] \quad (3.7.14)$$

The first operation that Alice performs is the application of a CNOT$_{\psi a}$ gate, which inverts the state of the second qubit, $|a\rangle$, one of the entangled qubits, if the first one, $|\psi\rangle$, is in the state $|\psi\rangle = |1\rangle$. Therefore, the system evolves to the state described by:

$$|\Phi_1\rangle = \text{CNOT}_{\psi a}|\Phi_0\rangle$$

$$= \frac{1}{\sqrt{2}}[\alpha|0\rangle(|00\rangle + |11\rangle) + \beta|1\rangle(|10\rangle + |01\rangle)] \quad (3.7.15)$$

The second operation is the application of the Hadamard gate to the first qubit, $|\psi\rangle$, leaving the system on state described by:

$$|\Phi_2\rangle = H_\psi |\Phi_1\rangle$$

$$= \frac{1}{2}\left[\alpha\big(|0\rangle + |1\rangle\big)\big(|00\rangle + |11\rangle\big) + \beta\big(|0\rangle - |1\rangle\big)\big(|10\rangle + |01\rangle\big)\right]$$

$$= \frac{1}{2}\big[\alpha\big(|000\rangle + |100\rangle + |011\rangle + |111\rangle\big)$$

$$+ \beta\big(|010\rangle - |110\rangle + |001\rangle - |101\rangle\big)\big] \tag{3.7.16}$$

At this point, the quantum feature of the teleport has already occurred, as can be seen after rewriting $|\Phi_2\rangle$:

$$|\Phi_2\rangle = \frac{1}{2}\big[|00\rangle\big(\alpha|0\rangle + \beta|1\rangle\big) + |01\rangle\big(\alpha|1\rangle + \beta|0\rangle\big)$$

$$+ |10\rangle\big(\alpha|0\rangle - \beta|1\rangle\big) + |11\rangle\big(\alpha|1\rangle - \beta|0\rangle\big)\big] \tag{3.7.17}$$

Clearly, Bob's qubit $|b\rangle$, is now in a superposition, which involves four different possible combinations, containing the coefficients α and β of the qubit state, that Alice wished to transmit at the beginning of the process. A measurement performed on her qubits will project the Bob's qubit to one of the possible combinations.

Therefore, the next step is a measurement, performed by Alice, in the qubits which are with her, $|\psi a\rangle$. After this measurement, Alice sends Bob a message by classical means, telling him the result that she found. Then Bob has to perform some operations, on his qubit, depending on the information sent by Alice. The operations, which are conditional to Alice's results, can be written as X^{M_a} and Z^{M_ψ}, being M_a and M_ψ the results found by Alice after the measurements on her qubit. For instance, if Alice's results were $M_a = 0$ and $M_\psi = 0$ (that is, $|00\rangle$ was the state measured), Bob does not have to do anything, since the state of his qubit is already the one Alice wished to transmit. On the other hand, if Alice finds $M_a = 1$ and $M_\psi = 1$ (which means that the state $|11\rangle$ was the measured one) Bob has to apply the operations X and Z, on his qubit. Hence, the final wave function depends on the result found by Alice:

$$|00\rangle \longrightarrow |\Phi_3\rangle = |\Phi_2\rangle = \alpha|0\rangle + \beta|1\rangle$$

$$|01\rangle \longrightarrow |\Phi_3\rangle = X|\Phi_2\rangle = X\big[\alpha|1\rangle + \beta|0\rangle\big] = \alpha|0\rangle + \beta|1\rangle$$

$$|10\rangle \longrightarrow |\Phi_3\rangle = Z|\Phi_2\rangle = Z\big[\alpha|0\rangle - \beta|1\rangle\big] = \alpha|0\rangle + \beta|1\rangle$$

$$|11\rangle \longrightarrow |\Phi_3\rangle = ZX|\Phi_2\rangle = ZX\big[\alpha|1\rangle - \beta|0\rangle\big] = \alpha|0\rangle + \beta|1\rangle \tag{3.7.18}$$

It is important to notice that for the teleport to be successfully implemented, a classical communication channel must be used, to send Alice's measurement results. Without that information Bob will never know that the state of his qubit has been changed.

3.8 QUANTUM ALGORITHMS

Perhaps, the most striking aspect of quantum computation are the quantum algorithms, which can compute states of bit sequences that are impossible for classical computers, such as superposition and entangled states. Here lies the power of the quantum algorithms.

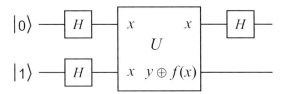

Figure 3.9 Quantum circuit to implement the Deutsch algorithm. Adapted with permission from [1].

Quantum algorithms can be divided into two classes, called A and B, the exponentially fast and the polynomially fast, respectively.

The first quantum algorithm, which demonstrated the power of quantum computation over the classical one, was discovered by David Deutsch [18]. There is no classical analogous to this algorithm. It uses the Hadamard logic gates to create superpositions, which is impossible for classical computers. The quantum search algorithm was created by Lov K. Grover [19], and it is polynomially faster than their classical versions, so it belongs to class B. The factorization algorithm, on another hand, created by Peter Shor [11,12] is exponentially faster than its classical version and it is therefore classified as an A-type algorithm.

In this section, a non-exhaustive discussion of these three algorithms, is made, showing the basic operations necessary for their implementation.

3.8.1 The Deutsch's algorithm

The Deutsch algorithm is used to test whether a binary function of one qubit is constant ($f(0) = f(1)$) or balanced ($f(0) \neq f(1)$), without the need of computing the two possible values $f(0)$ and $f(1)$, separately, and then comparing their results, as it would be made in a classical computer [18].

The quantum circuit that describes the Deutsch algorithm is illustrated in Figure 3.9, from which we can see that at the input the qubits are in a quantum state described by:

$$|\Phi_0\rangle = |0\rangle \otimes |1\rangle = |01\rangle \quad (3.8.1)$$

The first operation of the algorithm is a Hadamard logic gate applied to both qubits, yielding:

$$|\Phi_1\rangle = H_b H_a |\Phi_0\rangle = H_a |0\rangle H_b |1\rangle$$

$$= \frac{1}{2}(|0\rangle + |1\rangle)(|0\rangle - |1\rangle)$$

$$= \frac{1}{2}[|0\rangle(|0\rangle - |1\rangle) + |1\rangle(|0\rangle - |1\rangle)] \quad (3.8.2)$$

The next step is to perform an unitary operation U_f, which takes the two-qubit system from a generic state, $|x, y\rangle$ to the state $|x, y \oplus f(x)\rangle$. This transformation $|x, y\rangle \rightarrow |x, y \oplus f(x)\rangle$ is nothing but the sum of the second qubit, the bottom line of the circuit, with $f(x)$, that is the computed function of the first qubit. The binary function, $f(x)$, is the one to be

3.8. Quantum algorithms

verified. One can notice that any binary function, when applied to the system whose state is $|x, y\rangle = \frac{1}{\sqrt{2}}|x\rangle[|0\rangle - |1\rangle]$, yield the result $(-1)^{f(x)}\frac{1}{\sqrt{2}}|x\rangle[|0\rangle - |1\rangle]$:

$$U_f \frac{1}{\sqrt{2}}|x\rangle\big[|0\rangle - |1\rangle\big] = \frac{1}{\sqrt{2}}|x\rangle\big[|0 \oplus f(x)\rangle - |1 \oplus f(x)\rangle\big] = \cdots$$

$$= (-1)^{f(x)}\frac{1}{\sqrt{2}}|x\rangle\big[|0\rangle - |1\rangle\big] = \cdots \quad (3.8.3)$$

It is important to point that: $|0 \oplus f(x)\rangle = |0\rangle$ and $|1 \oplus f(x)\rangle = |1\rangle$ if $f(x) = 0$ or $|0 \oplus f(x)\rangle = |1\rangle$ and $|1 \oplus f(x)\rangle = |0\rangle$ if $f(x) = 1$. Therefore, after the operation U_f the state of the system is:

$$|\Phi_2\rangle = U_f|\Phi_1\rangle$$
$$= \frac{1}{2}\big[(-1)^{f(0)}|0\rangle(|0\rangle - |1\rangle) + (-1)^{f(1)}|1\rangle(|0\rangle - |1\rangle)\big] \quad (3.8.4)$$

from which one can see that, although both qubits are still in a superposition of states, the relative phase of the first one now depends on the result of the operations $f(0)$ and $f(1)$.

The state $|\Phi_2\rangle$ can be rewritten as:

$$|\Phi_2\rangle = \begin{cases} \pm\frac{1}{2}[(|0\rangle + |1\rangle)(|0\rangle - |1\rangle)] \text{ if } f(0) = f(1) \\ \pm\frac{1}{2}[(|0\rangle - |1\rangle)(|0\rangle - |1\rangle)] \text{ if } f(0) \neq f(1) \end{cases} \quad (3.8.5)$$

from which it is clear that the relative phase of the first qubit determines if the function is balanced or constant.

Next, a Hadarmad gate is applied to the first qubit, reminding that $H(|0\rangle + |1\rangle) = |0\rangle$ and $H(|0\rangle - |1\rangle) = |1\rangle$. Therefore, performing a measurement in the computational basis after the Hadamard operation, the system will be found in the state $|0\rangle$ if the function f is constant or balanced if the state is $|1\rangle$, as shown on Equation (3.8.6). Note that the measurement is performed on the first qubit only.

$$|\Phi_3\rangle = \begin{cases} \pm\frac{1}{\sqrt{2}}[|0\rangle(|0\rangle - |1\rangle)] \text{ if } f(0) = f(1) \\ \pm\frac{1}{\sqrt{2}}[|1\rangle(|0\rangle - |1\rangle)] \text{ if } f(0) \neq f(1) \end{cases} \quad (3.8.6)$$

There exists a variation of this algorithm for systems containing more than two qubits, which was derived by Deutsch and Jozsa, and it will not be discussed here. It uses the same principles as above, and is referred as the Deutsch–Jozsa algorithm [20].

3.8.2 The quantum search algorithm

A classical search algorithm needs about $O(N)$ operations in order to find a specified item in a disordered list containing N elements. The quantum search algorithm, created by Grover is quadratically faster than its classical analogous, since only $O(\sqrt{N})$ operations are needed [19]. In a quantum computer, the number of elements to be searched is the number of possible states of the system $N = 2^n$, where n is the number of qubit system. Grover's algorithm is then considered to be of B-type. For a two-qubit system, with $N = 2^2 = 4$

elements, the algorithm finds the solution with only one iteration. Searching algorithms play an important role in quantum computation, because they can also be used for searching solutions of an specified problem, which would take too long to calculate, or has too many operations to be performed.

The Grover's algorithm performs a search on the elements index, instead of searching the elements themselves. It uses two sets of distinct qubits, one containing the elements being searched ($|x\rangle$), and the other containing auxiliary qubits ($|q\rangle$). At the first stage of the algorithm the qubits in the state $|x\rangle$ are prepared in an uniform superposition. In such a state, any item has the same probability to be found after a measurement is performed. Such superposition can be achieved by applying a Hadamard gate to each qubit of the system, after preparing them on the state $|0\rangle$. This operation is represented as $H^{\otimes n}$, meaning $H \otimes H \otimes \cdots \otimes H$, n-times:

$$H^{\otimes n}|00..0\rangle = \frac{1}{\sqrt{2^n}} \sum_{x=0}^{2^n-1} |x\rangle \quad (3.8.7)$$

At the second stage, an operator, called Grover operator – G, is applied iteratively to the system, approximately $\sqrt{2^n}$ times, and after that the searched state will have a high probability of being found, when a measurement is performed. This operator is composed of four others, represented by: $G = H^{\otimes n} \cdot [2|0\rangle\langle 0| - 1] \cdot H^{\otimes n} \cdot O$, and they will be discussed below.

The first operator of G is an unitary controlled operation, represented by O. It inverts the phase of the state, which is being searched. This controlled operation is constructed by applying the transformation indicated on Equation (3.8.8), such as $f(x) = 1$ when x is the searched item and $f(x) = 0$ otherwise. Notice that the operation O only acts on the second set of qubits, leaving the first one intact.

$$O|x\rangle|q\rangle = |x\rangle|q \oplus f(x)\rangle \quad (3.8.8)$$

The operation O is considered to be a "black box" called an *Oracle*, whose construction has to be built individually for each item to be searched.

Reminding that the *Oracle* acts only on the second set of qubits, in order to invert the phase of the searched state, it is necessary to prepare the system in the state $|x\rangle[|0\rangle - |1\rangle]/\sqrt{2}$. In this case, when the *Oracle* is applied, the system will evolve as described on Equation (3.8.9), similarly to what happens in the Deutsch algorithm. As it can be seen, the solution gets marked after the *Oracle* operation, inverting the phase of the desired state, i.e. $|x\rangle \to -|x\rangle$, if $|x\rangle$ is the desired item.

$$O|x\rangle[|0\rangle - |1\rangle]/\sqrt{2} = (-1)^{f(x)}|x\rangle[|0\rangle - |1\rangle]/\sqrt{2} \quad (3.8.9)$$

Since the state of the second set of qubits does not change, it can be omitted from the notation:

$$O|x\rangle = (-1)^{f(x)}|x\rangle \quad (3.8.10)$$

After the application of the *Oracle*, three operations are necessary, a Hadamard on the $|x\rangle$ qubits, a phase shift $|x\rangle \to -(-1)^{\delta_{x0}}|x\rangle$ of all states, except for $|0\rangle$ ($\delta_{x0} = 1$ for $x = 0$,

3.8. Quantum algorithms

and $\delta_{x0} = 0$ for $x \neq 0$), and the Hadamard on the $|x\rangle$ qubits again. It is possible to demonstrate (see problems) that the operator which applies the conditional phase shift is:

$$[2|0\rangle\langle 0| - 1]|x\rangle = -(-1)^{\delta_{x0}}|x\rangle \qquad (3.8.11)$$

The set of operations $[H^{\otimes n} \cdot (2|0\rangle\langle 0| - 1) \cdot H^{\otimes n}]$ is called *inversion about the mean*, or *inversion about the average*. In order to illustrate this point let's consider the state:

$$|\psi\rangle = H^{\otimes n}|0\rangle = \frac{1}{\sqrt{2^n}} \sum_{x=0}^{2^n-1} |x\rangle \qquad (3.8.12)$$

An *inversion about the mean* is then given by:

$$H^{\otimes n} \cdot (2|0\rangle\langle 0| - I) \cdot H^{\otimes n} = [2|\psi\rangle\langle\psi| - \mathbf{1}] \qquad (3.8.13)$$

Its action on a generic state $|\varphi\rangle = \sum \alpha_k |k\rangle$ is:

$$[2|\psi\rangle\langle\psi| - \mathbf{1}]\sum_k \alpha_k |k\rangle = \sum_k [-\alpha_k + 2\langle\alpha\rangle]|k\rangle \qquad (3.8.14)$$

where $\langle\alpha\rangle \equiv \sum_k \alpha_k/N$, i.e. the mean value of α_k, hence the operator's name.

In summary, the Grover operator is given by the product of the *Oracle*, and the *inversion about the mean*:

$$G = [2|\psi\rangle\langle\psi| - \mathbf{1}] \cdot O \qquad (3.8.15)$$

In Figure 3.10, it is illustrated the action of the whole Grover operator in a generic state. The action of the *Oracle* inverts the phase of the searched state, selecting it, then a inversion about the mean is applied and the amplitude of the selected state is increased.

It can be shown that the number of times that the operator G has to be applied before the item is found is $\approx [\pi \sqrt{N/M}/4]$, where $N = 2^n$ is the number of states in the system, and M is the number of searched solutions [1]. For $N = 4$ and $M = 1$, which means looking for one item in 4, it is necessary to apply G only once. The inversion about the mean and *oracle* operators can also be described through their matrix representations. If, for instance,

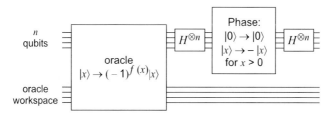

Figure 3.10 Quantum scheme to implement Grover operator. Adapted with permission from [1].

the searched state is $|11\rangle$:

$$[2|\psi\rangle\langle\psi| - \mathbf{1}] = \frac{1}{2}\begin{bmatrix} -1 & 1 & 1 & 1 \\ 1 & -1 & 1 & 1 \\ 1 & 1 & -1 & 1 \\ 1 & 1 & 1 & -1 \end{bmatrix}$$

$$O = \begin{bmatrix} 1 & 0 & 0 & 0 \\ 0 & 1 & 0 & 0 \\ 0 & 0 & 1 & 0 \\ 0 & 0 & 0 & -1 \end{bmatrix} \quad (3.8.16)$$

$$G = [2|\psi\rangle\langle\psi| - \mathbf{1}]O = \frac{1}{2}\begin{bmatrix} -1 & 1 & 1 & -1 \\ 1 & -1 & 1 & -1 \\ 1 & 1 & -1 & -1 \\ 1 & 1 & 1 & 1 \end{bmatrix}$$

As an example, let us consider the case of two qubits. The input state is $|\psi_0\rangle = |00\rangle$, from which an uniform superposition must be created, applying Hadamard gates on each qubit of the system. Therefore the following stage will be:

$$|\psi_1\rangle = \frac{1}{2}\big[|00\rangle + |01\rangle + |10\rangle + |11\rangle\big].$$

If the searched state is $|11\rangle$, the *oracle* must invert the phase of this particular state, hence selecting it, which takes the system to:

$$|\psi_2\rangle = O|\psi_1\rangle = \frac{1}{2}\big[|00\rangle + |01\rangle + |10\rangle - |11\rangle\big] \quad (3.8.17)$$

The next stage is the inversion about the mean operation, resulting in $|\psi_3\rangle$:

$$|\psi_3\rangle = [2|\psi\rangle\langle\psi| - \mathbf{1}]|\psi_2\rangle = |11\rangle \quad (3.8.18)$$

In Figure 3.11 the application of the Grover algorithm is illustrated, for (a) a two-qubit system and (b) a ten-qubit system ($N = 2^{10} = 1024$). Notice that the amplitude of the searched state oscillates with the number of times the G operator is applied. Thus, one must know in advance how many solutions exist and also the number of elements in the space where the search is being carried on, for there is a optimum number of runs of the algorithm. These numbers are approximately 1 and 25, for $n = 2$ and $n = 10$, respectively.

3.8.3 The quantum factorizing algorithm

The quantum factorizing algorithm was created by P. Shor [11], and is the most important application of quantum computing up to date. The main interest in this particular algorithm is because it can be used for breaking codes of cryptographic systems. The reason for this is that the best classical algorithm for factorizing a given number will run in a time proportional to $\exp((\log N)^{1/3}(\log(\log N))^{2/3})$, i.e. it runs in time, which grows exponentially with the number of digits of N ($\log N \approx$ the number of digits required to store N). Instead,

3.8. Quantum algorithms

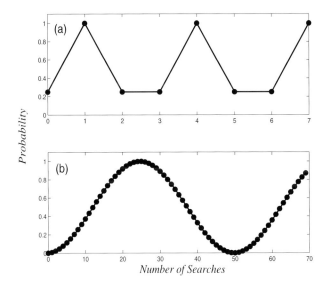

Figure 3.11 Action of Grover algorithm calculated for: (a) 2 qubits, (b) 10 qubits.

Shor's algorithm runs in time which grows only polynomially with log N. For instance, in order to factorize a number of 1024 bits, for instance, 100 thousand years are necessary, using present day classical computers. The same task could be made in less than 5 minutes, using a quantum computer running the Shor factorization algorithm.

One of the main features of the Shor algorithm is that it uses the QFT operation, which needs only $O(n^2)$ operations while its classical analogous, the FFT (Fast Fourier Transform) needs $O(n2^n)$ operations.

The factorization algorithm has 4 stages, but only the last one is quantum in nature. In fact, it turns out that the factorization problem can be reduced to an order finding problem, which can be implemented using basically the same quantum routine for phase estimation. Thus, phase estimation and order finding are "subroutines" to Shor algorithm, and they will be discussed in the next subsections.

Phase estimation

Let us first review an application of QFT: a quantum circuit to estimate the phase of a state, φ, which is eigenket of an operator U: $U|u\rangle = e^{2\pi i\varphi}|u\rangle$. The procedure requires two sets of qubits, called *registers*. The first register must have a number of qubits, sufficiently large to store the value of φ, with some precision; the larger this number, the better the precision. The second register must have enough qubits to represent the eigenstate $|u\rangle$. The initial state is prepared with the first register in the $|0\rangle$ state, whereas the second register must contain the eigenket $|u\rangle$. Thus the input state is:

$$|\psi_{in}\rangle = |0\rangle_t |u\rangle \quad (3.8.19)$$

where t is the number of qubits in the first register. At the first stage, a Hadamard gate is applied to each qubit in the first register, $H^{\otimes t}$, in order to create an uniform superposition

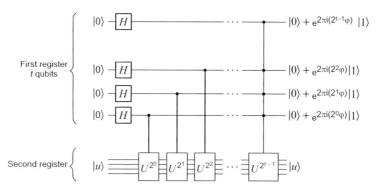

Figure 3.12 Quantum circuit to implement the routine of phase estimation. Adapted with permission from [1].

of states:

$$|\psi_1\rangle = H^{\otimes t}|\psi_{ini}\rangle = [H^{\otimes t}|0\rangle_t]|u\rangle = \frac{1}{\sqrt{2^t}}\left[\sum_{k=0}^{2^t-1}|k\rangle\right]|u\rangle = \cdots$$

$$= \frac{1}{\sqrt{2^t}}[(|0\rangle+|1\rangle)(|0\rangle+|1\rangle)\ldots(|0\rangle+|1\rangle)]|u\rangle \quad (3.8.20)$$

Then a series of logic gates are applied to the second register, controlled by the qubits of the first register, as can be seen from Figure 3.12.

These controlled operations apply the operator U on the second register, 2^{t-1-k} times, where k is the label of the qubit which is controlling the operation, noticing that $k = 0, 1, \ldots, t-1$. The U-controlled operation performs the transformation $|n\rangle|u\rangle \rightarrow |n\rangle U^n|u\rangle = e^{2\pi i\varphi n}|n\rangle|u\rangle$, taking the system to the state described by Equation (3.8.21). Notice that, although U is applied to the second register, it does not change it at all, since this register is storing $|u\rangle$, which is an eigenvector of U. Instead, the controlled U operation changes just the relative phase of every qubit in the first register. This operation takes the system to the state described on Equation (3.8.21).

$$|\psi_2\rangle = U^{t-k}|\psi_1\rangle$$

$$= \frac{1}{\sqrt{2^t}}[(|0\rangle + e^{2\pi i\varphi 2^{t-1}}|1\rangle)(|0\rangle + e^{2\pi i\varphi 2^{t-2}}|1\rangle)$$

$$\cdots(|0\rangle + e^{2\pi i\varphi 2^0}|1\rangle)]|u\rangle = \cdots = \frac{1}{\sqrt{2^t}}\left[\sum_{k=0}^{2^t-1}e^{2\pi i\varphi k}|k\rangle\right]|u\rangle \quad (3.8.21)$$

Therefore, the t qubits of the first register have become the QFT of the state $|\varphi\rangle$, i.e. QFT $|\varphi\rangle$. At the last stage, the inverse Quantum Fourier Transform QFT†, which can be obtained by reversing the operations required to implement the QFT, is applied to the first register. This last operation stores the value of the phase, φ, on the t qubits of the first register, leaving the system in the final state: $|\widetilde{\varphi}\rangle_t|u\rangle = |\varphi_1\varphi_2\cdots\varphi_t\rangle|u\rangle$. At this point, a measurement of the first register will yield an estimative of the phase φ. This value is only an estimative, because its precision depends on the number of qubits of the first register.

3.8. Quantum algorithms

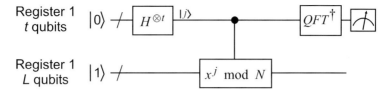

Figure 3.13 Quantum circuit to implement the order-finding protocol. Adapted with permission from [1].

Quantum order-finding

There is another quantum routine to be discussed before moving to Shor's factorization algorithm. The *order* of a number is a concept from Number Theory, which is beyond the scope of this book. However, some of its features, necessary to understand Shor's algorithm, will be discussed on this section.

The *order* of a number is the least integer r such as $x^r = 1 \pmod{N}$. This notation comes from modular arithmetic meaning that x^r leaves a remain 1, when divided by N. In this sense, $1 = 5 = 9 = 13 = 1 \pmod 4$, since they all leave the same remain, 1, when divided by 4. In modular arithmetic, which has some interesting features for understanding the properties of numbers, one is interested only in the remains, and this part of mathematics deals exclusively with integer numbers.

Finding the order of a number is a difficult task to classical computers, particularly if the number is large. Further below, it will be shown why this routine can be used to find the non-trivial factors of a number. Here we will be focused on the necessary quantum procedures for implementing this routine, that is illustrated on Figure 3.13. As one can see, this is just the phase estimation routine, with a different input, at the second register.

Again, the two registers have to be prepared in a specific quantum state, and the controlled-U operator, which performs the transformation, displayed on Equation (3.8.22), has to be applied, as may be seen from Figure 3.13.

$$U|y\rangle \equiv |xy \pmod{N}\rangle \qquad (3.8.22)$$

This function is periodic for $0 \leqslant y \leqslant N-1$, so that $xy \pmod N = x(y+N) \pmod N$.

Since the phase estimation routine will be used, it its necessary to construct the eigenstate of the operator U. As it can be observed, the ket described on Equation (3.8.23) is an eigenstate of the operator U, with eigenvalue $e^{2\pi is/r}$:

$$|u_s\rangle \equiv \frac{1}{\sqrt{r}} \sum_{k=0}^{r-1} \exp\left(\frac{-2\pi i s k}{r}\right) |x^k \pmod{N}\rangle \qquad (3.8.23)$$

$$U|u_s\rangle = \frac{1}{\sqrt{r}} \sum_{k=0}^{r-1} \exp\left(\frac{-2\pi i s k}{r}\right) |x^{k+1} \pmod{N}\rangle = \exp\left(\frac{2\pi i s}{r}\right)|u_s\rangle \qquad (3.8.24)$$

The preparation of $|u_s\rangle$ could be a problem, since the order r must be known beforehand, but that is exactly what one wishes to calculate. Fortunately, there is a way around it. By using an uniform superposition of the sates $|u_s\rangle$, with s running from 0 to $r-1$, given by Equation (3.8.25), being $\log(r)$ the number of qubits necessary for storing this set, it is possible to obtain a set of eigenkets of the operator U, that can be used in the order-finding routine:

$$\frac{1}{\sqrt{r}}\sum_{s=0}^{r-1}|u_s\rangle = |1\rangle \qquad (3.8.25)$$

With the eigenket of the operator U at hand, we can use the phase estimation routine for finding the order, r, such as $x^r = 1 \pmod{N}$. For that, the system must be prepared in the initial state $|\psi_{ini}\rangle = |0\rangle_t|1\rangle$, and a Hadamard gate applied on every t qubit of the first register, in order to create a uniform superposition. The quantum state of the system is then given by:

$$|\psi_1\rangle = H^{\otimes t}|\psi_{ini}\rangle = \frac{1}{\sqrt{2^t}}\left(\sum_{k=0}^{2^t-1}|k\rangle\right)|1\rangle \qquad (3.8.26)$$

Next, the controlled operation U, described by Equation (3.8.22), is applied to the second register. However in this case, this operation is controlled by the qubits on the first register such as $|k\rangle U|y\rangle = |k\rangle|yx^k \bmod N\rangle$, and the sate of the system will evolve to:

$$|\psi_2\rangle = U|\psi_1\rangle = \frac{1}{\sqrt{2^t}}\sum_{k=0}^{2^t-1}|k\rangle|x^k \bmod N\rangle \qquad (3.8.27)$$

Using the relation given by the Equation (3.8.24), this last expression can be rewritten as:

$$|\psi_2\rangle = \frac{1}{\sqrt{r2^t}}\sum_{s=0}^{r-1}\sum_{k=0}^{2^t-1}e^{2\pi i sk/r}|k\rangle|u_s\rangle \qquad (3.8.28)$$

It can be noticed that the first register is now the Fourier expansion of the $\widetilde{|s/r\rangle}$ state. It is important to point out that the value of s/r depends on the number of qubits of the first register.

At this stage, the $\text{QFT}^\dagger$ operator has to be applied, in order to transfer the approximate value of the phase $\widetilde{s/r}$ to the state of the first register:

$$\text{QFT}^\dagger|\psi_2\rangle = \frac{1}{\sqrt{r}}\sum_{s=0}^{r-1}\widetilde{|s/r\rangle}|u_s\rangle \qquad (3.8.29)$$

Therefore, after a measurement is performed on the first register, the value of $\widetilde{s/r}$ is determined. From the sequence of operations discussed above, it is clear that the problem of order finding was reduced to phase estimation. The value of s/r can then be efficiently determined, within an arbitrary precision, which depends on the number of qubits of the first register. Unfortunately, the first register is on a superposition of states, so that any value of s

can be measured. However, it is also known that s/r is a ratio between two integers, and if one could obtain the closest fraction of s/r the value of r will be determined. Fortunately, there is a classical algorithm, known as *continued fraction algorithm*, which was idealized for describing rational numbers in terms of integers, and that allows the determination of r (see Ref. [1] for more details).

The order finding routine do have some problems, which will be discussed as follows. As it was pointed out before, after the application of the inverse Quantum Fourier Transform ($\text{QFT}^\dagger$), the first register is in an uniform superposition of states. Therefore, several eigenkets with any value of s can be obtained after a measurement is performed on the first register. The continued fraction algorithm will only return the correct value if s is coprime of r. This is a problem, because if s is a factor of r, only a factor of r, and not the proper order, will be obtained after running the continued fraction algorithm. The easiest way around this is to verify if the found value of r satisfies the condition $x^r \bmod N$. If this condition is not satisfied, the order-finding routine should be run again, in order to obtain a different state $\widetilde{|s/r\rangle}$, and therefore another value of $\widetilde{s/r}$. Repeating the routine several times until the a value of s/r, that can be properly used, seems a waste of time and that may compromise the efficiency and speed of the algorithm. However, there is a high probability of finding a value of s that is not a factor of r, by picking up a value of this quantity randomly, and that is exactly what is done when an uniform superposition is measured, allowing the determination of the correct value of r. There are other ways to avoid many repetitions [1], which will not be discussed here. The other case in which this routine fails is if a bad estimate of s/r is obtained. In this case, it will be impossible to determine r. This problem can be avoided by increasing the number of qubits in the first register.

Shor's factoring algorithm

The factoring problem can be reduced to order-finding, and that is what is actually done in Shor's factorizing algorithm, which is discussed in this section. We begin this discussion by showing why the factorization problem can be reduced to order-finding, followed by the stages of the factoring a given number. We illustrate that with a simple example.

Let N, a positive integer, be the number we want to factorize. We start by selecting randomly a number x, such as $0 < x < N$ and $\gcd(x, N) = 1$ (where $\gcd(x, N)$ stands for Great Common Divisor between x and N). If $\gcd(x, N) > 1$ we are lucky, because x is already a factor of N. However, if $\gcd(x, N) = 1$, then there is a smallest positive integer $r \leqslant N$, such that $x^r = 1 \pmod{N}$. This number is also called the period of x in respect to N. In order this procedure to work, i.e. to determine the prime factors of N, the period must be even, so that:

$x^r - 1 = 0 \pmod{N}$ leading to $(x^{r/2} + 1)(x^{r/2} - 1) = 0 \pmod{N}$, which means that $(x^{r/2} + 1)(x^{r/2} - 1) = jN$, for a nonzero integer j. This, does not mean that $(x^{r/2} + 1)$ or $(x^{r/2} - 1)$ divide N, separately, but they $((x^{r/2} + 1)$ and $(x^{r/2} - 1))$ must have some common factors in respect to N. Therefore, the $\gcd(x^{r/2} + 1, N)$ and $\gcd(x^{r/2} - 1, N)$ might give some nontrivial factors of N. Even if only one prime factor of N is returned, this procedure can be used to find the others. However, if N is a power of a single prime number, this procedure fails. Fortunately, there is an efficient classical algorithm to test that possibility [1].

In order to illustrate the procedure described above, let's factorize 15, which is the smallest nontrivial odd integer. Let's pick the attempts 7, 9 and 13, for didactical purposes. Let us now calculate the order (period in respect to 15), which can done by testing one by one:

r	1	2	3	4	5	6	7	8	9	10
$7^r \bmod N$	7	4	13	1	7	4	13	1	7	4
$9^r \bmod N$	9	6	9	6	9	6	9	6	9	6
$13^r \bmod N$	13	4	7	1	13	4	7	1	13	4

For the first choice, $x = 7$, one can see clearly that the period is 4, which is even, and then calculate the $\gcd(7^{4/2} + 1, 15) = \gcd(50, 15) = 5$ and $\gcd(7^{4/2} - 1, 15) = \gcd(48, 15) = 3$, which will yield the trivial answer $3 \times 5 = 15$. For the second choice, $x = 9$, the period is 2, and $\gcd(9^{2/2} + 1, 15) = \gcd(10, 15) = 5$, but $\gcd(9^{2/2} - 1, 15) = \gcd(8, 15) = 1$. Here we a have case where the choice did not give all the prime factors of 15, but left us with at least one good answer, from which the other can be derived. For the third and final choice, $x = 13$, the period is also 4, giving $\gcd(13^{4/2} + 1, 15) = 5$ and $\gcd(13^{4/2} - 1, 15) = 3$.

It is clear that for large integers, this procedure is not efficient, since finding the order is a non-trivial procedure. The power of Shor's factorization algorithm lies in the fact that a quantum routine, which is extremely efficient, can be used to determine the order of a number.

At this point, we are finally ready to describe the quantum factoring algorithm, which has several tasks to be followed. However, some are simple and can run on a classical computer, with no harm to the performance of the algorithm.

The first task, which is simple, is to verify whether N is even, returning the value 2, for a positive result, and then restarting the algorithm. The second task is to determine if $N = a^b$ for $b \geq 2$, i.e. if it is a power of some single prime number, and return a, for a positive answer. The third task is to pick randomly a number x, such that $1 \leq x \leq N$, returning $\gcd(x, N)$ if $\gcd(x, N) > 1$, since a prime factor of N has already been found. The fourth task, if all of the previous ones have failed, is to find the *order*, r, of x ($x^r = 1 \pmod{N}$). If r has been found to be even, and $x^{r/2} \neq -1 \pmod{N}$, calculate $\gcd(x^{r/2} - 1, N)$ and $\gcd(x^{r/2} + 1, N)$, determining which are the non-trivial solutions, that could be the answer, i.e. the prime factors of N. If the answer has not been found, the third and fourth tasks have to be repeated, using a different x, also randomly picked.

For illustration purposes, we will show how to factorize the number 15, using the Shor's algorithm. This is also the number used in the first, and only, experimental implementation of this algorithm, utilizing the Nuclear Magnetic Resonance technique [21] (see Chapter 5). Since 15 is the product of two prime numbers, 3 and 5, the two first tasks will fail. The third task is to pick randomly a number x, let's choose $x = 7$, and check if $\gcd(7, 15) > 1$. This task is also going to fail, because 7 is not a factor of 15. Therefore, the next step is to find the order, r.

As explained above, at the first stage of the order-finding routine, the system is the state described by Equation (3.8.30), i.e. the first register, that contains t qubits, is in a uniform superposition, whereas the second register stores the state $|1\rangle$:

$$|\psi_1\rangle = \frac{1}{\sqrt{2^t}} \big[|0\rangle + |1\rangle + |2\rangle + \cdots + |t - 1\rangle\big]|1\rangle \qquad (3.8.30)$$

3.8. Quantum algorithms

At the second stage, the operation U, such as $|j\rangle U|y\rangle = |j\rangle|y7^j \bmod 15\rangle$, is applied to the second register, conditionally to the states of each individual qubit in the first register. The system is then lead to the following state $|\psi_2\rangle$:

$$|\psi_2\rangle = \frac{1}{\sqrt{2^t}}\big[|0\rangle|1\rangle + |1\rangle|7\rangle + |2\rangle|4\rangle + |3\rangle|13\rangle + \cdots$$
$$+ |4\rangle|1\rangle + |5\rangle|7\rangle + |6\rangle|4\rangle + |7\rangle|13\rangle + \cdots\big] \quad (3.8.31)$$

The next step is to apply the inverse Quantum Fourier Transform ($\text{QFT}^\dagger$) to the first register and measure it in order to obtain the value of s/r. This task can be performed by calculating the density matrix of the first register, through the partial trace operation, assuming that the density matrix of the whole system has been measured, and then applying the $\text{QFT}^\dagger$. However, there is a periodicity in the second register of $|\psi_2\rangle$, as can be seen on Equation (3.8.31), and an alternative can be used from this point. This alternative requires the use of the implicit measurement principle, which states that at the end of a quantum circuit, it can be assumed that all the qubits have been measured, even though they were not [1]. Since the qubits of the second register can only be in one of those states: $|1\rangle, |7\rangle, |4\rangle$ or $|13\rangle$, a projective measurement will make the wave-function collapse to one of these states. Supposing that such a measurement was performed and the second register was found to be on the state $|4\rangle$ (anyone will do), the system would now be on the state described by:

$$|\psi_3\rangle = \sqrt{\frac{4}{2^t}}\big[|2\rangle + |6\rangle + |10\rangle + |14\rangle + \cdots\big]|4\rangle \quad (3.8.32)$$

Now, the $\text{QFT}^\dagger$ must be applied. Assuming that $t = 11$ qubits were used on the first register, the value of s/r will be determined with a precision of $1/4$. This means that there are $2^{11} = 2048$ allowed states, and after the application of the inverse Quantum Fourier Transform the system will be in superposition of the following states: $|0\rangle, |512\rangle, |1024\rangle$ and $|1536\rangle$:

$$|\psi_4\rangle = \text{QFT}^\dagger|\psi_3\rangle = \Bigg[\sum_{k=0}^{2^t-1}\alpha_k|k\rangle\Bigg]|4\rangle$$
$$= \frac{1}{2}\big[|0\rangle + |512\rangle + |1024\rangle + |1536\rangle\big]|4\rangle \quad (3.8.33)$$

This last procedure, the application of the inverse Quantum Fourier Transform, is not straightforward, but can be easily calculated numerically, and the outcome of this result depends on the number of qubits in the first register.

The next step is to perform a measurement. Assuming that the system has been found on the state $|1536\rangle$, it is possible to determine the order of $x = 7$, using the continued fractions algorithm, to find $r = 4$, since $1536/2048 = 3/4$. Because 4 is even, it is then possible to calculate $\gcd(x^{r/2} \pm 1, N)$ and test if any of the answers are solutions of the problem. Since $\gcd(7^{4/2} - 1, 15) = \gcd(48, 15) = 3$ and $\gcd(7^{4/2} + 1, 15) = \gcd(50, 5) = 3$, we have that $15 = 3 \times 5$. The same result is obtained if the state $|512\rangle$ is measured instead. If the state $|0\rangle$ is the measured one, it will be impossible to determine the order of 7 (mod 15), and if the outcome is $|1024\rangle$, the order-finding routine has failed, since it provided the value $r = 2$.

This is due to $1024/2048 = 2/4 = 1/2$, which means that value of $s = 2$ is a factor of $r = 4$, and the continued fraction algorithm fails, as discussed earlier. For further discussion, see [22] and [23].

3.9 QUANTUM SIMULATIONS

There are various important applications of simulations, in physics, chemistry and many others areas. Simulations help in the construction of better cars, planes, buildings, etc. In this section, we discuss how quantum computers can be used to simulate quantum systems, a hard task for the classical computers. Examples of quantum simulations implemented by NMR are shown in Chapter 5.

One of the most remarkable applications of quantum computing is the ability to simulate others quantum systems. In fact, classical computers cannot be used to simulate a quantum system efficiently [24,25]. The basic problem is the dimension of the Hilbert's space, that is 2^n for a simple system containing n particles with only two degrees of freedom. It is obvious that as the number of particles increases, this problem becomes intractable, since it scales exponentially.

The second postulate of quantum mechanics dictates the rules for the time evolution of quantum systems:

$$|\psi(t)\rangle = \exp(-i\mathcal{H}t/\hbar)|\psi(0)\rangle \tag{3.9.1}$$

where $\mathcal{H}$ represents the system Hamiltonian. The trick of using a quantum system for simulating another one is to perform a set of unitary operations, $U_k(\Delta t_k)$, which altogether accomplish the operation $\exp(-i\mathcal{H}_s \Delta t/\hbar)$, where $\mathcal{H}_s$ is the Hamiltonian of the system one wishes to simulate:

$$\exp(-i\mathcal{H}_s \Delta t/\hbar) = \prod_k U_k(\Delta t_k) \tag{3.9.2}$$

where Δt is the total time for the duration of the simulation, noting that $\sum_k \Delta t_k = \Delta t$.

There are cases in which the exponentiation of $\mathcal{H}_s$ is difficult, for instance when the number of particles is large. Sometimes first order approximations can be made:

$$\exp(-i\mathcal{H}_s \Delta t/\hbar) \approx \mathbf{1} - i\mathcal{H}_s \Delta t/\hbar \tag{3.9.3}$$

but usually this will lead to an unsatisfactory result, and others procedures become necessary.

Some quantum systems have Hamiltonians that can be divided into two parts $\mathcal{H}_0$ and $\mathcal{H}'$, where the first one contains the main interactions and a second one that acts like a perturbation on the system and is controlled by an external agent. Examples of such systems are electrons bound by an atomic potential that can be induced to "jump" from an orbital to another by laser beams, and the orientation of the nuclear magnetic moment, along a strong magnetic field that can be manipulated by the weak radio-frequency pulses. These two parts should act on the system in which the simulation is to be run. The procedure only works if $\mathcal{H}_s$ can be efficiently described by $\mathcal{H}_0$ and $\mathcal{H}'$, i.e. the simulation also depends on the system in which it is to be run.

Many particles systems are usually described by a class of Hamiltonians that are composed of local interactions, i.e. between a small number of particles, like the Ising and Heisenberg models:

$$\mathcal{H} = \sum_k \mathcal{H}_k \qquad (3.9.4)$$

where $\mathcal{H}_k$ acts only on a small part of the system. Therefore, if $\mathcal{H}_k$ can be exponentiated, $\mathcal{H}$ will be efficiently simulated. However, in general $[\mathcal{H}_k, \mathcal{H}_l] \neq 0$, which means that $\exp(-i\mathcal{H}t/\hbar) \neq \prod_k \exp(-i\mathcal{H}_k t/\hbar)$. The solution to this problem lies in the so-called *Trotter formula*, also called the asymptotic approximation theorem, for Hermitian operators:

$$\lim_{n \to \infty} \left(\exp(iAt/n) \exp(iBt/n) \right)^n = \exp(i[A+B]t) \qquad (3.9.5)$$

This formula is valid even if A and B do not commute. In chapter 5 some implementations of quantum simulations through Nuclear Magnetic Resonance are discussed.

3.10 QUANTUM INFORMATION IN PHASE SPACE

Quantum mechanics, and consequently quantum computation and quantum information, can be formulated and analyzed in phase space. As it will be exemplified in Chapter 5, some quantum processes and algorithms can be represented in phase space and interpreted as quantum maps [26,27]. The bridge that connects the two distinct approaches is the Wigner function [28], which is a distribution function that enables the quantum states, and their time evolution to be represented in the classical phase-space scenario. In this section, the Wigner function representation of quantum mechanics is discussed for finite systems, with 2^n-dimensional Hilbert space, being n the number of qubits. We start by describing the definition of the Wigner function, discussing some of its basic properties, and exemplifying with some simple coherent quantum states. A discussion about the application of a quantum circuit, known as the scattering circuit, for determining directly the phase space map of quantum system is presented.

3.10.1 The Wigner function

The Wigner function, for a continuous system, is defined as [26]:

$$W(q, p) = \frac{1}{\pi \hbar} \int_{-\infty}^{+\infty} \exp(2ipy/\hbar) \langle q - y | \rho | q + y \rangle \, dy \qquad (3.10.1)$$

Therefore, the Wigner function is directly related to the density matrix operator, ρ, which characterizes the quantum system [28] in Hilbert space.

The Wigner function is a distribution for the position (q) and momentum (p) of a system. From the knowledge of the Wigner function of a system, its density matrix can be determined in a kind of quantum state tomography.

The Wigner function defined above describes continuous systems. Until a few yeas ago, the definition of Wigner functions for discrete systems, like spin systems, was an open question. The first attempt to describe a formalism in this direction was made by Wooters [29], for a system with prime dimensions. Later, Cohendet et al. [30] established the formalism for system of odd dimensions, and Leonhardt [31] generalized the problem for discrete systems, with an arbitrary dimension. Discrete systems have finite phase spaces, and their Wigner function possess some peculiarities, which have to be taken into account. For discrete systems, with eigenkets labeled by integers $|m\rangle$, such as $|m+d\rangle = |m\rangle$, where d is the system dimension, the Wigner function can be described by:

$$W(q, p) = \frac{1}{D} \sum_{n} \exp\left(\frac{4\pi i}{d} pn\right) \langle q - n|\rho|q + n\rangle \tag{3.10.2}$$

where $D = d$ for odd-dimensional systems (bosons) and $D = 2d$ for even-dimensional ones (fermions). Furthermore, the discrete Wigner function is real and normalized, and also periodic in phase space, with the period equal to the dimension [31], i.e.

$$W(q, p) = W(q + D, p) = W(q, p + D).$$

For bosons, n should run from $-(d-1)/2$ to $(d-1)/2$, taking integers values between them, and from $-d/2$ to $(d-1)/2$, taking half odds and integers values for fermions. However, in the fermionic systems, a convention establishes that the density matrix elements of half odd should be taken as zero.

For spin systems, the position eigenkets and eigenvalues play the role of the spin component along the z direction whereas the equivalent for momentum play the role of the quantum phases. They are connected through the discrete Fourier Transform:

$$|p\rangle = \frac{1}{\sqrt{d}} \sum_{p=0}^{d-1} \exp\left(\frac{2\pi i}{d} pq\right) |q\rangle \tag{3.10.3}$$

The systems dimension in this case is given by $d = 2j + 1$, where d is odd for bosons, and even for fermions. Because the arguments of the kets and bras have to be integers, in order to evaluate the Wigner function for spin systems, the following notation is used: $J_z|m\rangle = m|m\rangle$ being $(m = -j, \ldots, j)$ for bosons, and $J_z|m\rangle = (m - \frac{1}{2})|m\rangle$ being $(m = -j + \frac{1}{2}, \ldots, j + \frac{1}{2})$ for fermions [32]. In quantum computation, the Hilbert space has always an even dimension, equal to 2^n where n is the number of qubits. As a consequence, the Wigner function has a periodicity of $2d$.

An alternative way for describing the discrete Wigner function is the one given in terms of the discrete phase-point operator, given by [26]:

$$A(q, p) = \frac{1}{2N} U^q R V^{-p} \exp\left(\frac{2\pi i q p}{2^n}\right) \tag{3.10.4}$$

where U and V are the "translation" operators, in position $(U|q\rangle = |q+1\rangle)$ and momentum $(V|p\rangle = |p+1\rangle)$, being R the reflection operator $(R|n\rangle = |N-n\rangle)$, with $N = 2^n$.

The discrete Wigner function can be summarized as:

$$W(q, p) = \text{Tr}\left[A(q, p)\rho\right] \tag{3.10.5}$$

This procedure is particularly useful, as will be discussed below.

3.10. Quantum information in phase space

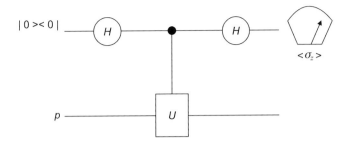

Figure 3.14 The scattering circuit. By measuring the output state of the fist qubit, one obtains information about either the input matrix ρ, or about the transformation U. Adapted with permission from [26].

3.10.2 Measuring the Wigner function

There is a generic quantum circuit that has been used in many algorithms and has various applications [1,23]. It is known as the "scattering circuit", because it resembles a scattering process, as illustrated on Figure 3.14. It uses an ancilla qubit which plays the role of the probe particle. By measuring the expectation values $\langle\sigma_z\rangle$ and $\langle\sigma_y\rangle$ of the probe, we obtain information about either the interaction (U), or the input state (ρ).

$$\langle\sigma_z\rangle = \text{Re}\big[\text{Tr}(\rho U)\big] \quad \text{and} \quad \langle\sigma_y\rangle = \text{Im}\big[\text{Tr}(\rho U)\big] \tag{3.10.6}$$

The scattering circuit can be used to measure the Wigner function of a quantum system, by setting the controlled operator to be $U = A(q, p)$, since $W(q, p) = \text{Tr}[A(q, p)\rho]$. Such a implementation has been performed by NMR [33], and will be discussed in Chapter 5.

3.10.3 Quantum states in phase space

Since the Wigner function is connected to the density matrix, it can be calculated from the knowledge of ρ, upon some repetitive applications of the operators U and V for each point of the phase space. Here the phase space of some interesting coherent quantum states is calculated numerically, for illustration purposes.

On Figure 3.15 the phase space relative to the four possible states of the computational basis – (a) $|00\rangle$, (b) $|01\rangle$, (c) $|10\rangle$ and (d) $|11\rangle$ is shown. The position q is plotted on the horizontal axis, and the momentum p along the vertical axis. As it can be seen from the figure, the "position" states are well defined, but the uncertainty on the momentum is maximum. We can also observe that the Wigner functions of the second state of the computational basis is equal to the first one, dislocated of two positions in the phase space, and so on.

By applying the QFT on the following combination states of the computational basis: (a) $\frac{1}{2}[|00\rangle + |01\rangle + |10\rangle + |11\rangle]$, (b) $\frac{1}{2}[|00\rangle + i|01\rangle - |10\rangle - i|11\rangle]$, (c) $\frac{1}{2}[|00\rangle - i|01\rangle - |10\rangle + i|11\rangle]$ and (d) $\frac{1}{2}[|00\rangle - |01\rangle + |10\rangle - |11\rangle]$ – we obtain states which are well defined in momentum, but undefined in position, as shown in Figure 3.16.

Other interesting Wigner functions for the following superpositions of states: (a) $\frac{1}{\sqrt{2}}[|00\rangle + |01\rangle]$, (b) $\frac{1}{\sqrt{2}}[|00\rangle + |10\rangle]$, (c) $\frac{1}{\sqrt{2}}[|01\rangle + |11\rangle]$ and (d) $\frac{1}{\sqrt{2}}[|10\rangle + |11\rangle]$, are

128   3. Fundamentals of Quantum Computation and Quantum Information

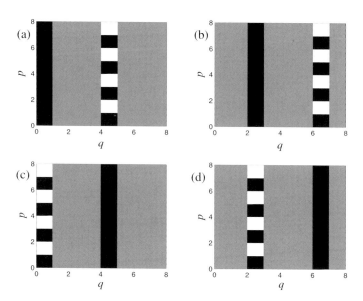

Figure 3.15 Phase space representation of the states of the computational basis of two qubits: (a) $|00\rangle$, (b) $|01\rangle$, (c) $|10\rangle$ and (d) $|11\rangle$. The states are well defined in position (horizontal axis) and undefined in momentum (vertical axis). Amplitudes vary from -0.125 (white) to $+0.125$ (black). Due to the periodic boundary conditions, an interference pattern appears as the black and white stripes.

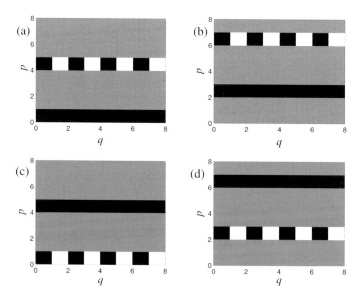

Figure 3.16 Phase space representation of the states with well defined momentum: (a) $\frac{1}{2}[|00\rangle + |01\rangle + |10\rangle + |11\rangle]$, (b) $\frac{1}{2}[|00\rangle + i|01\rangle - |10\rangle - i|11\rangle]$, (c) $\frac{1}{2}[|00\rangle - i|01\rangle - |10\rangle + i|11\rangle]$ and (d) $\frac{1}{2}[|00\rangle - |01\rangle + |10\rangle - |11\rangle]$. Amplitudes vary from -0.125 (white) to $+0.125$ (black).

3.10. Quantum information in phase space

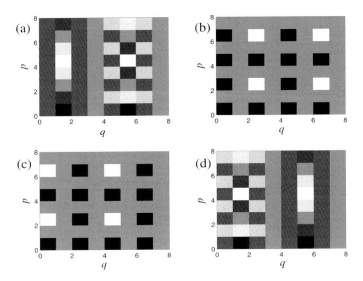

Figure 3.17 Wigner functions for some uniform superposition states: (a) $\frac{1}{\sqrt{2}}[|00\rangle + |01\rangle]$, (b) $\frac{1}{\sqrt{2}}[|00\rangle + |10\rangle]$, (c) $\frac{1}{\sqrt{2}}[|01\rangle + |11\rangle]$ and (d) $\frac{1}{\sqrt{2}}[|10\rangle + |11\rangle]$. Amplitudes vary from -0.125 (white) to $+0.125$ (black).

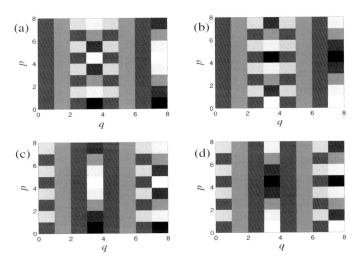

Figure 3.18 Wigner functions for four Bell states: (a) $\frac{1}{\sqrt{2}}[|00\rangle + |11\rangle]$, (b) $\frac{1}{\sqrt{2}}[|00\rangle - |11\rangle]$, (c) $\frac{1}{\sqrt{2}}[|01\rangle + |10\rangle]$ and (d) $\frac{1}{\sqrt{2}}[|01\rangle - |10\rangle]$. Amplitudes vary from -0.125 (white) to $+0.125$ (black).

displayed in Figure 3.17. In these cases, interference patterns between the quantum states can be observed, which is characteristic superposition of states.

Finally, Figure 3.18 exhibits the Wigner function of the Bell's basis: (a) $\frac{1}{\sqrt{2}}[|00\rangle + |11\rangle]$, (b) $\frac{1}{\sqrt{2}}[|00\rangle - |11\rangle]$, (c) $\frac{1}{\sqrt{2}}[|01\rangle + |10\rangle]$ and (d) $\frac{1}{\sqrt{2}}[|01\rangle - |10\rangle]$.

3.11 DETERMINING EIGENVALUES AND EIGENVECTORS

In quantum physics, one of the most relevant problem is to determine the eigenvalues and their respective eigenvectors for a certain Hamiltonian. This kind of problem requires an amount of time to be solved, which scales exponentially with the number of particles present in the system. This kind of problem is usually hard to be computed, even more complicated if *ab initio* calculations are involved. In this section, a description of an efficient algorithm that finds eigenvectors and eigenvalues for a local Hamiltonian in polynomial rather then exponential time is presented [34]. This algorithm makes the use of the Quantum Fourier Transform (QFT), and uses three sets of qubits, being m for the application of the QFT, l for the storing the Hilbert space of the Hamiltonian to be diagonalized and w extra working qubits, that may be necessary. The total number of qubits required by the implementation of the algorithm is then $m + l + w$.

In order to solve the problem it is necessary to apply the operator $U = e^{-i\mathcal{H}t/\hbar}$, which describes the evolution of the system represented by the Hamiltonian $\mathcal{H}$, whose eigenvalues and eigenvectors are to be obtained. Therefore, a simulation of the system represented by $\mathcal{H}$ is necessary. The algorithm requires that an approximate eigenvector $|V_a\rangle$ of U, and therefore of $\mathcal{H}$, to be generated in polynomial time, i.e. the system can be put in this state using a polynomial number of quantum operations. Defining $|\phi_k\rangle$ as the real eigenvector of $\mathcal{H}$, with λ_k eigenvalue, we have that: If $|\langle \phi_k | V_a\rangle|^2$ is not exponentially small, which means that the approximate eigenvector V_a has at least a small component of $|\phi_k\rangle$, then it is possible to find the λ_v with accuracy ϵ in time proportional to $1/|\langle \phi_k | V_a\rangle|^2$ and $1/\epsilon$. However, if the state is degenerated then the problem of determining its eigenvectors is more complicated.

The calculation starts by preparing the system in the initial state:

$$|\Psi_1\rangle = |0\rangle |V_a\rangle \qquad (3.11.1)$$

where the first m qubits are in the state $|0\rangle$ and l qubits are used to store $|V_a\rangle$.

The next step is to put the first m qubits in an uniform superposition:

$$|\Psi_2\rangle = \frac{1}{\sqrt{2^m}} \sum_{j=0}^{2^m-1} |j\rangle |V_a\rangle \qquad (3.11.2)$$

This can be achieved by applying a Hadamard gate to each qubit of the first set, as discussed earlier.

Then, a set of applications of the operator $U = e^{-i\mathcal{H}t/\hbar}$ is done to the approximate eigenvector $|V_a\rangle$, controlled by the m qubits of the first register:

$$|\Psi_3\rangle = \frac{1}{\sqrt{2^m}} \sum_{j=0}^{2^m-1} |j\rangle U^j |V_a\rangle \qquad (3.11.3)$$

It is worth pointing that U is applied j times, where j is the label of a qubit of the first register. Similar procedure is also used in the phase estimation and finding order subroutines. Here, the difference is the application of U^j, where j varies 0 to $2^m - 1$, whereas in the others subroutines required the application of even powers of U.

Rewriting the approximate eigenvector in terms of the real ones, the Equation (3.11.3) transforms to:

$$|\Psi_3\rangle = \frac{1}{\sqrt{2^m}} \sum_{j=0}^{2^m-1} |j\rangle U^j \sum_k c_k |\phi_k\rangle \qquad (3.11.4)$$

And after the application of U, the state $|\Psi_3\rangle$ is then described by:

$$|\Psi_3\rangle = \frac{1}{\sqrt{2^m}} \sum_{j=0}^{2^m-1} |j\rangle \sum_k \lambda_k^j c_k |\phi_k\rangle \qquad (3.11.5)$$

Writing the eigenvalues λ_k as $e^{i\omega_k}$, and exchanging the order of the two registers, the systems state can be rewritten as:

$$|\Psi_3\rangle = \frac{1}{\sqrt{2^m}} \sum_k c_k |\phi_k\rangle \sum_{j=0}^{2^m-1} e^{i\omega_k j} |j\rangle \qquad (3.11.6)$$

It is clear that the application of the QFT on the m register followed by a measurement will yield the values of the phases ω_k, which are the desired eigenvalues.

Furthermore, once a measurement is made and the eigenvalue λ_k is determined, the remaining l qubits will be projected onto the corresponding eigenvector. Therefore, others properties of the quantum system under study can be obtained by simply continuing the calculation, and many important physical information can then be extracted.

For the degenerate case, or situations in which the accuracy does not allows distinguishing among some eigenstates, this procedure can also be applied and the system will be projected onto the corresponding subspace. However, many measurements may be necessary in order to determine all the eigenvectors [34]. In summary, eigenvalues and eigenvectors can be obtained using the procedures discussed above, but the accuracy of the answer depends upon a number of variables of the system, that have to be precisely controlled.

PROBLEMS WITH SOLUTIONS

P3.1 - Prove the relation (3.2.7) $\sum_m M_m^\dagger M_m = 1$.

Solution
We have $p_m = \langle \psi | M_m^\dagger M_m | \psi \rangle$ and that:

$$\sum_m p_m = \sum_m \langle \psi | M_m^\dagger M_m | \psi \rangle = \langle \psi | \sum_m M_m^\dagger M_m | \psi \rangle = 1 \qquad (3.11.7)$$

But $\langle \psi || \psi \rangle = 1$, so that:

$$\langle \psi || \psi \rangle = \langle \psi | \sum_m M_m^\dagger M_m | \psi \rangle \Rightarrow \sum_m M_m^\dagger M_m = 1 \qquad (3.11.8)$$

P3.2 - Verify the relation given by Equation (3.6.7).

$$\rho = \begin{bmatrix} a & b+ic \\ b-ic & 1-a \end{bmatrix} \tag{3.11.9}$$

Solution
Evaluating the expected values we have: $\text{Tr}(\rho\sigma_x) = 2b$, $\text{Tr}(\rho\sigma_y) = -2c$ and $\text{Tr}(\rho\sigma_z) = 2a - 1$. We also have that: $\text{Tr}(\rho) = 1$. Therefore:

$$\frac{\text{Tr}(\rho) + \text{Tr}(\rho\sigma_x)\sigma_x + \text{Tr}(\rho\sigma_y)\sigma_y + \text{Tr}(\rho\sigma_z)\sigma_z}{2} = \cdots \tag{3.11.10}$$

$$\frac{1 + 2a\sigma_x - 2c\sigma_y + (2a-1)\sigma_z}{2} = \begin{bmatrix} a & b+ic \\ b-ic & 1-a \end{bmatrix} = \rho \tag{3.11.11}$$

P3.3 - Show that the QFT, given by Equation (3.5.5), is an unitary transformation.

Solution
We have that:

$$\text{QFT}|j\rangle = \frac{1}{\sqrt{2^n}} \sum_{k=0}^{2^n-1} e^{2i\pi jk/2^n} |k\rangle = \frac{1}{\sqrt{2^n}} \left(|0\rangle + \cdots + e^{i\pi j(2^n-1)}|2^n-1\rangle\right) \tag{3.11.12}$$

Hence:

$$\langle j|\text{QFT}^\dagger = \frac{1}{\sqrt{2^n}} \sum_{k=0}^{2^n-1} e^{-i2\pi jk/2^n} \langle k| = \frac{1}{\sqrt{2^n}} \left(\langle 0| + \cdots + e^{-i\pi j(2^n-1)}\langle 2^n-1|\right) \tag{3.11.13}$$

$$\langle j|\text{QFT}^\dagger \text{QFT}|j\rangle = \frac{1}{2^n} \left(\langle 0|(|0\rangle + \cdots + e^{i\pi j}|2^n-1\rangle) + \cdots + e^{-i\pi j(2^n-1)}\langle 2^n-1|)(|0\rangle + \cdots \right.$$
$$\left. + e^{i\pi j(2^n-1)}|2^n-1\rangle\right) \tag{3.11.14}$$

$$\langle j|\text{QFT}^\dagger \text{QFT}|j\rangle = \frac{1}{2^n} \left(\langle 0||0\rangle + \langle 1||1\rangle + \cdots + \langle 2^n-1||2^n-1\rangle\right) = 1 \tag{3.11.15}$$

then: $\quad \text{QFT}^\dagger \text{QFT} = \mathbf{1} \tag{3.11.16}$

Where we have used the fact $\langle k||k'\rangle = \delta_{kk'}$, since the set of kets $|k\rangle$ forms an orthonormal basis.

P3.4 - Demonstrate that the operator which applies the conditional phase shift can be described as shown on Equation (3.8.11), $[2|0\rangle\langle 0| - \mathbf{1}]|x\rangle = -(-1)^{\delta_{x0}}|x\rangle$, and give the general form of $\text{H} \cdot [2|0\rangle\langle 0| - \mathbf{1}] \cdot \text{H}$, where H represents the Hadamard.

Solution
As may be seen:

$$|0\rangle\langle 0| = \begin{bmatrix} 1 & 0 & \cdots & 0 \\ 0 & 0 & \cdots & 0 \\ \vdots & \vdots & \ddots & \vdots \\ 0 & 0 & \cdots & 0 \end{bmatrix} \quad \text{and} \quad 2|0\rangle\langle 0| - \mathbf{1} = \begin{bmatrix} 1 & 0 & \cdots & 0 \\ 0 & -1 & \cdots & 0 \\ \vdots & \vdots & \ddots & \vdots \\ 0 & 0 & \cdots & -1 \end{bmatrix}$$

which is equal to $-(-1)^{\delta_{x0}}$.

Problems with solutions

The `Hadamard` applied to both sides will result in:

$$[2H|0\rangle\langle 0|H - 1] = \frac{1}{2}\begin{bmatrix} -1 & 1 & \cdots & 1 \\ 1 & -1 & \cdots & 1 \\ \vdots & \vdots & \ddots & \vdots \\ 1 & 1 & \cdots & -1 \end{bmatrix}$$

P3.5 - Show that the `QFT` can be described by Equation (3.5.6).

Solution

$$\frac{1}{\sqrt{2^n}} \sum_{k=0}^{2^n-1} e^{2\pi i jk/2^n} |k\rangle = \cdots$$

$$= \frac{1}{\sqrt{2^n}} \sum_{k_1=0}^{1} \cdots \sum_{k_n=0}^{1} \exp\left(2\pi i j \sum_{l=1}^{n} k_l/2^l\right) |k_1 k_2 \ldots k_n\rangle = \cdots$$

$$= \frac{1}{\sqrt{2^n}} \sum_{k_1=0}^{1} \cdots \sum_{k_n=0}^{1} \bigotimes_{l=1}^{n} \exp(2\pi i j k_l/2^l)|k_l\rangle = \cdots$$

$$= \frac{1}{\sqrt{2^n}} \bigotimes_{l=1}^{n} \sum_{k_l=0}^{1} \exp(2\pi i j k_l/2^l)|k_l\rangle = \frac{1}{\sqrt{2^n}} \bigotimes_{l=1}^{n} [|0\rangle + e^{2\pi i j/2^l}|1\rangle] = \cdots$$

$$= \frac{1}{\sqrt{2^n}} \left(|0\rangle + e^{2\pi i 0.j_n}|1\rangle\right)\left(|0\rangle + e^{2\pi i 0.j_{n-1} j_n}|1\rangle\right) \cdots \left(|0\rangle + e^{2\pi i 0.j_1 j_2 \cdots j_n}|1\rangle\right) \quad (3.11.17)$$

P3.6 - Show that $[2|\psi\rangle\langle\psi| - 1] \sum_k \alpha_k |k\rangle = \sum_k [2\langle\alpha\rangle - \alpha_k]|k\rangle$.

Solution

$$|\psi\rangle = \frac{1}{\sqrt{N}} \sum_{x=0}^{2^n-1} |x\rangle \quad \text{and} \quad \langle\psi| = \frac{1}{\sqrt{N}} \sum_{y=0}^{2^n-1} \langle y|$$

$$[2|\psi\rangle\langle\psi| - 1] \sum_{k=0}^{2^n-1} \alpha_k |k\rangle$$

$$= \sum_k \left[\frac{2}{N} \sum_{x=0}^{2^n-1} \sum_{y=0}^{2^n-1} |x\rangle\langle y||k\rangle\alpha_k - \alpha_k|k\rangle\right] = \cdots$$

$$= \sum_k^{2^n-1} \left[\frac{2}{N} \sum_{x=0}^{2^n-1} \sum_{y=0}^{2^n-1} |x\rangle\alpha_k \delta_{ky} - \alpha_k|k\rangle\right] = \sum_k^{2^n-1} \left[\frac{2}{N} \sum_{x=0}^{2^n-1} |x\rangle\alpha_k - \alpha_k|k\rangle\right] = \cdots$$

$$= \sum_{x=0}^{2^n-1} |x\rangle \frac{2}{N} \sum_{k=0}^{2^n-1} \alpha_k - \frac{2}{N} \sum_{k=0}^{2^n-1} \alpha_k |k\rangle = \sum_{x=0}^{2^n-1} |x\rangle\langle\alpha\rangle - \sum_{k=0}^{2^n-1} \alpha_k |k\rangle = \cdots = \sum_k [2\langle\alpha\rangle - \alpha_k]|k\rangle$$

P3.7 - Show that the controlled operations of the phase estimation algorithm implicates in the following transformation: $|k\rangle|u\rangle \to |k\rangle U^k |u\rangle$.

Solution

In the binary basis, an arbitrary number k is written as $k_0 k_1 k_2 \ldots k_{n-1}$, using n digits to represent the number, so that $k = k_0 \times 2^{n-1} + k_1 \times 2^{n-2} + \cdots + k_{n-1} \times 2^0$. Following the operations shown on the Figure 3.12, one can see that the last qubit of the first register, k_{n-1}, controls the application of the operator U to the second register, i.e. U is applied only if $k_{n-1} = 1$. Another way of representing this is $U^{2^0 k_{n-1}}$, which is equal to 1 only when $k_{n-1} = 1$.

Therefore, the sequence of controlled operations can be written as:

$$U^{2^{n-1} k_0} \ldots U^{2^2 k_{n-3}} U^{2^1 k_{n-2}} U^{2^0 k_{n-1}} = U^{2^{n-1} \times k_0 + \cdots + 2^2 \times k_{n-3} + 2^1 \times k_{n-2} + 2^0 \times k_{n-1}} = U^k$$

Which gives $|k\rangle|u\rangle \to |k\rangle U^k|u\rangle$.

P3.8 - Show that $\frac{1}{\sqrt{r}} \sum_{s=0}^{r-1} |u_s\rangle = |1\rangle$.

Solution

Using the fact: $\sum_{s=0}^{r-1} \exp(-2\pi i s k / r) = r \delta_{k0}$, which can be easily demonstrated, we have:

$$\frac{1}{\sqrt{r}} \sum_{s=0}^{r-1} |u_s\rangle = |1\rangle = \frac{1}{\sqrt{r}} \sum_{s=0}^{r-1} \frac{1}{\sqrt{r}} \sum_{k=0}^{r-1} \exp(-2\pi i s k / r) |x^k \bmod N\rangle = \cdots$$

$$= \frac{1}{r} \sum_{k=0}^{r-1} |x^k \bmod N\rangle \sum_{s=0}^{r-1} \exp(-2\pi i s k / r) = \cdots$$

$$= \frac{1}{r} \sum_{k=0}^{r-1} |x^k \bmod N\rangle r \delta_{k0} = \cdots$$

$$= |x^0 \bmod N\rangle = |1\rangle$$

P3.9 - Show that the components $\langle \sigma_z \rangle = \mathrm{Re}[\mathrm{tr}(U\rho)]$ and $\langle \sigma_y \rangle = \mathrm{Im}[\mathrm{tr}(U\rho)]$, of the measured qubit of the scattering circuit.

Solution

At the beginning of the quantum circuit, the system is the state $|\psi\rangle = |0\rangle|\phi\rangle$, i.e. the first qubit is in the others qubits of the system are in the state $|\phi\rangle$, with ρ being given by $\rho = |\phi\rangle\langle\phi|$. The operation is to apply the Hadamard gate to the first qubit creating an uniform superposition. The system sate is the given by:

$$|\psi_1\rangle = (H|0\rangle)|\phi\rangle = \frac{1}{\sqrt{2}}(|0\rangle + |1\rangle)|\phi\rangle = \frac{1}{\sqrt{2}}(|0\rangle|\phi\rangle + |1\rangle|\phi\rangle)$$

The next operation is controlled one, which applies U to $|\phi\rangle$, if the first qubit is in the sate $|1\rangle$. The system will then be in the state:

$$|\psi_2\rangle = \frac{1}{\sqrt{2}}(|0\rangle|\phi\rangle + |1\rangle U|\phi\rangle)$$

The last operation is to apply the `Hadamard` again to the first qubit, leaving the system at the state:

$$|\psi_3\rangle = \frac{1}{2}\big[(|0\rangle + |1\rangle)|\phi\rangle + (|0\rangle - |1\rangle)U|\phi\rangle\big] = \frac{1}{2}\big[|0\rangle(1+U)|\phi\rangle + |1\rangle(1-U)|\phi\rangle\big]$$

As $\langle \sigma_z \rangle = \langle \psi_3 | \sigma_z | \psi_3 \rangle$, we have:

$$\langle \sigma_z \rangle = \frac{1}{4}\big\{[\langle\phi|(1+U^\dagger)\langle 0| + \langle\phi|(1-U^\dagger)\langle 1|]\sigma_z[|0\rangle(1+U)|\phi\rangle + |1\rangle(1-U)|\phi\rangle]\big\}$$

$$\langle \sigma_z \rangle = \frac{1}{4}\big\{[\langle\phi|(1+U^\dagger)\langle 0| + \langle\phi|(1-U^\dagger)\langle 1|] \cdot [|0\rangle(1+U)|\phi\rangle - |1\rangle(1-U)|\phi\rangle]\big\}$$

$$\langle \sigma_z \rangle = \frac{1}{4}\{\langle\phi|(1+U+U^\dagger+1)|\phi\rangle - \langle\phi|(1-U-U^\dagger+1)|\phi\rangle\}$$

$$\langle \sigma_z \rangle = \frac{1}{4}\{\langle\phi|(U+U^\dagger)|\phi\rangle + \langle\phi|(U+U^\dagger)|\phi\rangle\}$$

We then have:

$$\langle \sigma_z \rangle = \frac{1}{2}\langle\phi|(U+U^\dagger)|\phi\rangle = \langle\phi|\operatorname{Re} U|\phi\rangle = \operatorname{Re}[\operatorname{tr}(U\rho)]$$

REFERENCES

[1] M.A. Nielsen, I.L. Chuang, *Quantum Computation and Quantum Information* (Cambridge Press, 2001).
[2] D. Bouwmeester, A. Ekert, A. Zeilinger, *The Physics of Quantum Information* (Springer Verlag, 2001).
[3] C.P. Williams, S.H. Clearwater, *Explorations in Quantum Computing* (Springer Verlag, 1998).
[4] A.O. Pittenger, *An Introduction to Quantum Computing Algorithms* (Birkhäuser, 1999).
[5] C. Cohen-Tannoudji, B. Diu, F. Laloë, *Quantum Mechanics* (Hermann and Jonh Wiley & Sons, 1977).
[6] E. Knill, R. Laflamme, R. Martinez, C.-H. Tseng, An algorithmic benchmark for quantum information processing, *Nature* **404** (2000) 368.
[7] C. Bennett, Logical reversibility of computation, *IBM J. Res. Develop.* **17** (1973) 525.
[8] P. Benioff, The computer as a physical system: A microscopic quantum mechanical Hamiltonian model of computers as represented by Turing machines, *J. Stat. Phys.* **22** (1980) 563.
[9] D.P. DiVincenzo, Quantum gates and circuits, *Proc. Royal Soc. London A* **454** (1998) 261.
[10] R. Jozsa, Quantum algorithms and the Fourier transform, *Proc. Royal Soc. London A* **454** (1998) 323.
[11] P. Shor, Algorithms for quantum computation: Discrete logarithms and factoring, in: *Proc. 35th Ann. Symp. Found. Comp. Science*, 1994, p. 124.
[12] P. Shor, Polynomial-time algorithms for prime factorization and discrete logarithms on a quantum computer, *SIAM J. Comput.* **26** (1997) 1484.
[13] A. Einstein, B. Podolsky, N. Rosen, Can quantum-mechanical description of physical reality be considered complete?, *Phys. Rev.* **47** (1935) 777.
[14] J.S. Bell, On the Einstein–Podolsky–Rosen paradox, *Physics 1* **195** (1964).
[15] F. Mintert, A.R.R. Carvalho, M. Kús, A. Buchleitner, Measures and dynamics of entangled states, *Phys. Rep.* **415** (2005) 207.
[16] A. Aspect, P. Grangier, G. Roger, Experimental realization of Einstein–Podolsky–Rosen–Bohm Gedanken-experiment: A new violation of Bell's inequalities, *Phys. Rev. Lett.* **49** (1982) 91.
[17] G. Brassard, S.L. Braunstein, R. Cleve, Teleportation as a quantum computation, *Physica D* **120** (1998) 43.
[18] D. Deutsh, Quantum theory, the Church-Turing principle and the universal quantum computer, *Proc. Royal Soc. London A* **400** (1985) 97.
[19] L. Grover, Quantum mechanics helps in searching for a needle in a haystack, *Phys. Rev. Lett.* **79** (1997) 325.
[20] D. Deutsh, R. Jozsa, Rapid solution of problems for quantum computation, *Proc. Royal Soc. London A* **439** (1992) 553.
[21] L.M.K. Vandersypen, M. Steffen, G. Breyta, C.S. Yannoni, M.H. Sherwood, I.L. Chuang, Experimental realization of Shor's quantum factoring algorithm using nuclear magnetic resonance, *Nature* **414** (2001) 883.
[22] A. Ekert, R. Jozsa, Quantum computation and Shor's algorithm, *Rev. Mod. Phys.* **68** (1996) 733.
[23] R. Cleve, A. Ekert, C. Macchiavello, M. Mosca, Quantum algorithms revisited, *Proc. Royal Soc. London A* **454** (1998) 339.
[24] R. Feynman, Simulating physics with computers, *Int. J. Theor. Phys.* **21** (1982) 467.
[25] C. Zalka, Simulating quantum systems on a quantum computer, *Proc. Royal Soc. London A* **454** (1998) 313.
[26] C. Miquel, J.P. Paz, M. Saraceno, Quantum computers in phase space, *Phys. Rev. A* **65** (2002) 062309.
[27] C. Miquel, J.P. Paz, M. Saraceno, E. Knill, R. Laflamme, C. Negrevergne, Interpretation of tomography and spectroscopy as dual forms of quantum computation, *Nature* **418** (2002) 59.
[28] M. Hillery, R.F. O'Connell, M.O. Scully, E.P. Wigner, Distribution functions in physics: Fundamentals, *Phys. Rep.* **106** (1984) 121.

[29] W.K. Wooters, A Wigner-function formulation of finite-state quantum mechanics, *Ann. Phys.* **176** (1987) 1.
[30] O. Cohendet, Ph. Combe, M. Sirugue, M. Sirugue-Collin, A stochastic treatment of the dynamics of an integer spin, *J. Phys. A* **21** (1988) 2875.
[31] U. Leonhardt, Quantum-state tomography and discrete Wigner function, *Phys. Rev. Lett.* **74** (1995) 4101.
[32] U. Leonhardt, Discrete Wigner function and quantum-state tomography, *Phys. Rev. A* **53** (1996) 2998.
[33] J.P. Paz, A.J. Roncaglia, M. Saraceno, Quantum algorithms for phase-space tomography, *Phys. Rev. A* **69** (2004) 032312.
[34] D. Abraham, S. Lloyd, Quantum algorithm providing exponential speed increase for finding eigenvalues and eigenvectors, *Phys. Rev. Lett.* **83** (1999) 5162.

– 4 –

Introduction to NMR Quantum Computing

> ... Such a computer would look nothing like the machine that sits on your desk; surprisingly, it might resemble the cup of coffee at its side. – N. Gershenfeld and I.L. Chuang [Scientific American, June, 1998]

4.1 THE NMR QUBITS

We have seen that qubits can be accomplished by different quantum properties of a system. The basic requirement is that they must be well characterized and susceptible to manipulation by an external perturbation, so that the input states can be adequately prepared and controlled to produce the desired calculation. Besides, the physical representation of the qubit in quantum information processing must be unequivocal. This is certainly a requirement that NMR systems fulfill. In fact, a natural implementation of a qubit is an isolated spin 1/2 in a magnetic field [1]. In the I_z operator basis, the general state of this spin can be represented by $|\psi\rangle = \alpha|+1/2\rangle + \beta|-1/2\rangle$ (Figure 4.1). Labeling the states $|+1/2\rangle$ as $|0\rangle$ and $|-1/2\rangle$ as $|1\rangle$, each state of the system can be represented by a single label, $|0\rangle$ or $|1\rangle$, which means one-qubit of information.

The association of the spin states with logical labeling can be done in real systems. For example, let us consider a solution of ^{13}C enriched chloroform, $^{13}CHCl_3$. This system can be well approximated by two coupled spins I_1, I_2 with spin I_1 being the ^{13}C and spin I_2 the ^1H. The relevant NMR interactions are the Zeeman, chemical shifts and the J-coupling, which are represented by the following secular Hamiltonian (see Chapter 2) [2]:

$$\mathcal{H} = -\hbar\omega_{01} I_{1z} - \hbar\omega_{02} I_{2z} + \hbar 2\pi I_{1z} I_{2z} \tag{4.1.1}$$

where ω_{01} and ω_{02} are the resonance frequencies of each nucleus, including the chemical shift contribution. Considering the weak coupling limit, $|\omega_{01} - \omega_{02}| \gg 2\pi J_{12}$, the energy eigenvalues can be straightforwardly calculated, yielding the energy levels shown in Figure 4.2.[1] Using the same logical labeling as for single spin 1/2 states, we represent the eigenstates as $|0\rangle \otimes |0\rangle \equiv |00\rangle, |0\rangle \otimes |1\rangle \equiv |01\rangle, |1\rangle \otimes |0\rangle \equiv |10\rangle, |1\rangle \otimes |1\rangle \equiv |11\rangle$, which represents the computational basis of a two-qubit system. Because for spin 1/2 systems each qubit is associated to a spin, a n-qubit system can be implemented by a system of n spins.

Another aspect of NMR for quantum information processing concerns the coupling between the spins. Many quantum operations involve conditioning the states of two different

---
[1] See corresponding energy calculation in Section 2.8.

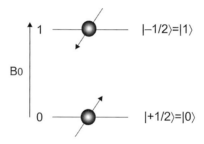

Figure 4.1 Representation of the possible configurations of one spin 1/2 in a magnetic field.

qubits (controlled gates). Since this can be easily achieved by evolution under coupling, the J-coupling between the spins is also essential. Thus, a system of n qubits can be implemented by a system on n coupled spins 1/2, with the following Hamiltonian:

$$\mathcal{H} = -\sum_{i}^{n} \hbar \omega_i I_z^i + \hbar 2\pi \sum_{i \neq j}^{n} J_{ij} I_{iz} I_{jz} \qquad (4.1.2)$$

In practical situations, many qubits NMR liquid-state systems are not easy to be achieved. This is because they require a sample with n NMR distinguishable spins in a single molecule, which can be very difficult to obtain for large values of n. Besides, such an approach is not scalable (see Chapter 6). To implement the 7-qubit NMR system used to demonstrate the Shor's algorithm, Vandersypen et al. [3] designed a special molecule where ^{13}C and ^{19}F nuclei were used as qubits (see Chapter 5). However, since a NMR system is not constituted by a single molecule, but by an ensemble of identical molecules, neglecting the intermolecular interactions, we can think of a NMR liquid sample as constituted by a huge number ($\approx 10^{23}$) of molecular quantum processors executing a kind of parallel processing.[2]

Another NMR system that has been used for QIP implementations is constituted of quadrupolar spins ($I > 1/2$) in oriented media. In this case, the quadrupolar nuclei are part of the structure of a solid or liquid crystal. An oriented media is always required in this case, otherwise, as described previously, the random orientation of the electric field gradient tensors along the sample would lead to a distribution of quadrupolar couplings and consequently to broadening of the NMR spectrum. In a crystalline system all the quadrupolar nuclei experience basically the same electric field gradient and can be characterized by a single quadrupolar coupling constant (which, actually, corresponds to a valve average in the case of liquid crystalline media). Besides, in typical systems used for NMR QIP the quadrupolar coupling is the dominant internal NMR interaction, allowing to describe the NMR system with the following Hamiltonian:

$$\mathcal{H} = -\hbar \omega_0 I_z + \hbar \omega_Q (3 I_z^2 - \mathbf{I}^2) \qquad (4.1.3)$$

where ω_Q is the quadrupolar frequency, defined in Chapter 2. The energy levels obtained for a spin 3/2 system are shown in Figure 4.2b. This Hamiltonian gives rise to four un-

---

[2] However, this parallelism does not increase the power of the computer, since it represents only redundancy.

4.1. The NMR qubits

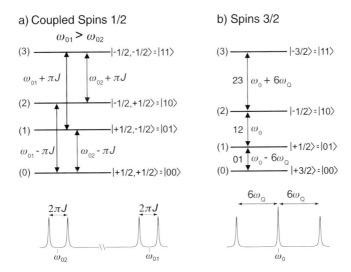

Figure 4.2 Energy levels in a two-qubit NMR system: (a) two J-coupled spin 1/2 system. Direct dipolar coupled systems in the weak coupling limit have similar energy configuration replacing the J coupling by the dipolar coupling constant. (b) Quadrupolar spin 3/2 system. The three allowed transitions labeled 01, 12, and 23 are indicated.

equally spaced energy states, originating an NMR spectrum containing three lines, corresponding to transitions between adjacent levels. The states,

$$|+3/2\rangle \text{ or } 0, \quad |+1/2\rangle \text{ or } 1, \quad |-1/2\rangle \text{ or } 2, \quad |-3/2\rangle \text{ or } 3$$

can be labelled as

$$|00\rangle, \quad |01\rangle, \quad |10\rangle, \quad |11\rangle$$

so they can represent a two-qubit system. This "rule" can be extended to other quadrupolar spin systems: generally, an ensemble of I spin nuclei experiencing quadrupolar interaction can be used to represent an n-qubit system provided that $2^n = 2I + 1$.

The physical implementation of the qubit in quadrupolar systems is not so obvious as in spin 1/2 systems, since one single quadrupolar spin carries more than one bit of information. Despite that, many experiments have shown that logic gates and even quantum algorithms can be performed in such systems [4,5]. There are also reports on the use of spin 7/2 (three-qubits system) to implement logic gates [6]. The main disadvantage of using quadrupolar systems for NMR QIP concerns relaxation effects. As mentioned before, quadrupolar coupling are usually much stronger than the other NMR interactions. Then, the relaxation times are also much faster. For example, while a typical T_2 of spin 1/2 in a isotropic solution is usually of order of a few seconds, for quadrupolar nuclei in liquid crystalline matrices it is typically of order of a few milliseconds [7].

In order to conclude this brief discussion on the NMR implementations of qubits, let us mention the direct dipolar coupled spins. This spin system is becoming quite important in NMR QIP, since many of the candidate approaches to produce scalable NMR quantum computers (see Chapter 7), are based on spin 1/2 systems in solid-state materials, where the dominant internal NMR interaction is the direct dipolar coupling [9]. Actually, the secular part of the dipolar coupling Hamiltonian has, in the weak coupling limit, similar form as the J-coupling, which allows an easy adaptation of most of the pulse schemes already developed for solution NMR. Other typical solid-state NMR methods, such as CP, spin-locking, decoupling, multiple pulse techniques, etc., can also be used for QIP implementations [9].

4.2 QUANTUM LOGIC GATES GENERATED BY RADIOFREQUENCY PULSES

In Chapter 3 we stated that a quantum computer is based on a set of universal logic gates, just like its classical counterpart. Thus, any technique capable of executing such logic gates would be a natural candidate to be used for quantum information processing. It was also discussed that quantum logic gates are nothing but unitary operations whose precise control is of fundamental importance for quantum processing. This requirement is found in NMR, which has a long tradition in manipulating spin states through unitary transformations using RF pulses or evolutions under internal nuclear spin interactions. From this point of view, NMR has the appropriate tools for implementing quantum logic gates. In this section we will present some of the NMR implementations of Hadamard, $\pi/8$, phase shift, CNOT, SWAP gates for coupled spin $1/2$ system, as well as quadrupolar spin $3/2$ systems.

4.2.1 Elementary single-qubit gates and their implementations using RF pulses

Single-qubit unitary operations play an important role in QIP. Using the nuclear spins as qubits, the most elementary single-qubit operations are those that perform a rotation of a single spin and can be represented by the following rotation operator:

$$R_{\hat{n}}(\theta) = \exp(-i\theta \mathbf{n} \cdot \mathbf{I}) \qquad (4.2.1)$$

where $\mathbf{n}$ is the unitary vector that defines the rotation axis, $\mathbf{I} = I_x\mathbf{i} + I_y\mathbf{j} + I_z\mathbf{k}$ is the nuclear spin operator, and θ is the rotation angle. In Chapter 2 we saw that RF pulses are direct experimental implementations of unitary rotations. To illustrate that, let us consider an isolated spin $1/2$ in a magnetic field as an idealization of an NMR system with a single-qubit. The action of an on-resonance RF pulse with arbitrary phase ϕ and duration t_p is described by the following pulse propagator:

$$(\theta)_\phi^I = \exp(-i\omega_1 t_p I_\phi) = \exp(-i\theta I_\phi) \qquad (4.2.2)$$

where $I_\phi = I_x \cos(\phi) + I_y \sin(\phi)$, and $\theta = \omega_1 t_p$. The equivalence between this propagator and the rotation operator of Equation (4.2.1) is readily recognized, showing that any spin rotation in the xy plane can be generated by a single RF pulse with proper phase, amplitude and duration. A consequence of this statement is that gates such X (equivalent to the NOT gate) and Y are obtained directly from the corresponding RF pulse, as shown in Equation (3.4.1) i.e.,

$$(\pi)_x^I = e^{-i\frac{\pi}{2}} \begin{pmatrix} 0 & 1 \\ 1 & 0 \end{pmatrix}; \qquad (\pi)_y^I = \begin{pmatrix} 0 & -1 \\ 1 & 0 \end{pmatrix} \qquad (4.2.3)$$

4.2. Quantum logic gates generated by radiofrequency pulses

It is obvious that rotations around the z-axis cannot be implemented by single RF pulses. A way of achieving that is introducing a resonance offset to shift the reference frame by an angle θ. Thus, all subsequent pulses and also the receiver will be seen in the new reference frame, which is equivalent to a rotation of θ around the z-axis. Another way of implementing z-rotations is to make use of the property that allows us to write a rotation around z as rotations about the x or y axes. For example, a rotation of an angle θ around the z-axis can be written as:

$$R_z(\theta) = R_x\left(\frac{\pi}{2}\right) R_y(\theta) R_{-x}\left(\frac{\pi}{2}\right) \qquad (4.2.4)$$

Thus, because rotations around the axis x and y can be trivially implemented using RF pulses, rotations around z can be generated using (4.2.4). Actually, the pulse sequence $\left(\frac{\pi}{2}\right)_{-x} - (\theta)_y - \left(\frac{\pi}{2}\right)_x$ is well known in NMR, being usually named as composite z-pulse [2]. From the QIP point of view, one spin rotations can be seen as single-qubit gates, which means that we might produce a whole set of single-qubit operations. The question is which RF pulses, or set of them, generate gates necessary for QIP, such as NOT, Hadamard (H), Phase Shift (S), $\pi/8$ or T, Z, etc. [15].

The answer is simple if we compare the matrix representation for the RF pulse $(\theta)^I_\phi$ and the gate operator. For example, consider π-pulses around the x and y axis, respectively. The matrix representation of the corresponding pulse operators are:

$$(\pi)^I_x = \exp(-i\pi I_x) = \begin{pmatrix} 0 & -i \\ -i & 0 \end{pmatrix} = e^{-i\frac{\pi}{2}} \begin{pmatrix} 0 & 1 \\ 1 & 0 \end{pmatrix} = e^{-i\frac{\pi}{2}} \times \text{NOT}$$

$$(\pi)^I_y = \exp(-i\pi I_y) = \begin{pmatrix} 0 & -1 \\ 1 & 0 \end{pmatrix} \qquad (4.2.5)$$

We can see that a π-pulse along the x axis has the same matrix as the NOT gate, times a global phase factor. Since global phase factors do not affect unitary rotations, we can say that $(\pi)^I_x$ is equivalent to a NOT gate. If the pulse is applied along the y-axis it no longer represents a NOT gate, which shows that a good control of the pulse phase is essential

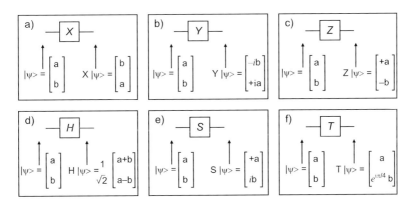

Figure 4.3 Representation of operators corresponding to some single qubit logic gates. The result of the gates application to some selected states is also shown.

in NMR QIP. Fortunately in conventional NMR spectrometers such control is excellent, achieving an accuracy of less than 0.1 degrees.

Other single-qubit gates that can be directly implemented by RF pulses are the phase gate, S, the $\pi/8$ or T gate, and the Z gate. These gates can be obtained from a z-rotation of arbitrary flip angle, which is implemented by the pulse sequence equivalent to the following propagator [10],

$$R(\theta) = \left(\frac{\pi}{2}\right)^I_x (\theta)^I_y \left(\frac{\pi}{2}\right)^I_{-x} = e^{-i\frac{\theta}{2}} \begin{pmatrix} 1 & 0 \\ 0 & e^{i\theta} \end{pmatrix} \quad (4.2.6)$$

Therefore, setting $\theta = \frac{\pi}{4}$, $\theta = \frac{\pi}{2}$, and, $\theta = \pi$ in Equation (4.2.6) we obtain T, S, and Z gates, respectively.

Another fundamental single-qubit gate is the Hadamard gate. As it can be observed from Equation (4.2.7), a $\left(\frac{\pi}{2}\right)$ pulse along the y-axis closely resembles a Hadamard matrix H.

$$\left(\frac{\pi}{2}\right)^I_y = \exp\left(-i\frac{\pi}{2}I_y\right) = \frac{1}{\sqrt{2}} \begin{pmatrix} 1 & -1 \\ 1 & 1 \end{pmatrix}$$

$$\mathrm{H} = \frac{1}{\sqrt{2}} \begin{pmatrix} 1 & 1 \\ 1 & -1 \end{pmatrix} \quad (4.2.7)$$

Despite the similarity between the two matrices, a $\left(\frac{\pi}{2}\right)_y$ pulse can only be classified as pseudo-Hadamard gate. This is so, because a single $\left(\frac{\pi}{2}\right)_y$ pulse is not self-reversible (that is, when applied twice to a given quantum state it does not recover the original state), which is a fundamental property of a true Hadamard gate. However, for applications where self reversibility is not required, it is common to use a $\left(\frac{\pi}{2}\right)_y$ pulse in place of a Hadamard gate. To produce a true Hadamard operation in NMR it is necessary to introduce an extra pulse to perform the necessary phase correction in the pulse matrix. This can be achieved by adding a π-pulse around the z- or x-axis after the $\left(\frac{\pi}{2}\right)_y$ pulse, i.e.,

$$\mathrm{H} = \frac{1}{\sqrt{2}} \begin{pmatrix} 1 & 1 \\ 1 & -1 \end{pmatrix} \equiv (\pi)^I_z \left(\frac{\pi}{2}\right)^I_{-y} = \left(\frac{\pi}{2}\right)^I_x (\pi)^I_y \left(\frac{\pi}{2}\right)^I_{-x} \left(\frac{\pi}{2}\right)^I_{-y}$$

$$= (\pi)^I_x \left(\frac{\pi}{2}\right)^I_y \quad (4.2.8)$$

Since RF pulses are capable of producing general spin rotations, it looks obvious that any single-qubit gate can be implemented by RF pulses [11].

Until now we have considered that the NMR system is composed by isolated spins, which might look quite unrealistic. However, if each qubit is represented by a distinct kind of nucleus this can be a reasonable approximation. For example, let us consider two J-coupled heteronuclear spins, like the ^{13}C, ^{1}H spin pair in ^{13}C labeled chloroform. Because they have distinct NMR frequencies, we can apply a resonant pulse to one of them without affecting the other. If the pulse amplitude is much higher than the magnitude of the J-coupling, we can also neglect the effect of the coupling during the pulse. In other words,

4.2. Quantum logic gates generated by radiofrequency pulses

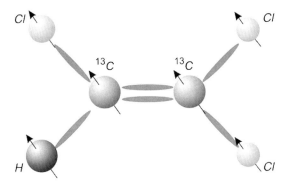

Figure 4.4 A scheme for the 3-qubit molecule of trichloroethylene. The qubits are represented by the two ^{13}C nuclei and the ^{1}H nucleus.

we can apply a pulse that acts as a rotation to each individual nuclear spin. These RF pulses act as a selective pulse to each spin and from now on we will call them as a spin selective pulse, regardless that in the heteronuclear system they are actually hard pulses. This is valid for both spins, which means that single-qubit gates can be performed independently for each one of them.

The situation is considerably more complicated when we consider coupled homonuclear spins, such as the case of the two ^{13}C nuclei in the molecule of trichloroethylene, shown in Figure 4.4.[3] In these cases, the resonance frequencies are close and to perform a single-qubit rotation to one of the spins it is necessary to use RF pulses capable of acting selectively only in a narrow range of frequency. This is achieved by narrow band selective pulses, which have been widely exploited in NMR spectroscopy and imaging [12]. To provide narrow excitation profiles their durations are much longer than in non selective pulses, and they are usually amplitude and/or phase modulated by different functions. Typical amplitude modulation functions for selective pulses are sinc ($\text{sinc}(x) \equiv \sin(x)/x$), Gaussian, Hermite, sine bell, a class of numerically optimized pulses known as BURP (Band-selective, Uniform Response, Pure-phase), etc. [12] (see Figure 4.5). Each of these pulses have particular advantages such narrow band excitation, phase accuracy, uniform response, good refocusing of J-coupling, etc. Because the maximum RF power used in selective pulses is low, they are also know as soft pulses.

Besides the excitation profiles, there are other features that can compromise the ability of selective pulses to implement single-qubit rotations. For example, selective pulses are usually long and it is not possible to neglect the evolution under the J-coupling during them. This might be critical because the simultaneous evolution under RF pulses and J-couplings makes the rotation induced by the pulse dependent on the coupling constant, which is certainly not a desirable feature. The solution for that is the use of self-refocusing pulses, which are able to refocus the evolution under the J-coupling at the end of the pulse [2,12], or setting the pulse duration to a multiple of the J-coupling evolution period. Thus, only the effective rotation around the xy plane is performed [12]. Another important feature of narrow band selective pulses concerns the central frequency of the excitation

---
[3]This molecule actually has three coupled spin 1/2 nuclei (the two ^{13}C nuclei plus the ^{1}H nucleus). The chlorine nuclei have effectively negligible couplings with all other nuclei. Such molecule can be used to implement 3-qubit operations, as it has been done in a lot of NMR QIP experiments.

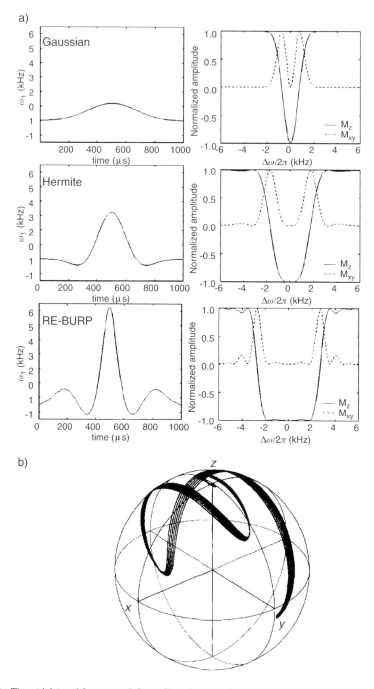

Figure 4.5 Time (right) and frequency (left) profiles of some soft π pulses used for narrow band excitation in NMR spectroscopy. (a) Gaussian pulse; Hermite pulse; and RE-BURP pulse. It is supposed that before the pulse the magnetization was in the z direction. (b) Typical magnetization trajectories for a inversion RE-BURP pulse with distinct frequency offsets. Adapted with permission from References [12,15] (Copyright 2007 American Physical Society and Elsevier).

profile. It can be controlled by shifting the pulse carrier frequency, but in some situations it is advantageous to keep the carrier frequency constant and set the central frequency using linearly phase modulated pulses [12]. To produce a frequency shift by linear phase modulation we need to increase the pulse phase at a constant rate $\frac{\Delta\phi}{\Delta t}$ during the RF pulse. The Hamiltonian of the RF pulse with initial phase ϕ, carrier frequency ω_{RF}, and amplitude modulation described by $\omega_1(t)$, can be written as:

$$\mathcal{H}_{RF}(t) = \hbar\omega_1(t)\left\{\cos\left[\omega_{RF}t + \left(\phi + \frac{\Delta\phi}{\Delta t}t\right)\right]I_x\right.$$
$$\left. + \sin\left[\omega_{RF}t + \left(\phi + \frac{\Delta\phi}{\Delta t}t\right)\right]I_y\right\}$$

$$\mathcal{H}_{RF}(t) = \hbar\omega_1(t)\left\{\cos\left[\left(\omega_{RF} + \frac{\Delta\phi}{\Delta t}\right)t + \phi\right]I_x\right.$$
$$\left. + \sin\left[\left(\omega_{RF} + \frac{\Delta\phi}{\Delta t}\right)t + \phi\right]I_y\right\}$$

(4.2.9)

It is clear from the bottom equation in (4.2.9) that a linear phase modulation at rate $\frac{\Delta\phi}{\Delta t}$ is equivalent to a frequency shift of $\Delta\omega = \frac{\Delta\phi}{\Delta t}$.

Another class of selective pulses that deserves a comment refers to the so called multi-frequency pulses. These are amplitude or phase modulated pulses capable of acting simultaneously at different frequencies. To illustrate how this can be done, let us consider the pulse described by (4.2.9) without phase modulation, and add an extra amplitude modulation of $\cos(\Delta\omega t)$,

$$\mathcal{H}_{RF}(t) = \cos(\Delta\omega t)\{\hbar\omega_1(t)[\cos(\omega_{RF}t + \phi)I_x + \sin(\omega_{RF}t + \phi)I_y]\} \quad (4.2.10)$$

In the framework of the Fourier transform we see that the frequency profile of the pulse described by (4.2.10) will be given by the convolution between the Fourier transform of the $\cos(\Delta\omega t)$ modulation and the frequency profile of the pulse. Since the Fourier transform of $\cos(\Delta\omega t)$ corresponds to two delta functions at frequencies $-\Delta\omega$ and $+\Delta\omega$, we conclude that with the $\cos(\Delta\omega t)$ modulation the selective pulse can be seen as two identical pulses acting simultaneously at $-\Delta\omega$ and $+\Delta\omega$. With this technique, multi-frequency selective pulses can be obtained just by modulating the pulse by a sum of cosine functions, i.e., $\sum_{i=1}^{n} a_i \cos(\Delta\omega_i t)$, which produces $2n$ selective pulses acting simultaneously at $[\Delta\omega_i, -\Delta\omega_i]$. Notice that by using distinct weighting factors, a_i, it is possible to vary the amplitude of the pulses at each pair of frequencies. Other approach to produce multi-frequency excitation is replacing the cosine amplitude modulation by a phase modulation [12]. This can provide a shaped pulse that is equivalent to an odd number of selective pulses acting simultaneously at different frequencies.

Selective pulses can also be used to excite a single transition between two energy levels of a spin system, as shown in Figure 4.2. Because J-couplings are usually small, long selective pulses (ultra-soft pulses) are need to achieve a single transition excitation (or equivalently a single line excitation in the NMR spectrum). Transition selective pulses have been used in some spin 1/2 NMR QIP applications [13,14], but, depending on the coupling constants, the pulse length can be too long, and pulse imperfections and relaxation effects limit their use. However, for quadrupolar spins transition selective pulses are fundamental for implementing logic gates. In typical oriented media used for NMR QIP, the quadrupolar interaction is at least two orders of magnitude stronger that J-couplings in typical isotropic solutions. Because of that, transition selective pulses can be made much shorter in quadrupolar systems, making advantageous to use them, even in the presence of fast relaxation.

To provide some examples of single-qubit gates implemented in quadrupolar systems, let us consider the spin 3/2 system described in Section 4.1. Because this system has two-qubits, single-qubit gates will act only in one of the them, keeping the other untouched. The single transition selective pulses operators corresponding to the pulse sequences that implement NOT_A, NOT_B, H_A, H_B single-qubit gates, are shown in (4.2.11). Using the pulse operators for transition selective pulses mentioned in Chapter 2, the matrix form of the corresponding gates operators can be obtained (Problem P4.1):

a) $\text{NOT}_A = (\pi)_x^{23}(\pi)_x^{01}(\pi)_x$

b) $\text{NOT}_B = (\pi)_x^{23}(\pi)_x^{01}$

c) $\text{H}_A^{|00\rangle,|10\rangle} = (2\pi)_{-x}^{01}(\pi)_y^{01}\left(\dfrac{\pi}{2}\right)_y^{12}(\pi)_{-y}^{01}$

d) $\text{H}_A^{|01\rangle,|11\rangle} = (2\pi)_{-x}^{01}(\pi)_y^{12}\left(\dfrac{\pi}{2}\right)_y^{23}(\pi)_{-y}^{12}$

e) $\text{H}_B^{|00\rangle,|01\rangle} = (\pi)_x^{01}\left(\dfrac{\pi}{2}\right)_{-y}^{01}$

f) $\text{H}_B^{|10\rangle,|11\rangle} = (\pi)_x^{23}\left(\dfrac{\pi}{2}\right)_{-y}^{23}$

(4.2.11)

The upper indexes, $|00\rangle, |01\rangle, |10\rangle, |11\rangle$, are to emphasize that the corresponding operators only execute a true Hadamard gate when they act on the indicated states. The indexes 01, 12, and 23 indicate the pulse transition as indicated in Figure 4.2. However, the operators c), d) and e), g) can be implemented by a single pulse sequence if we use two-frequency pulses to excite simultaneously two transitions. For example, $U_{H_B} = (\pi)_x^{01\text{-}23}\left(\dfrac{\pi}{2}\right)_{-y}^{01\text{-}23}$, where 01-23 indicates a two-frequency selective pulse that act simultaneously on the transitions 01 and 23 see Figure 4.2, will implement a H_B operation independently of the initial state. All the Hadamard transformations indicated in (4.2.11) are self-reversible.

4.2.2 Elementary two-qubit gates and their implementation in NMR

As discussed in Chapter 3, two-qubit gates such as CNOT and Hadamard are fundamental[4] for quantum information processing; any experimental method that aims to be used as

---
[4] Many algorithms start with an uniform superposition of states, produced by a multi-qubit Hadamard gate. Although such an operation can be formally written as a tensor product of one-qubit Hadamard gate, sometimes we have to regard it as a multi-qubit gate. This is the case, for instance, of quadrupole nuclei.

4.2. Quantum logic gates generated by radiofrequency pulses

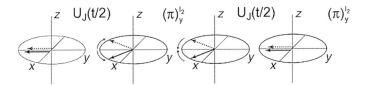

Figure 4.6 Vector representation of the evolution of two coupled spins. The figure shows the evolution of one of the spins during the refocusing pulse sequence.

a quantum hardware must be able to implement such operations. In NMR this can be achieved either for coupled spins 1/2 systems or for quadrupolar systems. In this section we discuss both cases, starting from coupled spins 1/2 and then discussing the case of quadrupolar spins.

Before discussing the pulse sequence used for implementing 2-qubit gates in J-coupled spin 1/2 systems, let us present a particularly useful method in NMR QIP implementations. The method consists of a pulse sequence that can momentarily "turn off" the J-coupling between two selected spins. It is named *refocusing* and has been used since the early days of multidimensional NMR [2]. To illustrate the idea of refocusing, let us first consider a system of two J-coupled spins ($\mathbf{I_1}$ and $\mathbf{I_2}$) in the weak coupling regime. We consider that spin $\mathbf{I_1}$ is initially rotated to point along the x-axis after a $\left(\frac{\pi}{2}\right)_y^{I_1}$ pulse. Then, it evolves under the J-coupling during a time interval of duration $t/2$ and after that a $(\pi)_y^{I_2}$ inverts the state of the $\mathbf{I_2}$ spin. This operation inverts the local field at the spin $\mathbf{I_1}$ position and, as a consequence, the J-coupling evolution of spin $\mathbf{I_1}$ is reversed. Then, after another $t/2$ evolution period under J-coupling it returns to its initial position. After that, the spin $\mathbf{I_2}$ is also taken to its initial state by another $(\pi)_y^{I_2}$ pulse. This second $(\pi)_y^{I_2}$ pulse ensures that both spins return to their initial situation regardless the initial state. Thus, after the refocusing pulse sequence both spins are in the same situation as they were before the evolution, which is equivalent to no overall evolution under the J-coupling during this period. The same effect can be produced by applying the refocusing pulses to the spin $\mathbf{I_1}$. A pictorial view of the spin evolution in the Bloch sphere during the refocusing pulse sequence is shown in Figure 4.6. A formal treatment of the refocusing can be done using the corresponding pulse matrices and the evolution operator under J-coupling U_J described in Chapter 2 (see Problem P4.2). The unitary operation corresponding to the refocusing sequence can be written as:

$$U_{refocusing} = U_J\left(\frac{t}{2}\right)(\pi)_y^{I_2} U_J\left(\frac{t}{2}\right)(\pi)_{-y}^{I_2} = U_J\left(\frac{t}{2}\right) U_J\left(-\frac{t}{2}\right) = 1 \quad (4.2.12)$$

where the matrix representations of the operator were described in the Section 2.8. Hence, the unitary operation that represents the whole pulse sequence is the identity operator, meaning that there is no overall evolution under the J-coupling at the end of the pulse sequence. The sequence can be applied at different instants during the evolution. This scheme can be used either for heteronuclear or homonuclear spins using the appropriated selective pulses. In multispin systems the refocusing techniques can be applied to turn off the J-coupling between any two specific pair of spins. This is an important feature that is used in many controlled gates to keep only the interaction between some selected spins [15]. Another usual application of refocusing pulse sequence is to make the coupling of a certain spin with different spins effectively equal. For example, the pulse sequence corresponding to the operation bellow can be applied to a four spin system (lets say spins 1, 2, 3 and 4) to make the effective coupling between all spins equal after a period of $\frac{1}{4J_{13}}$ [18],

$$U\left(\frac{1}{8J_{13}}-\frac{1}{8J_{12}}\right)\pi_y^{I_2}U\left(\frac{1}{8J_{12}}-\frac{1}{8J_{14}}\right)\pi_{-y}^{I_4}U\left(\frac{1}{8J_{13}}+\frac{1}{8J_{14}}\right) \quad (4.2.13)$$

Now, let us return to the implementation of two-qubit gates. In Chapter 3 we saw that the action of the CNOT gate is: "invert one of the qubits (the target qubit) provided the other (the control qubit) is in the state $|1\rangle$". In a two-qubit $|AB\rangle$ system this is accomplished by following operators:

$$\text{CNOT}_A = \begin{pmatrix} 1 & 0 & 0 & 0 \\ 0 & 1 & 0 & 0 \\ 0 & 0 & 0 & 1 \\ 0 & 0 & 1 & 0 \end{pmatrix}$$

$$\text{CNOT}_B = \begin{pmatrix} 1 & 0 & 0 & 0 \\ 0 & 0 & 0 & 1 \\ 0 & 0 & 1 & 0 \\ 0 & 1 & 0 & 0 \end{pmatrix}$$

(4.2.14)

where the indexes A and B indicate the control qubit. The pulse sequence corresponding to the NMR implementation of a CNOT gate in a 2-spin system, with $\mathbf{I}_1$ and $\mathbf{I}_2$ as the control or target qubits has the following time ordered sequence of events: (i) a $\pi/2$ pulse with y phase applied to the spin representing the target qubit; (ii) an evolution period of $t = 1/2J$ under the J-coupling; (iii) a $\pi/2$ pulse with x phase applied to the target qubit $\mathbf{I}_2$; (iv) two $\pi/2$ composite z-pulses applied at both spins. This can be represented by the following operators:

$$\text{CNOT}_A = \left(\frac{\pi}{2}\right)_z^{I_1}\left(-\frac{\pi}{2}\right)_z^{I_2}\left(\frac{\pi}{2}\right)_x^{I_2} U_J\left(\frac{1}{2J}\right)\left(\frac{\pi}{2}\right)_y^{I_2}$$

$$= \left(\frac{1-i}{\sqrt{2}}\right)\begin{pmatrix} 1 & 0 & 0 & 0 \\ 0 & 1 & 0 & 0 \\ 0 & 0 & 0 & 1 \\ 0 & 0 & 1 & 0 \end{pmatrix}$$

(4.2.15)

$$\text{CNOT}_B = \left(\frac{\pi}{2}\right)_z^{I_2}\left(-\frac{\pi}{2}\right)_z^{I_1}\left(\frac{\pi}{2}\right)_x^{I_1} U_J\left(\frac{1}{2J}\right)\left(\frac{\pi}{2}\right)_y^{I_1}$$

$$= \left(\frac{1-i}{\sqrt{2}}\right)\begin{pmatrix} 1 & 0 & 0 & 0 \\ 0 & 1 & 0 & 1 \\ 0 & 0 & 1 & 0 \\ 0 & 1 & 0 & 0 \end{pmatrix}$$

where the subscript indicates the control qubit. Observe that these sequences of events are closely related to the experiment leading to conditional inversion of one spin described in

4.2. Quantum logic gates generated by radiofrequency pulses

the Section 2.8. The resulting matrix corresponds exactly to the operator CNOT with the control in the first qubit, times an irrelevant global phase. For systems of higher number of qubits the same pulse sequences can be used for implementing the CNOT gates just by pulsing in the spins involved in the operation and considering the J-coupling between them. Notice that for in this case it might be necessary keeping only the interaction between two spins, which can be done by using refocusing schemes.

Although transition selective pulses have been used in spin 1/2 systems [13,14], they are much more frequent in applications involving quadrupolar nuclei. For example, the CNOT gate can be performed by a single transition selective π-pulse applied to an allowed transition of the 2-spin system. The CNOT gate has also been implemented in two-qubit spin 3/2 quadrupolar systems using transition selective pulses [5]. The resulting matrices correspond to the pulse sequence that executes a CNOT gate with control in the first (CNOT_A) and second (CNOT_B) qubits are:

$$\text{CNOT}_A = (\pi)_x^{23} = \begin{pmatrix} 1 & 0 & 0 & 0 \\ 0 & 1 & 0 & 0 \\ 0 & 0 & 0 & i \\ 0 & 0 & i & 0 \end{pmatrix}$$

$$\text{CNOT}_B = (\pi)_x^{12}(\pi)_x^{23}(\pi)_x^{12} = \begin{pmatrix} 1 & 0 & 0 & 0 \\ 0 & 0 & 0 & -1 \\ 0 & 0 & -1 & 0 \\ 0 & -1 & 0 & 0 \end{pmatrix}$$

(4.2.16)

Note that these operators are not exactly equal to CNOT operators, but they act as CNOT gates for most of two qubit states.

A two qubit-gate very used in quantum algorithms is the SWAP gate. It can be directly implemented by the pulses corresponding to three successive CNOT gates,

$$\text{SWAP} = \text{CNOT}_A \text{CNOT}_B \text{CNOT}_A = -\frac{1+i}{\sqrt{2}} \begin{pmatrix} 1 & 0 & 0 & 0 \\ 0 & 0 & 1 & 0 \\ 0 & 1 & 0 & 0 \\ 0 & 0 & 0 & 1 \end{pmatrix} \quad (4.2.17)$$

As a last example of two-qubit gates let us consider and important case where cascaded gates are used to produce the four states of the Bell basis. As discussed in Chapter 3, such states can be created from the computational basis states $|00\rangle, |01\rangle, |10\rangle, |11\rangle$ by the application of the so called EPR generator operator (see Problems 4.3 and 4.4), which is implemented by the pulses corresponding to a Hadamard followed by a CNOT gate:

$$\text{EPR} = \text{CNOT}_A H_A = -\frac{1+i}{2} \begin{pmatrix} 1 & 0 & 1 & 0 \\ 0 & 1 & 0 & 1 \\ 0 & 1 & 0 & -1 \\ 1 & 0 & -1 & 0 \end{pmatrix} \quad (4.2.18)$$

In the derivation of (4.2.18) and (4.2.17) the operators corresponding to the full pulse sequences for the CNOT and Hadamard gates in two J-coupled spins 1/2 (Equations (4.2.15) and (4.2.8)) were used.

In the case of long sequences of logical operations, such as quantum algorithms, it is usual to suppress some of the pulses that implement the gates in order to decrease the size of the pulse sequence and minimize decoherence effects. For example, in the CNOT gate of (4.2.15) the two z-rotations are only to perform a relative phase correction. If such correction is not really necessary, they can be omitted, decreasing the number of pulses in the sequence. In fact, this kind of pulse simplification has been used in experimental implementations of complex algorithms and some strategies have been developed for avoiding introducing errors or loosing information when this procedure is used [15].

4.2.3 Multi-qubit gates

In the last sections we restricted our discussion to NMR implementations of one- and two-qubit gates. This relies on the idea that a universal set of logical gates can be constructed only using one and two-qubit gates. [15,17]. It means that one can, in principle, use only individual pair of qubits to implement all necessary logic gates for QIP. This is very convenient, because the absence of three body interactions in nature prohibits the implementation of true multi-qubit gates (logic gates that act in many qubits simultaneously). Despite that, Deustch [16] has demonstrated the existence of a set of universal three-qubit gates and it would be interesting at least to simulate such multi-qubit gates. This can be achieved if we have a many particle system coupled by means of two-body interactions acting simultaneously. This is the case of NMR multi-spin system, where there are many spins interacting in pairs through J or direct dipolar couplings. The presence of such simultaneous two-body interactions can be further exploited to construct simulations of multi-qubit gates that are more efficient than sets of one and two-qubit gates [17]. One example of such gates is the NMR implementation of the Toffoli gate, which uses much less pulses than the same operation constructed by combination of two-qubit operations [1]. To illustrate that, let us consider a Toffoli gate where the target qubit is the first qubit. Lets also assume that the coupling constant J between spins 1 and 2 is equal to that between 1 and 3. If they are not, they can be done effectively equal by using partial refocusing similar to sequence of (4.2.13). With this assumptions the Toffoli gate can be implemented by the pulse sequence corresponding to the following operator [17–19],

$$U_{\text{Toffoli-}I_1} = \left(\frac{\pi}{2}\right)^{I_1}_{-x} U_J\left(\frac{1}{4J}\right) \left(\frac{\pi}{2}\right)^{I_1}_{-x} U_J\left(\frac{1}{4J}\right) \left(\frac{\pi}{2}\right)^{I_1}_{y} U_J\left(\frac{1}{4J}\right) \left(\frac{\pi}{2}\right)^{I_1}_{y}$$

$$U_{\text{Toffoli-}I_1} = e^{i\frac{3\pi}{8}} \begin{pmatrix} 1 & 0 & 0 & 0 & 0 & 0 & 0 & 0 \\ 0 & i & 0 & 0 & 0 & 0 & 0 & 0 \\ 0 & 0 & i & 0 & 0 & 0 & 0 & 0 \\ 0 & 0 & 0 & 0 & 0 & 0 & 0 & -1 \\ 0 & 0 & 0 & 0 & i & 0 & 0 & 0 \\ 0 & 0 & 0 & 0 & 0 & -i & 0 & 0 \\ 0 & 0 & 0 & 0 & 0 & 0 & -i & 0 \\ 0 & 0 & 0 & i & 0 & 0 & 0 & 0 \end{pmatrix} \qquad (4.2.19)$$

which is the same as the Toffoli operator, multiplied by an irrelevant global factor. It acts on the state basis as a Toffoli gate (Problem P4.5). Besides Toffoli gates, there are many NMR implementations of effective multi-qubit gates, such as Hadamard and generalized controlled CNOT, and even Quantum Fourier Transform, using the same idea of multi-qubit simplifications [17]. Furthermore, there are many other methods to implement such effective multi-qubit operations, including the use of selective pulses [18]. An interesting approach is producing specially shape designed, phase and frequency modulated pulses that simultaneously use all the convenience of RF pulses and NMR interactions. Such pulses are called Strongly Modulated Pulses (SMP) and will be the topic of the next section.

4.2.4 Use of strongly modulated RF pulses for quantum gate implementation in NMR QIP

In the above discussion we treated a logic gate as a set of independent RF pulses and evolutions. This approach is convenient because we can regard a logic gate as a set of individual rotations. However, the excellent control of amplitude and phases provided by RF pulses and the exact knowledge of the internal NMR interactions are features that can be further exploited in NMR QIP. An example of that is the method named Strongly Modulated Pulses (SMP) [21]. To illustrate the method, let us consider the Hamiltonian of a nuclear spin system in a strong magnetic filed, under the action of the an internal spin interaction, and a RF field. In the rotating frame defined by ω_{RF}, this Hamiltonian can be written as:

$$\mathcal{H} = \hbar \Delta \omega I_z + \mathcal{H}_{int} + \hbar \omega_1 I_\phi \quad (4.2.20)$$

were $\Delta \omega = (\omega_{RF} - \omega_0)$ is the resonance offset, $\mathcal{H}_{int}$ is an internal spin interaction (usually J-coupling, dipolar, or quadrupolar interaction), and $\hbar \omega_1 I_\phi$ represents a constant RF field with amplitude ω_1 and phase ϕ. The evolution of a spin system under the Hamiltonian (4.2.20) during a time interval t can be described by the unitary operator,

$$U(\Delta\omega, \omega_1, \phi, t) = \exp\left[-i\left(\Delta\omega I_z + \frac{\mathcal{H}_{int}}{\hbar} + \omega_1 I_\phi\right)t\right] \quad (4.2.21)$$

Notice that the externally controllable parameters in this unitary operator are $\Delta\omega$, ω_1, and ϕ. Now, let us consider another unitary operator composed by a product of M operators of the same kind, i.e.,

$$U = \prod_{i=1}^{M} U_i(\Delta\omega_i, \omega_{1i}, \phi_i, t_i) \quad (4.2.22)$$

With this, we have a set of $4M$ controllable parameters to define U. The idea behind the SMP method is to find the right set of parameters that make the operator U correspond to a desired logic gate. This is done using an optimization procedure. The first optimization step is to establish an objective operator U_{obj} that represents the logic gate. During the optimization, the operator U is compared with U_{obj} through a measure of fidelity, F, define as:

$$F = \frac{\text{Tr}(U_{obj}^{-1} U)}{N} \quad (4.2.23)$$

where N is the dimension of the corresponding Hilbert space. Thus, the fidelity parameter tends to 1 as U and U_{obj}^{-1} become close to each other. Besides U_{obj} and F, a penalty parameter ξ is also introduced in order to keep the pulse parameter within meaningfully values or introduce any experimental restriction. This is done by making ξ very large if any pulse parameter reaches not accessible experimental values. An optimization routine is used to minimize the function $f(P) = 1 - \sqrt{F(P)} + \xi(P)$, where P represent the hyperspace composed by the parameters $\Delta\omega_i, \omega_{1i}, \phi_i, t_i$. The optimization procedure stops when $F(P) \simeq 1$, which mens $U \simeq U_{obj}$. Therefore, the RF pulse that generates U_{obj} will be composed by M consecutive square shaped pulses each one defined by $\Delta\omega_i, \omega_{1i}, \phi_i, t_i$, i.e., it is a pulse modulated in amplitude, frequency and phase (see Figure 4.7a). Notice that there is no restriction about the duration of each block, which means that much shorter pulses compared with usual selective pulses can be obtained. Strongly Modulated Pulses are an efficient way of generating logic gates in NMR, mainly for systems where selec-

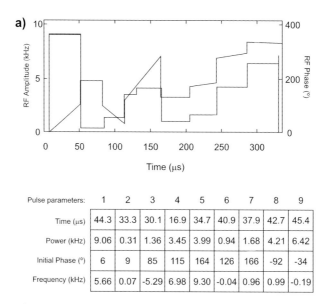

Pulse parameters:	1	2	3	4	5	6	7	8	9
Time (μs)	44.3	33.3	30.1	16.9	34.7	40.9	37.9	42.7	45.4
Power (kHz)	9.06	0.31	1.36	3.45	3.99	0.94	1.68	4.21	6.42
Initial Phase (°)	6	9	85	115	164	126	166	-92	-34
Frequency (kHz)	5.66	0.07	-5.29	6.98	9.30	-0.04	0.96	0.99	-0.19

b)

Quantum operation	Gate fidelity (F)	Unitary propagator gate fidelity (F_U)	Process duration (Δt_i) (μs)
Identify	0.68	0.998	(11.9)
C-NOT B	0.72	0.994	60.4
Had. A	0.68	0.993	46.2
Deutsch	0.81	0.986	55.1
Grover	0.76	0.980	94.8
Had. B	0.70	0.978	62.8
QFT	0.85	0.972	89.6

Figure 4.7 (a) Schematic representation of an strongly modulated pulses (SMP), with the respective table of amplitudes and phases. (b) Table of fidelity parameter obtained for several logical operation implemented using SMP pulses. Adapted with permission from References [15,22] (Copyright 2007 American Physical Society and American Institute of Physics).

tive pulses are needed. Figure 4.7b shows a set of operations designed by Kampermann and Veeman [22] using SMP pulses and tested experimentally in a system of Na nuclei (spin 3/2 – two-qubit system) in a solid crystal. The accuracy of the implementation was checked by reconstructing the operator U from the experimental data using the method called quantum process tomography and calculating the fidelity with respect to the theoretically expected operator. This fidelity is referred as gate fidelity in Figure 4.7.

4.3 PRODUCTION OF PSEUDO-PURE STATES

Conventional NMR deals with a large ensemble of spins. It means that the state of the system is in a statistical mixture, which is obviously inadequate for QIP. However, the NMR ability for manipulating spins states worked out by Cory et al. [24] and Chuang et al. [23] resulted in elegant methods for creating the so called *effectively pure* or *pseudo-pure* states. Behind the idea of the pseudo-pure states is the fact that NMR experiments are only sensitive to the traceless deviation density matrix. Thus, we might search for transformations that, applied to the thermal equilibrium density matrix, produce a deviation density matrix with the same form as a pure state density matrix. Once such state is created, all remaining unitary transformations will act only on such a deviation density matrix, which will transform as a true pure state.

Let us start with a short discussion about the kind of transformation we are seeking for. As described in the Chapter 2, the density matrix corresponding to a pure state is a projector, which satisfies the following properties (Chapter 3): $\rho = \rho^n$ and $\text{Tr}(\rho^2) = 1$. On the other hand, for a statistically mixed state, $\rho \neq \rho^n$ and $\text{Tr}(\rho^2) < 1$. Now, let us look at a density operator that is obtained from a mixed state operator ρ by a unitary transformation, $\rho' = U\rho U^\dagger$. The question is whether this operator can or cannot be a pure state operator. The trace and idempotency properties for the transformed operator become:

$$\text{Tr}(\rho'^2) = \text{Tr}([U\rho U^\dagger]^2) = \text{Tr}(U\rho U^\dagger U\rho U^\dagger)$$
$$= \text{Tr}(U\rho^2 U^\dagger) = \text{Tr}(\rho^2) < 1 \qquad (4.3.1)$$

$$\rho'^n = (U\rho U^\dagger)^n = \prod^n (U\rho U^\dagger)$$
$$= U\rho^n U^\dagger \neq U\rho U^\dagger. \quad \text{Therefore } \rho'^n \neq \rho \qquad (4.3.2)$$

Equations (4.3.1) and (4.3.2) show that if ρ is a mixed state operator so is ρ'. In other words, it is not possible to obtain a pure state from a mixed state only using unitary transformations. Therefore the creation of pseudo-pure states must involve not only a set of unitary transformations, but also non-unitary rotations or some kind of averaging over different mixed states. There are some different ways of creating such pseudo-pure states in NMR, but the most common methods are based on temporal or spacial averaging and logical labeling, which will be discussed in the next section. In the following discussion we will restrict to the production of pseudo-pure states for the computational basis $|00\rangle, |01\rangle, |10\rangle, |11\rangle$, but from these states the Bell basis pseudo-pure states can be directly obtained by applying the EPR generator operator of (4.2.18) (see Problem P4.4).

4.3.1 Temporal averaging

In the temporal averaging method a set of states, prepared by applying unitary transformations to a common initial state (usually the thermal equilibrium), are combined to produce an average state that behaves like a pure state in NMR experiments. If fact, this method do not really create a pure logical state, but allows to simulate the execution of a logical operation by analyzing the average result obtained after applying the operation to each preparation step individually.

To illustrate the general idea, let us consider a diagonal two-qubit density matrix with real diagonal elements a, b, c and, d, corresponding to the populations of the states $|00\rangle, |01\rangle, |10\rangle, |11\rangle$ (note that such matrix can be seen as a general representation of an equilibrium density matrix).

$$\rho_{ini} = \begin{pmatrix} a & 0 & 0 & 0 \\ 0 & b & 0 & 0 \\ 0 & 0 & c & 0 \\ 0 & 0 & 0 & d \end{pmatrix} \tag{4.3.3}$$

Lets then apply the unitary transformation U_0, U_1, and U_2 given by (4.3.4) to ρ_{ini}:

$$U_0 = \begin{pmatrix} 1 & 0 & 0 & 0 \\ 0 & 1 & 0 & 0 \\ 0 & 0 & 1 & 0 \\ 0 & 0 & 0 & 1 \end{pmatrix} \quad U_1 = \begin{pmatrix} 1 & 0 & 0 & 0 \\ 0 & 0 & 1 & 0 \\ 0 & 0 & 0 & 1 \\ 0 & 1 & 0 & 0 \end{pmatrix}$$

$$U_2 = \begin{pmatrix} 1 & 0 & 0 & 0 \\ 0 & 0 & 0 & 1 \\ 0 & 1 & 0 & 0 \\ 0 & 0 & 1 & 0 \end{pmatrix} \tag{4.3.4}$$

where one can identify U_0 as the identity matrix, U_1 can be constructed by two successive CNOT gates ($U_1 = \text{CNOT}_B \text{CNOT}_A$) and $U_2 = U_1^\dagger$. The resulting density matrices ρ_0, ρ_1 and ρ_2 are:

$$\rho_{10} = U_0 \rho_{ini} U_0^\dagger = \begin{pmatrix} a & 0 & 0 & 0 \\ 0 & b & 0 & 0 \\ 0 & 0 & c & 0 \\ 0 & 0 & 0 & d \end{pmatrix} \tag{4.3.5}$$

$$\rho_1 = U_1 \rho_{ini} U_1^\dagger = \begin{pmatrix} a & 0 & 0 & 0 \\ 0 & c & 0 & 0 \\ 0 & 0 & d & 0 \\ 0 & 0 & 0 & b \end{pmatrix} \tag{4.3.6}$$

$$\rho_2 = U_2 \rho_{ini} U_2^\dagger = \begin{pmatrix} a & 0 & 0 & 0 \\ 0 & d & 0 & 0 \\ 0 & 0 & b & 0 \\ 0 & 0 & 0 & c \end{pmatrix} \tag{4.3.7}$$

4.3. Production of pseudo-pure states

Taking the average over ρ_0, ρ_1 and ρ_2 we obtain the effective state,

$$\rho_{00} = \rho_0 + \rho_1 + \rho_2 = \begin{pmatrix} 3a & 0 & 0 & 0 \\ 0 & b+c+d & 0 & 0 \\ 0 & 0 & b+c+d & 0 \\ 0 & 0 & 0 & b+c+d \end{pmatrix} \quad (4.3.8)$$

But, since $b+c+d = 1-a$, we have:

$$\rho_{00} = \begin{pmatrix} 1-a & 0 & 0 & 0 \\ 0 & 1-a & 0 & 0 \\ 0 & 0 & 1-a & 0 \\ 0 & 0 & 0 & 1-a \end{pmatrix} + \begin{pmatrix} 4a-1 & 0 & 0 & 0 \\ 0 & 0 & 0 & 0 \\ 0 & 0 & 0 & 0 \\ 0 & 0 & 0 & 0 \end{pmatrix} \quad (4.3.9)$$

Or,

$$\rho_{00} = (1-a)\mathbf{1} + (4a-1)|00\rangle\langle 00| \quad (4.3.10)$$

The first term on the right side of the above expression of ρ_{00} is proportional to the identity and is not detected in a NMR experiment (neither is affected by RF pulses). However, the second part, which has exactly the same form as the density matrix corresponding to the pure state $|00\rangle$, does transform under the action of RF pulses and also contribute to detected signal. Therefore, once the average state ρ_{00} is created it behaves under any unitary transformation just like a pure state. Note that if we wish to test the implementation of a given logical operation, three full experiments must be done and combined to obtain the averaged answer, which will be the output to that logical operation. The other pseudo-pure states corresponding to the sates $|01\rangle, |10\rangle, |11\rangle$ can be created from $|00\rangle$ just by applying NOT gates implemented by π pulses i.e.,

$$\rho_{01} = \text{NOT}_B\big((1-a)\mathbf{1} + 4a - 1|00\rangle\langle 00|\big)\text{NOT}_B^\dagger = (1-a)\mathbf{1} + 4a - 1|01\rangle\langle 01|$$

$$\rho_{10} = \text{NOT}_A\big((1-a)\mathbf{1} + 4a - 1|00\rangle\langle 00|\big)\text{NOT}_A^\dagger = (1-a)\mathbf{1} + 4a - 1|10\rangle\langle 10|$$
(4.3.11)

$$\rho_{11} = \text{NOT}_B\text{NOT}_A\big((1-a)\mathbf{1} + 4a - 1|00\rangle\langle 00|\big)\text{NOT}_A^\dagger\text{NOT}_B^\dagger$$

$$= (1-a)\mathbf{1} + 4a - 1|11\rangle\langle 11|$$

As a specific example of generating pseudo-pure states in NMR spin systems, lets consider a case of two J-coupled spin $1/2$ (see Section 4.1). The equilibrium density matrix of this system can be written as (4.3.3). Remember that the pulse sequence for the CNOT gate, the operations U_0, U_1, and U_2 can be written as,

$$U_0 = 1$$

$$U_1 = \left(\frac{\pi}{2}\right)_x^{I_1} U_J\left(\frac{1}{2J}\right) \left(\frac{\pi}{2}\right)_y^{I_1} \left(\frac{\pi}{2}\right)_x^{I_2} U_J\left(\frac{1}{2J}\right) \left(\frac{\pi}{2}\right)_y^{I_2} \quad (4.3.12)$$

$$U_2 = \left(\frac{\pi}{2}\right)_y^{I_2} U_J\left(\frac{1}{2J}\right) \left(\frac{\pi}{2}\right)_x^{I_2} \left(\frac{\pi}{2}\right)_y^{I_1} U_J\left(\frac{1}{2J}\right) \left(\frac{\pi}{2}\right)_x^{I_1}$$

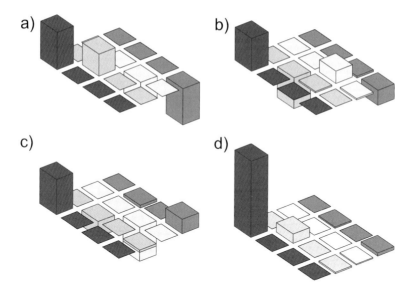

Figure 4.8 Experimental deviation density matrices: (a), (b) and (c) were obtained after the application of the operations U_0, U_1, and U_2 to the equilibrium density matrix. (d) Represents the average deviation density matrix. Adapted with permission from Reference [27] (Copyright 2007 American Chemical Society).

Notice that the z rotations in the CNOT gates are not necessary for producing U_1, and U_2. This is a typical example of the pulse simplification discussed in the last section.

Figure 4.8 shows experimental results for the deviation density matrix obtained after applying each operation for a 2-qubit system U_0, U_1, and U_2 as well as the average state (see also Problems P4.3 and P4.4). The deviation density matrices were obtained using the quantum state tomography process, which will be described in the next section. As it can be seen, the final averaged deviation density matrix is very similar to that of the pure state $|00\rangle$.

For coupled spins 1/2, the temporal averaging method described above can be generalized to systems with larger number of spins. In these cases, it is necessary to combine $2^n - 1$ prepared states to create a pseudo-pure state in a system of n spins. The operations for preparing the individual states can be obtained based on CNOT and SWAP gates. For example, for three spins systems the quantum circuits of these operations are shown in Figure 4.9.

The pseudo-pure state preparation by temporal averaging in quadrupolar nuclei can be done in a similar way. To illustrate the procedure lets take a two-qubit system implemented by spin 3/2 nuclei. The corresponding thermal equilibrium density matrix is given by:

$$\rho_{eq} = \frac{1}{4}\begin{pmatrix} 1 & 0 & 0 & 0 \\ 0 & 1 & 0 & 0 \\ 0 & 0 & 1 & 0 \\ 0 & 0 & 0 & 1 \end{pmatrix} + \frac{\hbar\omega_0}{4k_B T}\begin{pmatrix} 3/2 & 0 & 0 & 0 \\ 0 & 1/2 & 0 & 0 \\ 0 & 0 & -1/2 & 0 \\ 0 & 0 & 0 & -3/2 \end{pmatrix} \quad (4.3.13)$$

4.3. Production of pseudo-pure states

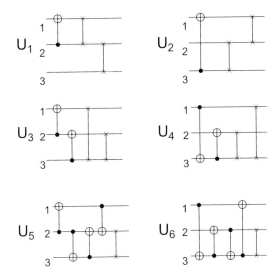

Figure 4.9 Quantum circuit used to create pseudo-pure states in a three qubit system by temporal averaging using two-qubit CNOT and SWAP gates. The pseudo-pure state $|000\rangle$ is obtained after combining the results of the seven (add the identity operator) U_i operations. Adapted with permission from Reference [27] (Copyright 2007 American Physical Society).

Here only two unitary operations are necessary to create the pseudo-pure states. Such operations are implemented by the following single-transition selective pulses [5]:

$$U_1 = \left(\frac{\pi}{2}\right)_x^{23}(\pi)_x^{12} = \frac{1}{\sqrt{2}}\begin{pmatrix} \sqrt{2} & 0 & 0 & 0 \\ 0 & 0 & i\sqrt{2} & 0 \\ 0 & i & 0 & i \\ 0 & -1 & 0 & 1 \end{pmatrix}$$

$$U_2 = \left(\frac{\pi}{2}\right)_{-x}^{23}(\pi)_x^{12} = \frac{1}{\sqrt{2}}\begin{pmatrix} \sqrt{2} & 0 & 0 & 0 \\ 0 & 0 & i\sqrt{2} & 0 \\ 0 & i & 0 & -i \\ 0 & 1 & 0 & 1 \end{pmatrix}$$

(4.3.14)

The density matrices after each pulse sequence as well as their addition become:

$$\rho_1 = \frac{1}{4}\begin{pmatrix} 1 & 0 & 0 & 0 \\ 0 & 1 & 0 & 0 \\ 0 & 0 & 1 & 0 \\ 0 & 0 & 0 & 1 \end{pmatrix} + \frac{\hbar\omega_0}{4k_BT}\begin{pmatrix} 3/2 & 0 & 0 & 0 \\ 0 & -1/2 & 0 & 0 \\ 0 & 0 & -1/2 & -i \\ 0 & 0 & i & -1/2 \end{pmatrix}$$

$$\rho_2 = \frac{1}{4}\begin{pmatrix} 1 & 0 & 0 & 0 \\ 0 & 1 & 0 & 0 \\ 0 & 0 & 1 & 0 \\ 0 & 0 & 0 & 1 \end{pmatrix} + \frac{\hbar\omega_0}{4k_BT}\begin{pmatrix} 3/2 & 0 & 0 & 0 \\ 0 & -1/2 & 0 & 0 \\ 0 & 0 & -1/2 & i \\ 0 & 0 & -i & -1/2 \end{pmatrix}$$

(4.3.15)

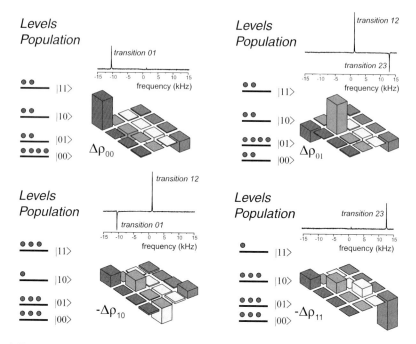

Figure 4.10 Experimental deviation density matrices and NMR spectra corresponding to the pseudo-pure states of a spin 3/2 system: (a): |00⟩, (b): |01⟩, (c): |10⟩, (d): |11⟩. The negative sign indicates that the corresponding level has a population deficit in respect to the other levels.

$$\rho_{00} = \frac{1}{4}(1-\varepsilon)\begin{pmatrix} 1 & 0 & 0 & 0 \\ 0 & 1 & 0 & 0 \\ 0 & 0 & 1 & 0 \\ 0 & 0 & 0 & 1 \end{pmatrix} + \varepsilon \begin{pmatrix} 1 & 0 & 0 & 0 \\ 0 & 0 & 0 & 0 \\ 0 & 0 & 0 & 0 \\ 0 & 0 & 0 & 0 \end{pmatrix}$$

where $\varepsilon = \frac{\hbar\omega_0}{2k_BT}$. The matrix ρ_{00} has the same form as (4.3.10). The other pseudo-pure states representing the other elements of the two qubits computational basis can be created in a similar manner:

$$\rho_{01} = \left[(\pi)_x^{01}\left(\frac{\pi}{2}\right)_{\pm x}^{23}(\pi)_x^{12}\right]\rho_{eq}\left[(\pi)_x^{01}\left(\frac{\pi}{2}\right)_{\pm x}^{23}(\pi)_x^{12}\right]^\dagger = \alpha\mathbf{1} + \varepsilon|01\rangle\langle 01|$$

$$\rho_{10} = \left[\left(\frac{\pi}{2}\right)_{\pm x}^{01}(\pi)_x^{23}(\pi)_x^{12}\right]\rho_{eq}\left[\left(\frac{\pi}{2}\right)_{\pm x}^{01}(\pi)_x^{23}(\pi)_x^{12}\right]^\dagger \qquad (4.3.16)$$
$$= \alpha\mathbf{1} + \varepsilon|10\rangle\langle 10|$$

$$\rho_{01} = \left[\left(\frac{\pi}{2}\right)_{\pm x}^{01}(\pi)_x^{12}\right]\rho_{eq}\left[\left(\frac{\pi}{2}\right)_{\pm x}^{01}(\pi)_x^{12}\right]^\dagger = \alpha\mathbf{1} - \varepsilon|11\rangle\langle 11|$$

Figure 4.10 shows the deviation density matrices and corresponding NMR spectra obtained for each of the above pseudo-pure states. As it can be observed, the NMR spectra of the different pseudo-pure states are clearly distinguishable. Thus, if the output state of a

4.3. Production of pseudo-pure states

give logical operation corresponds to a pseudo-pure state of the computational basis, it can be identified just by looking at the NMR spectrum.

The main advantage of the temporal averaging procedure is the facility of implementation and interpretation. The main disadvantage is the exponential increase in the number of states that must be combined upon increasing the number of qubits.

4.3.2 Spatial averaging

The use of spatial averaging for producing pseudo-pure states was first introduced by Cory et al. [24] and is based on dividing the system in spatially separated sub-ensembles. These sub-ensembles can be accessed independently in NMR by using a combination of RF pulses and pulsed magnetic gradients, which is equivalent to applying different unitary operations to each sub-ensemble. The pseudo-pure state is the average over all sub-ensembles. The main advantage of this method is that the pseudo-pure state is obtained after a single application of the pulse sequence, i.e., it is not necessary to combine different outputs to get the result of the computation.

A simple spatial averaging scheme can be used for creating pseudo-pure states in a system of quadrupolar nuclei. It can be illustrated using the above example of spins 3/2. Applying the operation U_1 of Equation (4.3.14) to the thermal equilibrium density matrix of Equation (4.3.13), the density matrix ρ_1 in Equation (4.3.15) is obtained. That matrix already has the same population distribution as the pseudo-pure state $|00\rangle$, but it does not correspond to a pseudo-pure state due to the presence of off-diagonal elements (coherences). The effect of a pulsed magnetic field gradient is to introduce a dephasing for the coherences associated with different spatial locations along the sample. Because this dephasing is proportional to the gradient strength, a high gradient strength pulse makes the coherences vary from 0 to 2π along the sample. In other words, one can consider the macroscopic sample as being constituted by a set of sub-ensembles each one represented by a density matrix with the same distribution of populations, but with off-diagonal elements out of phase. Therefore, the average density matrix over a reasonable number of sub-ensemble looks exactly like ρ_{00} in Equation (4.3.15), i.e., it is a pseudo-pure state density matrix. An example of creating a pseudo-pure state by spatial averaging in coupled spin 1/2 system is the pulse sequence corresponding to the operation (4.3.17) (see also Problem P4.6). The transformation that takes the equilibrium density matrix to the one corresponding to the pseudo-pure state $|00\rangle$, ρ_{00}, is:

$$\rho_{00} = \left[G_z(\tau) \left(\frac{\pi}{4}\right)^{I_1}_{-y} U_J\left(\frac{1}{2J}\right) \left(\frac{\pi}{4}\right)^{I_1}_{x} G_z(\tau) \left(\frac{\pi}{3}\right)^{I_2}_{x} \right] \rho_{eq}$$

$$\times \left[G_z(\tau) \left(\frac{\pi}{4}\right)^{I_1}_{-y} U_J\left(\frac{1}{2J}\right) \left(\frac{\pi}{4}\right)^{I_1}_{x} G_z(\tau) \left(\frac{\pi}{3}\right)^{I_2}_{x} \right]^{\dagger} \quad (4.3.17)$$

where $G_z(\tau)$ represents a gradient pulse of duration τ. The pseudo-pure state corresponding to the other eigenstates can be obtained from $|00\rangle$, as described in (4.3.11).

For many-qubit systems this method is very useful for obtaining pseudo-pure states. This is so, because temporal averaging would require too many repetitions for phase cancellation, which can be done at once by gradient pulses. The application of gradient pulses intercalated with RF pulses, as shown in (4.3.17), can also decrease the number of pulses.

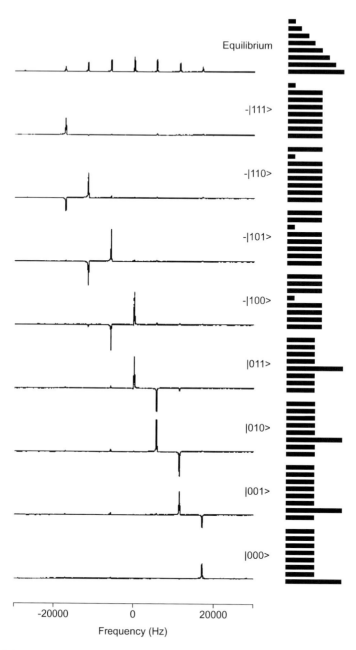

Figure 4.11 NMR spectra and schematic representation of the level populations in different states of a spin 7/2 system. Adapted with permission from Reference [25] (Copyright 2007 American Physical Society).

Gradient pulses can also be used together with multi-frequency pulses to create pseudo-pure states in systems with large number of qubits. An example of that is the case of a ^{133}Cs nuclei (spin 7/2 system) in a liquid crystal of cesium pentadecafluorooctanoate in D_2O [25]. This system has 8 different levels with linearly increasing populations, re-

4.3. Production of pseudo-pure states

sulting in 7 single-quantum transitions with different frequencies and transition rates, 7:12:15:16:15:12:7. Using multi-frequency selective pulses it is possible to simultaneously saturate all transitions except one, which is related to the desired pseudo-pure state. For example, to prepare the pseudo-pure state $|000\rangle$ we can simultaneously saturate six of the seven single-quantum transitions, i.e., irradiate them all with frequencies $\omega_{i,i+1}$, except the transition $\omega_{0,1}$. This can be achieved by a six-frequency pulse with central frequency at $(\omega_{3,4} + \omega_{4,5})/2$. This pulse produces the equalization of the populations of levels 1 to 7, keeping unchanged the population of the level 0. Thus, we obtain a state where all level populations are equal, except the population of the level 0. However, the application of multi-frequency pulses also creates undesired coherences in the deviation density matrix, making necessary to use a gradient pulse to remove those coherences. Figure 4.11 shows the NMR spectra and the schematic representation of the levels populations in different three-qubits pseudo-pure states created in a spin 7/2 system. Spatial averaging has also been performed in quadrupolar nuclei in solid crystals [8]. However, in these cases the T_2 relaxation times are in the order of microseconds, which allowed the use of single delays instead of gradient pulses to dephase the off-diagonal coherences.

4.3.3 State labeling

Another usual technique for creating pseudo-pure states is called *logical* or *state labeling*. This method was first introduced by Gershenfeld and Chuang [23] and does not make used of sub-ensembles of spins neither of averaging procedures. Considering a n-qubit system, in the state labeling method one of the qubits is used to label the state, while the others $n-1$ qubits are put in a pseudo-pure configuration. To illustrate how this is done, let us take a three qubits system formed by three homonuclear weakly J-coupled spins 1/2. The corresponding thermal equilibrium deviation density matrix and the relative populations (normalized by the factor $\hbar\omega_L/16K_BT$ and measured relative to the population of the sate $|11\rangle$) of the eigenstates $|\alpha\beta\gamma\rangle$ are:

$$\begin{array}{lcccccccc} \text{State:} & |000\rangle & |001\rangle & |010\rangle & |011\rangle & |100\rangle & |101\rangle & |110\rangle & |111\rangle \\ \text{Relative population:} & 6 & 4 & 4 & 2 & 4 & 2 & 2 & 0 \end{array} \quad (4.3.18)$$

$$\Delta\rho_{eq} = \frac{\hbar\omega_0}{16k_BT} \begin{pmatrix} 3 & 0 & 0 & 0 & 0 & 0 & 0 & 0 \\ 0 & 1 & 0 & 0 & 0 & 0 & 0 & 0 \\ 0 & 0 & 1 & 0 & 0 & 0 & 0 & 0 \\ 0 & 0 & 0 & -1 & 0 & 0 & 0 & 0 \\ 0 & 0 & 0 & 0 & 1 & 0 & 0 & 0 \\ 0 & 0 & 0 & 0 & 0 & -1 & 0 & 0 \\ 0 & 0 & 0 & 0 & 0 & 0 & -1 & 0 \\ 0 & 0 & 0 & 0 & 0 & 0 & 0 & -3 \end{pmatrix} \quad (4.3.19)$$

Now, let us look at this three qubit system as being composed by two independent subspaces, labeled by the first qubit. In other words, the first four states are seen as a two-qubit system with labeling qubit equal to 0, while the other remaining states form another two-qubit system with the labeling qubit equal to 1. Imagine now that we apply two consecutive CNOT gates, one with target in the first qubit, and control in the third, followed by another CNOT with the first qubit as the target and the second as the control, i.e., $\text{CNOT}_{1-2}\text{CNOT}_{1-3}$. The first CNOT operation simultaneously swaps the populations of the states $|001\rangle$ with $|101\rangle$ and $|011\rangle$ with $|111\rangle$ and the second CNOT gate swaps the populations of the states $|010\rangle$ with $|110\rangle$ and $|011\rangle$ with $|111\rangle$. Thus, after applying both gates, the following state is obtained (see Problem P4.7):

$$\begin{array}{lcccccccc} \text{State:} & |000\rangle & |001\rangle & |010\rangle & |011\rangle & |100\rangle & |101\rangle & |110\rangle & |111\rangle \\ \text{Population:} & 6 & 2 & 2 & 2 & 4 & 4 & 4 & 0 \end{array} \tag{4.3.20}$$

$$\Delta\rho_{pps} = \frac{\hbar\omega_0}{16 k_B T} \begin{pmatrix} 3 & 0 & 0 & 0 & 0 & 0 & 0 & 0 \\ 0 & -1 & 0 & 0 & 0 & 0 & 0 & 0 \\ 0 & 0 & -1 & 0 & 0 & 0 & 0 & 0 \\ 0 & 0 & 0 & -1 & 0 & 0 & 0 & 0 \\ 0 & 0 & 0 & 0 & 1 & 0 & 0 & 0 \\ 0 & 0 & 0 & 0 & 0 & 1 & 0 & 0 \\ 0 & 0 & 0 & 0 & 0 & 0 & 1 & 0 \\ 0 & 0 & 0 & 0 & 0 & 0 & 0 & -3 \end{pmatrix} \tag{4.3.21}$$

Starting from the equilibrium state, the CNOT gates do not introduce any off-diagonal elements into the deviation density matrix, so we can regard (4.3.21) as two subsets of pseudo-pure states, the state $|00\rangle_0$ with label state $|0\rangle$ and the state $|11\rangle_1$ with state label $|1\rangle$, i.e,

$$|\psi\rangle \propto |0\rangle \otimes |00\rangle_0 + |1\rangle \otimes |11\rangle_1 \tag{4.3.22}$$

Here the first qubit plays an important role because it defines the working subspace during the execution of a logical operation. This means that the label state must index the output in such a way that, after executing a logical operation, the output associated to the working subspace can be identified. For example, if we apply a two-qubit gate to the pseudo-pure state $|11\rangle$, the state label qubit $|1\rangle$ can be used to search for (or isolate) the output signal from the corresponding subspace.

The logical labeling scheme was also generalized for a n-qubit [26] system with the advantage of using a smaller number of operations than the temporal or spatial averaging approaches. The disadvantage of this method is that the label spin cannot be used for the computation.

There are other less usual methods for creating pseudo-pure states, for example using flip and swap logic gates, randomization of group of spins, or using entanglement [27]. However, the most important are the ones described above.

4.4 RECONSTRUCTION OF DENSITY MATRICES IN NMR QIP: QUANTUM STATE TOMOGRAPHY

The ability of an experimental technique for preparing initial states and implementing an universal set of logic gates are two important features for its use in quantum information processing. Another equally important requirement is the characterization of the output state. In many cases we wish more than a simple readout, but a full characterization of the system state. This can achieved by determining all elements of the density matrix of the system. In quantum information processing, the reconstruction of the density matrix allows many interesting applications. For example: (a) testing the preparation of quantum states; (b) estimating the experimental errors and calculating the fidelity of the implementation of a quantum gate; (c) monitoring the implementation of quantum gates in intermediate steps; (d) characterizing decoherence and dissipation effects in quantum systems, etc.

The reconstruction of the density matrix of a system usually involves performing a series of measurements (readouts) and combining the results to obtain the density matrix elements. As stated before, the density matrix of a n qubits system has $(2^n - 1)(2^{n-1} + 1)$ elements, but not all element are accessible by a single measurement. To see that, let us consider a quantum system represented by a density matrix ρ in a $|u\rangle$ basis. A set of measurements in the system via the measurement operator $A = |u\rangle\langle u|$ provides the probability distribution of finding the system in a given state $|u\rangle$. Thus, applying such procedure for all basis states its possible to obtain the full set of probabilities, which indeed represent the diagonal elements of the density matrix. To obtain the other elements of ρ it is necessary to perform a set of measurements in the same state at different bases until all elements of ρ are fully determined. This procedure is known as density matrix tomography or quantum state tomography (QST). Next section address how to achieve that in NMR systems.

4.4.1 NMR Quantum State Tomography

In standard NMR experiments the readout is made in the basis of the Zeeman interaction, i.e, the I_z basis. Then, the basic ingredient for NMR QST would be to know how to execute the readout at different bases. However, this is completely equivalent to rotate the qubits and execute the measurements in a fixed base. Indeed:

$$\mathrm{Tr}\left[\rho U |m\rangle\langle m| U^\dagger\right] \equiv \mathrm{Tr}\left[U^\dagger \rho U |m\rangle\langle m|\right] \tag{4.4.1}$$

Using this property, it is possible to design a procedure for NMR QST based on a rotation of the qubits via RF pulses, followed by measurements in the I_z base, although there are peculiarities of each method according to the spin system of interest. In general, NMR QST methods rely on the same idea, i.e., to execute specially designed unitary rotation through RF pulses and reconstructing the density matrix from the intensities of the resulting NMR spectra. In the following section we will describe some of these methods for J-coupled spins 1/2 in diluted solution and quadrupolar spins in oriented media or solid-state. For direct dipolar coupled spins 1/2 the same method as for J-coupled spins can be used.

4.4.2 NMR Quantum State Tomography in coupled spin 1/2 systems

The first NMR QST method was developed by Chuang et al. [26] for systems of coupled spins 1/2. It consists of preparing the state to be tomographed and applying RF pulses (tomography pulses) to execute selective unitary rotations in the qubits. Then, the NMR spectrum (readout) is acquired and the line intensities are recorded. This procedure is repeated varying the phases of the tomography pulses and, after a certain numbers of readouts, it is possible to construct a set of equations involving the line intensities and the elements of the original density matrix. To illustrate it in more details, let us consider a general deviation density matrix of two coupled spins:

$$\Delta\rho = \begin{pmatrix} x_{11} & x_{12}+iy_{12} & x_{13}+iy_{13} & x_{14}+iy_{14} \\ x_{12}-iy_{12} & x_{22} & x_{23}+iy_{23} & x_{24}+iy_{24} \\ x_{13}-iy_{13} & x_{23}-iy_{23} & x_{33} & x_{34}+iy_{34} \\ x_{14}-iy_{14} & x_{24}-iy_{24} & x_{34}-iy_{34} & x_{44} \end{pmatrix} \quad (4.4.2)$$

where $\Delta\rho$ is Hermitian and traceless, that is, $x_{11}+x_{22}+x_{33}+x_{44}=0$. It is straightforward to show that the spectrum corresponding to spin $\mathbf{I_1}$ depends only on the elements $x_{12}+iy_{12}$ and $x_{34}+iy_{34}$, while the spectrum corresponding to spin $\mathbf{I_2}$ depends only on the elements $x_{13}+iy_{13}$ and $x_{24}+iy_{24}$ (see Problem P4.8). Since the NMR spectrum of this system has four lines (two for each nucleus), taking the real and imaginary parts of these lines (Re(S_i), Im(S_i)) one can obtain a system of eight equations relating the spectral line intensities with these eight elements. From the solution of such system of equations the unknown elements can be determined. Therefore, preparing the state to be tomographed and recording the NMR spectra corresponding to both spins allow us to determine the elements corresponding to the 12, 34, 13 and, 24 entries of the corresponding deviation density matrix (we will call these entries as "reading positions"). To obtain the other elements it is necessary to execute a rotation in the spin system that brings the desired element to one of these reading positions. A general set of RF pulses that perform such task (i.e., the complete density matrix tomography) is,

$$\Delta\rho_{read} = \mathbf{1}\Delta\rho\mathbf{1}$$

$$\Delta\rho_{read} = \left[\left(\frac{\pi}{2}\right)_x^{I_1}\right]\Delta\rho\left[\left(\frac{\pi}{2}\right)_x^{I_1}\right]^\dagger$$

$$\Delta\rho_{read} = \left[\left(\frac{\pi}{2}\right)_x^{I_2}\right]\Delta\rho\left[\left(\frac{\pi}{2}\right)_x^{I_2}\right]^\dagger$$

$$\Delta\rho_{read} = \left[\left(\frac{\pi}{2}\right)_y^{I_1}\right]\Delta\rho\left[\left(\frac{\pi}{2}\right)_y^{I_1}\right]^\dagger$$

$$\Delta\rho_{read} = \left[\left(\frac{\pi}{2}\right)_y^{I_2}\right]\Delta\rho\left[\left(\frac{\pi}{2}\right)_y^{I_2}\right]^\dagger \quad (4.4.3)$$

$$\Delta\rho_{read} = \left[\left(\frac{\pi}{2}\right)_x^{I_2}\left(\frac{\pi}{2}\right)_x^{I_1}\right]\Delta\rho\left[\left(\frac{\pi}{2}\right)_x^{I_1}\left(\frac{\pi}{2}\right)_x^{I_2}\right]^\dagger$$

$$\Delta\rho_{read} = \left[\left(\frac{\pi}{2}\right)_x^{I_2}\left(\frac{\pi}{2}\right)_y^{I_1}\right]\Delta\rho\left[\left(\frac{\pi}{2}\right)_x^{I_1}\left(\frac{\pi}{2}\right)_y^{I_2}\right]^\dagger$$

$$\Delta\rho_{read} = \left[\left(\frac{\pi}{2}\right)_y^{I_2}\left(\frac{\pi}{2}\right)_x^{I_1}\right]\Delta\rho\left[\left(\frac{\pi}{2}\right)_y^{I_1}\left(\frac{\pi}{2}\right)_x^{I_2}\right]^\dagger$$

$$\Delta\rho_{read} = \left[\left(\frac{\pi}{2}\right)_y^{I_2}\left(\frac{\pi}{2}\right)_y^{I_1}\right]\Delta\rho\left[\left(\frac{\pi}{2}\right)_y^{I_1}\left(\frac{\pi}{2}\right)_y^{I_2}\right]^\dagger$$

where $\Delta\rho_{read}$ indicates the deviation density matrix that originates the NMR spectra. Since the pulse matrices are known, the reading positions of $\Delta\rho_{read}$ can be calculated as a function the elements of $\Delta\rho$. Consequently, the elements of $\Delta\rho$ can be directed associated to the line intensities in the corresponding NMR spectra. The net result of this procedure is a system of 72 equations (each line in (4.4.3) give rise to eight equations – four complex spectral intensities) that can be solved to determine the 16 elements of $\Delta\rho$. The equation set (4.4.3) is redundant and can be simplified to decrease the number of readouts [28]. Such feature is important for systems with many qubits, where the application of the method will require much more readouts. This procedure for QST can also be adapted to coupled homonuclear spins just by using the appropriate selective pulses to perform the desired rotations.

Another useful QST method was developed using the two-dimensional Fourier transform technique [29]. In this method the diagonal elements of the density matrix are obtained in a 1D experiment where a short pulse, similar to that discussed above, is applied to retrieve the populations. The quantum coherences, including those not directly observable, are codified in a 2D spectrum $S(\omega_1, \omega_2)$ so that the 2D intensities depend on these coherences. This allows to extract the quantum coherences by fitting the 2D spectrum. The main advantage of such method is its scalability, as the 2D acquisition provides an efficient way of extracting the coherences even for a large number of qubits (>5).

4.4.3 NMR Quantum State Tomography of quadrupole nuclei

The first QST procedure dedicated to quadrupolar systems was developed by Kampermann and Veeman for spin 3/2 nuclei [8]. It is a direct adaptation from the method used for spin 1/2, but uses transition selective pulses instead of spin selective pulses. Similar to the spin 1/2 method, the transition selective pulses are used to bring an specific set of populations and coherences to the reading position of the deviation density matrix (single quantum coherences) and after that the NMR spectrum is acquired. The transformations (transition selective pulses) that are applied to the deviation density matrix $\Delta\rho$ in order to bring the unknown elements to the reading positions are,

$$\Delta\rho_{read} = \mathbf{1}\Delta\rho\mathbf{1}$$

$$\Delta\rho_{read} = \left[\left(\frac{\pi}{2}\right)_x^{01}\right]\Delta\rho\left[\left(\frac{\pi}{2}\right)_x^{01}\right]^\dagger$$

$$\Delta\rho_{read} = \left[\left(\frac{\pi}{2}\right)_x^{12}\right]\Delta\rho\left[\left(\frac{\pi}{2}\right)_x^{12}\right]^\dagger \quad (4.4.4)$$

$$\Delta\rho_{read} = \left[\left(\frac{\pi}{2}\right)_x^{23}\right]\Delta\rho\left[\left(\frac{\pi}{2}\right)_x^{23}\right]^\dagger$$

$$\Delta\rho_{read} = \left[(\pi)_x^{23}(\pi)_x^{01}\right]\Delta\rho\left[(\pi)_x^{01}(\pi)_x^{23}\right]^\dagger$$

where $\Delta\rho_{read}$ indicates the deviation density matrix that originates the NMR signal. Since $\Delta\rho_{read}$ can be calculated from $\Delta\rho$ using the single transition pulse operators described in Chapter 2, the NMR signal as a function of the elements of $\Delta\rho$ can be determined. This allows to relate the spectra line intensities with the desired elements of $\Delta\rho$. Then, the normalized real and imaginary integrals of the spectra line intensities are used to build a set of linear equations that has the elements of $\Delta\rho$ as unknowns. This set of equations is also solved by standard methods in order to obtain the tomographed deviation density matrix.

Another method for performing quantum state tomography of a quadrupolar spin 3/2 NMR system is based in the combination of transition selective and hard readout pulses [30]. Basically, the method consists in determining the diagonal elements of the deviation density matrix, after performing operations on the system, which selectively drag the off-diagonal elements into the main diagonal. The following steps are executed:

(1) The diagonal elements of the deviation density matrix, x_{11}, x_{22}, x_{33}, and x_{44}, are determined from the intensities of the three lines measured in an averaged spectrum obtained after the application of a hard $\pi/20$ readout pulse under the CYCLOPS phase cycling scheme (see Chapter 2). Using the pulse operators and the CYCLOPS readout scheme to calculate the average spectrum, it is possible to obtain a set of three equations relating the matrix elements of the $\pi/20$ pulse, e_{ij}, the elements x_{11}, x_{22}, x_{33}, x_{44} of the deviation density matrix, and the three line intensities, A_1, A_2, and A_3. A fourth equation is obtained from the trace relation for the deviation density matrix. Thus, the following set of equations is found,

$$A_1 = \sqrt{3}(e_{11}e_{12}x_{11} - e_{12}e_{22}x_{22} - e_{23}e_{13}x_{33} - e_{13}e_{14}x_{44})$$
$$A_2 = 2(e_{13}e_{12}x_{11} + e_{22}e_{23}x_{22} - e_{23}e_{22}x_{33} - e_{13}e_{12}x_{44})$$
$$A_3 = \sqrt{3}(e_{13}e_{14}x_{11} + e_{13}e_{23}x_{22} + e_{12}e_{22}x_{33} - e_{11}e_{12}x_{44})$$
$$x_{11} + x_{22} + x_{33} + x_{44} = 0$$
(4.4.5)

Since the pulse matrix elements e_{ij} are known and the line intensities, A_1, A_2, and A_3 can be measured, x_{11}, x_{22}, x_{33}, x_{44} can be determined from the solution of (4.4.5).

(2) To obtain the off-diagonal elements of the deviation density matrix, $\pi/2$ transition selective pulses with proper phases are applied to the system prior to the readout pulse. The effect of the application of such pulses is to bring the off-diagonal elements of the density matrix to the main diagonal, as illustrated in (4.4.6), where only the main diagonal elements are displayed for simplicity,

$$\left(\frac{\pi}{2}\right)_x^{01} \Delta\rho \left(\frac{\pi}{2}\right)_x^{01\dagger} = \begin{pmatrix} x_{11} + 2y_{12} + x_{22} & & & \\ & x_{11} - 2y_{12} + x_{22} & & \\ & & 2x_{33} & \\ & & & 2x_{44} \end{pmatrix}$$

$$\left(\frac{\pi}{2}\right)_x^{23} \Delta\rho \left(\frac{\pi}{2}\right)_x^{23\dagger} = \begin{pmatrix} 2x_{11} & & & \\ & 2x_{22} & & \\ & & x_{33} + 2y_{34} + x_{44} & \\ & & & x_{33} - 2y_{34} + x_{44} \end{pmatrix}$$
(4.4.6)

4.4. Reconstruction of density matrices in NMR QIP: Quantum State Tomography

Then, the same procedure of item 1 is applied to determine the new diagonal elements of the transformed matrix and, since the elements x_{11}, x_{22}, x_{33}, x_{44} of original matrix were already determined, each element dragged to the diagonal is obtained. Using transition selective pulses with distinct frequency and phases, either the real or imaginary part of each element are determined. The elements in the first diagonal (single quantum coherences) are obtained using a single selective pulse, while for elements in the second (double quantum coherences) and third (triple quantum coherences) diagonal two and three pulses are used, respectively. A complete set of equations that allows the determination of all off-diagonal elements is shown in (4.4.7). In this cases $\omega_Q t_p$ (t_p is the pulse length) is chosen to be a multiple of 2π in order to avoid the effect of the quadrupolar evolution during the pulse [31].

$$x_{12} = \left\{ \rho_{11}\left[\left(\frac{\pi}{2}\right)_y^{01}\right] - \rho_{22}\left[\left(\frac{\pi}{2}\right)_y^{01}\right] \right\}/2$$

$$y_{12} = \left\{ \rho_{11}\left[\left(\frac{\pi}{2}\right)_x^{01}\right] - \rho_{22}\left[\left(\frac{\pi}{2}\right)_x^{01}\right] \right\}/2$$

$$x_{23} = \left\{ \rho_{22}\left[\left(\frac{\pi}{2}\right)_y^{12}\right] - \rho_{33}\left[\left(\frac{\pi}{2}\right)_y^{12}\right] \right\}/2$$

$$y_{23} = \left\{ \rho_{22}\left[\left(\frac{\pi}{2}\right)_x^{12}\right] - \rho_{33}\left[\left(\frac{\pi}{2}\right)_x^{12}\right] \right\}/2$$

$$x_{34} = \left\{ \rho_{33}\left[\left(\frac{\pi}{2}\right)_y^{23}\right] - \rho_{44}\left[\left(\frac{\pi}{2}\right)_y^{23}\right] \right\}/2$$

$$y_{34} = \left\{ \rho_{33}\left[\left(\frac{\pi}{2}\right)_x^{23}\right] - \rho_{44}\left[\left(\frac{\pi}{2}\right)_x^{23}\right] \right\}/2$$

$$x_{13} = \left\{ \rho_{22}\left[\left(\frac{\pi}{2}\right)_x^{12}\left(\frac{\pi}{2}\right)_x^{01}\right] - \rho_{33}\left[\left(\frac{\pi}{2}\right)_x^{12}\left(\frac{\pi}{2}\right)_x^{01}\right] - \sqrt{2}y_{23} \right\}/\sqrt{2} \quad (4.4.7)$$

$$y_{13} = \left\{ \rho_{33}\left[\left(\frac{\pi}{2}\right)_y^{12}\left(\frac{\pi}{2}\right)_x^{01}\right] - \rho_{22}\left[\left(\frac{\pi}{2}\right)_y^{12}\left(\frac{\pi}{2}\right)_x^{01}\right] + \sqrt{2}x_{23} \right\}/\sqrt{2}$$

$$x_{24} = \left\{ \rho_{33}\left[\left(\frac{\pi}{2}\right)_x^{23}\left(\frac{\pi}{2}\right)_x^{12}\right] - \rho_{44}\left[\left(\frac{\pi}{2}\right)_x^{23}\left(\frac{\pi}{2}\right)_x^{12}\right] - \sqrt{2}y_{34} \right\}/\sqrt{2}$$

$$y_{24} = \left\{ \rho_{44}\left[\left(\frac{\pi}{2}\right)_y^{23}\left(\frac{\pi}{2}\right)_x^{12}\right] - \rho_{33}\left[\left(\frac{\pi}{2}\right)_y^{23}\left(\frac{\pi}{2}\right)_x^{12}\right] + \sqrt{2}x_{34} \right\}/\sqrt{2}$$

$$x_{34} = \left\{ \rho_{33}\left[\left(\frac{\pi}{2}\right)_y^{23}\left(\frac{\pi}{2}\right)_y^{12}\left(\frac{\pi}{2}\right)_y^{01}\right] - \rho_{44}\left[\left(\frac{\pi}{2}\right)_y^{23}\left(\frac{\pi}{2}\right)_y^{12}\left(\frac{\pi}{2}\right)_y^{01}\right] \right.$$
$$\left. - \sqrt{2}x_{34} + x_{24} \right\}/\sqrt{2}$$

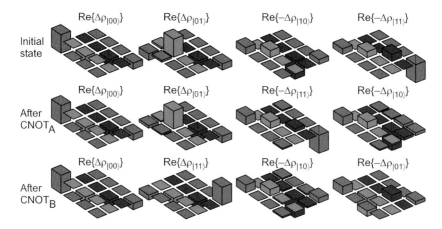

Figure 4.12 Deviation density matrix truth table for CNOT gates implemented in a quadrupolar spin 3/2 system. Adapted with permission from Reference [31] (Copyright 2007 Elsevier).

$$y_{34} = \left\{ \rho_{44} \left[\left(\frac{\pi}{2}\right)_x^{23} \left(\frac{\pi}{2}\right)_x^{12} \left(\frac{\pi}{2}\right)_x^{01} \right] - \rho_{33} \left[\left(\frac{\pi}{2}\right)_x^{23} \left(\frac{\pi}{2}\right)_x^{12} \left(\frac{\pi}{2}\right)_x^{01} \right] \right. $$
$$\left. + \sqrt{2} y_{34} + x_{24} \right\} \Big/ \sqrt{2}$$

The expression $\rho_{kk}\left[\left(\frac{\pi}{2}\right)_\phi^{ij} \cdots \left(\frac{\pi}{2}\right)_\phi^{mn}\right]$ stands for the diagonal element (kth line and kth column) of the density matrix after the application of the respective $\left(\left(\frac{\pi}{2}\right)_\phi^{ij} \cdots \left(\frac{\pi}{2}\right)_\phi^{mn}\right)$ pulse sequence, i.e.,

$$\rho_{kk}\left[\left(\frac{\pi}{2}\right)_\phi^{ij} \cdots \left(\frac{\pi}{2}\right)_\phi^{mn}\right] = \left[\left(\frac{\pi}{2}\right)_\phi^{ij} \cdots \left(\frac{\pi}{2}\right)_\phi^{mn} \Delta\rho \left(\frac{\pi}{2}\right)_\phi^{ij\dagger} \cdots \left(\frac{\pi}{2}\right)_\phi^{mn\dagger}\right]_{kk}$$

An illustration of the use of the density matrix tomography process is shown in Figure 4.12. It is the measured truth table for a CNOT gate with control on qubits A and B, obtained from experimentally determined density matrices.

The methods for QST of spin 3/2 systems as well as for coupled homonuclear spin 1/2 describe above can, in principle, be adapted for higher spin systems. However, this will certainly imply in the use of more selective pulses.

4.5 EVOLUTION OF BLOCH VECTORS AND OTHER QUANTITIES OBTAINED FROM TOMOGRAPHED DENSITY MATRICES

An appropriate and useful approach to follow the evolution of a quantum state is the Bloch sphere representation, introduced in Chapter 3. This is a geometrical scheme in which the quantum state and its evolution is represented by the trajectory of a vector over the so-called Bloch sphere (Figure 4.13). In the Bloch sphere, the poles represent the two eigenstates of the system, whereas the equatorial plane corresponds to an uniform superposition of these two eigenstates.

4.5. Evolution of Bloch vectors and other quantities obtained from tomographed density matrices

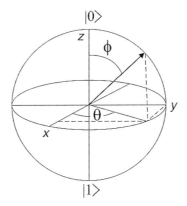

Figure 4.13 Schematic representation of a state vector in the Bloch sphere.

To illustrate the Bloch sphere representation of an NMR system, let us consider the density matrix of an ensemble of spins 1/2 nuclei. Because the high temperature deviation density matrix is proportional to I_z, the effect of an RF pulse is to induce rotations that transform the initial density matrix into a linear combination of the spin operator components,

$$\Delta\rho = \xi_x(t)I_x + \xi_y(t)I_y + \xi_z(t)I_z \qquad (4.5.1)$$

Thus, the evolution of such density matrix is contained in the coefficients $\xi_i(t)$, which represent the component of the vector $\mathbf{I}(t) = \mathbf{i}\xi_x(t)I_x + \mathbf{j}\xi_y(t)I_y + \mathbf{k}\xi_z(t)I_z$. Notice that the thermal average of $\mathbf{I}(t)$ is proportional to the macroscopic magnetization, which means that for one-qubit systems both vectors have the same dynamics. Considering a spherical representation with $\sigma = \mathbf{i}I_x + \mathbf{j}I_y + \mathbf{k}I_z$ and

$$\mathbf{r} = \mathbf{i}\xi_1 \sin[\theta(t)]\cos[\phi(t)] + \mathbf{j}\xi_2 \sin[\theta(t)]\sin[\phi(t)] + \mathbf{k}\sin[\theta(t)]$$

we can write $\Delta\rho = \mathbf{r}\cdot\sigma$, and the full density matrix of the system becomes,

$$\rho(t) = \frac{1+\mathbf{r}\cdot\sigma}{2} \qquad (4.5.2)$$

Using this equation it is possible to obtain the components of the Bloch vector $\mathbf{r}$ from the density matrix $\rho(t)$ as $\xi_x(t) = \text{Re}[2\rho_{21}(t) - 1]$, $\xi_y(t) = \text{Im}[2\rho_{21}(t) - 1]$, $\xi_z(t) = \text{Re}[2\rho_{11}(t) - 1]$. As a result, the evolution of the Bloch vector can be accompanied if the experimental density matrix, or the deviation density matrix, is determined. One example of such procedure is shown in Figure 4.14. Figure 4.14a shows the evolution of the Bloch vector in a Bloch sphere during a Hadamard operation implemented by the pulses $(\pi)_x^I\left(\frac{\pi}{2}\right)_y^I$. The initial pseudo-pure state $|0\rangle$, corresponding to the top of the sphere, it transformed into the state $(|0\rangle + |1\rangle)/\sqrt{2}$, the equatorial plane of the sphere, by a Hadamard transformation. The Bloch vector starts its evolution at z and it is transferred to the x direction by the pulse $\left(\frac{\pi}{2}\right)_y^I$ where it remains during the application of the pulse $(\pi)_x^I$, which is applied only to provide the phase correction. For many qubit systems a similar approach allows the determination of the Bloch vectors corresponding to both qubits. The method

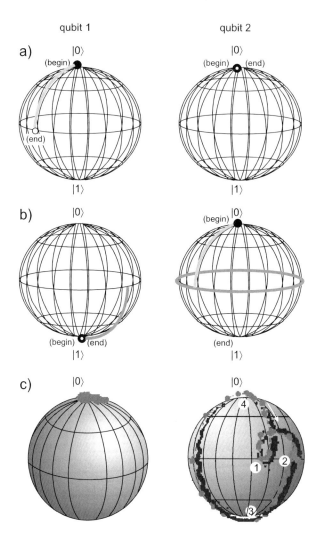

Figure 4.14 Representation in the Bloch Sphere of one qubit evolution during the execution of some logic gates. (a) One qubit Hadamard Operation; (b) CNOT$_A$|10⟩ operation; (c) Experimental and simulated evolution of both qubits during the double application of a Hadamard Gate. Adapted with permission from Reference [30] (Copyright 2007 Elsevier).

consists in calculating the partial trace of the density matrix over the subspace corresponding to each qubit, and relating it to the Bloch vector by the following expression,

$$\mathrm{Tr}_A\big[\rho(t)\big] = \frac{\mathbf{1} + \mathbf{r}_A \cdot \boldsymbol{\sigma}}{2} \qquad \mathrm{Tr}_B\big[\rho(t)\big] = \frac{\mathbf{1} + \mathbf{r}_B \cdot \boldsymbol{\sigma}}{2} \qquad (4.5.3)$$

Figure 4.14b shows the simulated evolution of the Bloch vectors in a system of two coupled spins 1/2, during the execution of a CNOT$_A$|10⟩ operation. As a result of the $\left(\frac{\pi}{2}\right)_z^{I_1}$ rotation, the qubit A, initially at the south pole of the sphere (state |1⟩), it is transferred to the equatorial plane, but returns to the south pole at the end of the gate. Meanwhile, due to all the pulses applied to the $\mathbf{I_2}$ spin, the qubit B, initially at the state |0⟩, execute a complex trajectory in the Bloch sphere before going to the state $|1\rangle$ at the end of the gate. An example of an experimental monitoring of the Bloch vector during the double application of an H_B logic gate is shown in Figure 4.14c. This was achieved for a spin 3/2 quadrupolar system and the experimental Bloch vector trajectories were determined from a set of tomographed density matrices, obtained at each step of the evolution [30]. The qubit A remains almost unchanged during all the operation, while the qubit B initially at the state $|0\rangle$, is changed to the state $(|0\rangle + |1\rangle)/\sqrt{2}$ at the end of a single application of the H_B gate, and returns to the state $|0\rangle$ after applying the gate twice. The full line is a calculation of the qubits evolution performed using the RF and quadrupolar Hamiltonians according to the selective Gaussian shaped RF pulses used to implement the H_B gate.

Important information can also be extracted from the entire system density matrix. For example, the distance $D(t) = \text{Tr}[\rho(0) - \rho(t)]$ and the fidelity

$$F(t) = \text{Tr}\left\{\sqrt{\rho(t)^{1/2} \cdot \rho(0)^{1/2} \cdot \rho(t)^{1/2}}\right\}$$

between any two quantum states can be calculated [32]. These parameters were obtained after the application of the double-Hadamard gate of Figure 4.14c. Comparing the initial and final states, one finds $D(\rho_1, \rho_4) = 0.3 \pm 0.2$ and $F(\rho_1, \rho_4) = 0.9 \pm 0.2$, confirming the self reversibility of the implemented H_B logic gate.

PROBLEMS WITH SOLUTIONS

P4.1 – Using the pulse matrix for the transition selective pulses in quadrupolar systems, find the pulse operators for the single-qubit gates NOT_A, NOT_B, H_A, H_B. Apply such pulse operators to the state $|00\rangle$ and show the they execute the expected actions. Then, show that the two Hadamard gates are self-reversible.

Solution
The relevant pulse operator matrices obtained from the general equation of Chapter 2 are:

$$\left(\frac{\pi}{2}\right)_y^{01} = \begin{pmatrix} \frac{1}{\sqrt{2}} & \frac{1}{\sqrt{2}} & 0 & 0 \\ -\frac{1}{\sqrt{2}} & \frac{1}{\sqrt{2}} & 0 & 0 \\ 0 & 0 & 1 & 0 \\ 0 & 0 & 0 & 1 \end{pmatrix} \quad \left(\frac{\pi}{2}\right)_y^{12} = \begin{pmatrix} 1 & 0 & 0 & 0 \\ 0 & \frac{1}{\sqrt{2}} & \frac{1}{\sqrt{2}} & 0 \\ 0 & -\frac{1}{\sqrt{2}} & \frac{1}{\sqrt{2}} & 0 \\ 0 & 0 & 0 & 1 \end{pmatrix}$$

$$\left(\frac{\pi}{2}\right)_y^{23} = \begin{pmatrix} 1 & 0 & 0 & 0 \\ 0 & 1 & 0 & 0 \\ 0 & 0 & \frac{1}{\sqrt{2}} & \frac{1}{\sqrt{2}} \\ 0 & 0 & -\frac{1}{\sqrt{2}} & \frac{1}{\sqrt{2}} \end{pmatrix} \quad (\pi)_x^{01} = \begin{pmatrix} 0 & i & 0 & 0 \\ i & 0 & 0 & 0 \\ 0 & 0 & 1 & 0 \\ 0 & 0 & 0 & 1 \end{pmatrix}$$

$$(\pi)_x^{12} = \begin{pmatrix} 1 & 0 & 0 & 0 \\ 0 & 0 & i & 0 \\ 0 & i & 0 & 0 \\ 0 & 0 & 0 & 1 \end{pmatrix} \quad (\pi)_x^{23} = \begin{pmatrix} 1 & 0 & 0 & 0 \\ 0 & 1 & 0 & 0 \\ 0 & 0 & 0 & i \\ 0 & 0 & i & 0 \end{pmatrix}$$

$$(\pi)_x = \begin{pmatrix} 0 & 0 & 0 & 1 \\ 0 & 0 & 1 & 0 \\ 0 & 1 & 1 & 0 \\ 1 & 0 & 0 & 1 \end{pmatrix} \quad (\pi)_y = \begin{pmatrix} 0 & 0 & 0 & 1 \\ 0 & 0 & -1 & 0 \\ 0 & -1 & 0 & 0 \\ 1 & 0 & 0 & 0 \end{pmatrix}$$

Performing the multiplication of the operators corresponding to the pulse sequence that implement the NOT gates we find:

$$NOT_A = (\pi)_x^{23}(\pi)_x^{01}(\pi)_x = \begin{pmatrix} 0 & 0 & -1 & 0 \\ 0 & 0 & 0 & -1 \\ -1 & 0 & 0 & 0 \\ 0 & -1 & 0 & 0 \end{pmatrix}$$

$$NOT_B = (\pi)_x^{23}(\pi)_x^{01} = \begin{pmatrix} 0 & i & 0 & 0 \\ i & 0 & 0 & 0 \\ 0 & 0 & 0 & i \\ 0 & 0 & i & 0 \end{pmatrix}$$

To obtain the matrix representation of the pulse operators of the Hadamard gates we can use that:

$$\left(\frac{\pi}{2}\right)_{-y}^{12} = \left[\left(\frac{\pi}{2}\right)_y^{12}\right]^\dagger \quad \text{and} \quad (2\pi)_x^{01} = (\pi)_x^{01}(\pi)_x^{01}$$

thus the Hadamard pulse operator becomes:

$$H_A^{|00\rangle,|01\rangle} = (2\pi)_{-x}^{01}(\pi)_y^{01}\left(\frac{\pi}{2}\right)_y^{12}(\pi)_{-y}^{01} = \begin{pmatrix} -\frac{1}{\sqrt{2}} & 0 & -\frac{1}{\sqrt{2}} & 0 \\ 0 & -1 & 0 & 0 \\ -\frac{1}{\sqrt{2}} & 0 & \frac{1}{\sqrt{2}} & 0 \\ 0 & 0 & 0 & 1 \end{pmatrix}$$

$$H_A^{|10\rangle,|11\rangle} = (2\pi)_{-x}^{01}(\pi)_y^{12}\left(\frac{\pi}{2}\right)_y^{23}(\pi)_{-y}^{12} = \begin{pmatrix} -1 & 0 & 0 & 0 \\ 0 & -\frac{1}{\sqrt{2}} & 0 & -\frac{1}{\sqrt{2}} \\ 0 & 0 & 1 & 0 \\ 0 & -\frac{1}{\sqrt{2}} & 0 & \frac{1}{\sqrt{2}} \end{pmatrix}$$

$$H_B^{|00\rangle,|01\rangle} = (\pi)_x^{01}\left(\frac{\pi}{2}\right)_{-y}^{01} = \begin{pmatrix} \frac{1}{\sqrt{2}} & \frac{1}{\sqrt{2}} & 0 & 0 \\ \frac{1}{\sqrt{2}} & -\frac{1}{\sqrt{2}} & 0 & 0 \\ 0 & 0 & 1 & 0 \\ 0 & 0 & 0 & 1 \end{pmatrix}$$

$$H_B^{|01\rangle,|11\rangle} = (\pi)_x^{23}\left(\frac{\pi}{2}\right)_{-y}^{23} = \begin{pmatrix} 1 & 0 & 0 & 0 \\ 0 & 1 & 0 & 0 \\ 0 & 0 & \frac{1}{\sqrt{2}} & \frac{1}{\sqrt{2}} \\ 0 & 0 & \frac{1}{\sqrt{2}} & -\frac{1}{\sqrt{2}} \end{pmatrix}$$

In order to see the action of these pulse operators, let us consider the state $|00\rangle$ of the computational basis:

$$|00\rangle = \begin{pmatrix} 1 \\ 0 \\ 0 \\ 0 \end{pmatrix}; \quad |01\rangle = \begin{pmatrix} 0 \\ 1 \\ 0 \\ 0 \end{pmatrix}; \quad |10\rangle = \begin{pmatrix} 0 \\ 0 \\ 1 \\ 0 \end{pmatrix}; \quad |11\rangle = \begin{pmatrix} 0 \\ 0 \\ 0 \\ 1 \end{pmatrix}$$

Hence,

$$NOT_A |00\rangle = \begin{pmatrix} 0 & 0 & -1 & 0 \\ 0 & 0 & 0 & -1 \\ -1 & 0 & 0 & 0 \\ 0 & -1 & 0 & 0 \end{pmatrix}\begin{pmatrix} 1 \\ 0 \\ 0 \\ 0 \end{pmatrix} = -\begin{pmatrix} 0 \\ 0 \\ 1 \\ 0 \end{pmatrix} = -|10\rangle$$

$$NOT_B |00\rangle = \begin{pmatrix} 0 & i & 0 & 0 \\ i & 0 & 0 & 0 \\ 0 & 0 & 0 & i \\ 0 & 0 & i & 0 \end{pmatrix} \begin{pmatrix} 1 \\ 0 \\ 0 \\ 0 \end{pmatrix} = -\begin{pmatrix} 0 \\ -i \\ 0 \\ 0 \end{pmatrix} = i|01\rangle$$

$$H_A^{|00\rangle,|01\rangle} |00\rangle = \begin{pmatrix} -\frac{1}{\sqrt{2}} & 0 & -\frac{1}{\sqrt{2}} & 0 \\ 0 & -1 & 0 & 0 \\ -\frac{1}{\sqrt{2}} & 0 & \frac{1}{\sqrt{2}} & 0 \\ 0 & 0 & 0 & 1 \end{pmatrix} \begin{pmatrix} 1 \\ 0 \\ 0 \\ 0 \end{pmatrix} = -\frac{1}{\sqrt{2}} \begin{pmatrix} 1 \\ 0 \\ 1 \\ 0 \end{pmatrix} = -\frac{|00\rangle + |10\rangle}{\sqrt{2}}$$

$$H_B^{|00\rangle,|01\rangle} |00\rangle = \begin{pmatrix} 1 & 0 & 0 & 0 \\ 0 & 1 & 0 & 0 \\ 0 & 0 & \frac{1}{\sqrt{2}} & \frac{1}{\sqrt{2}} \\ 0 & 0 & \frac{1}{\sqrt{2}} & -\frac{1}{\sqrt{2}} \end{pmatrix} \begin{pmatrix} 1 \\ 0 \\ 0 \\ 0 \end{pmatrix} = -\frac{i}{\sqrt{2}} \begin{pmatrix} 1 \\ 1 \\ 0 \\ 0 \end{pmatrix} = -i \frac{|00\rangle + |01\rangle}{\sqrt{2}}$$

Where, regardless a global phase, we see that the operators act as expected. To exemplify the reversibility of the Hadamard gates let us apply them twice to the state $|00\rangle$:

$$H_A^{|00\rangle,|01\rangle} H_A^{|00\rangle,|01\rangle} |00\rangle = H_A^{|00\rangle,|01\rangle} \left(-\frac{|00\rangle + |10\rangle}{\sqrt{2}} \right)$$

$$= -\frac{1}{\sqrt{2}} \begin{pmatrix} -\frac{1}{\sqrt{2}} & 0 & -\frac{1}{\sqrt{2}} & 0 \\ 0 & -1 & 0 & 0 \\ -\frac{1}{\sqrt{2}} & 0 & \frac{1}{\sqrt{2}} & 0 \\ 0 & 0 & 0 & 1 \end{pmatrix} \begin{pmatrix} 1 \\ 0 \\ 1 \\ 0 \end{pmatrix} = \begin{pmatrix} 1 \\ 0 \\ 0 \\ 0 \end{pmatrix} = |00\rangle$$

$$H_B^{|00\rangle,|01\rangle} H_B^{|00\rangle,|01\rangle} |00\rangle = H_B^{|00\rangle,|01\rangle} \left(-i \frac{|00\rangle + |01\rangle}{\sqrt{2}} \right)$$

$$= -\frac{i}{\sqrt{2}} \begin{pmatrix} 1 & 0 & 0 & 0 \\ 0 & 1 & 0 & 0 \\ 0 & 0 & \frac{1}{\sqrt{2}} & \frac{1}{\sqrt{2}} \\ 0 & 0 & \frac{1}{\sqrt{2}} & -\frac{1}{\sqrt{2}} \end{pmatrix} \begin{pmatrix} 1 \\ 1 \\ 0 \\ 0 \end{pmatrix} = -\begin{pmatrix} 1 \\ 0 \\ 0 \\ 0 \end{pmatrix} = -|00\rangle$$

In both case the initial state $|00\rangle$ is recovered, regardless a global phase.

P4.2 – Use the proper pulse and evolution operators to show (4.2.12).

Solution
The relevant pulse operators matrices are:

$$(\pi)_y^{I_2} = \begin{pmatrix} 0 & 0 & -1 & 0 \\ 0 & 0 & 0 & -1 \\ 1 & 0 & 0 & 0 \\ 0 & 1 & 0 & 0 \end{pmatrix} \quad (\pi)_{-y}^{I_2} = \begin{pmatrix} 0 & 0 & 1 & 0 \\ 0 & 0 & 0 & 1 \\ -1 & 0 & 0 & 0 \\ 0 & -1 & 0 & 0 \end{pmatrix}$$

$$U_J(t) = \begin{pmatrix} e^{-i\pi Jt/2} & 0 & 0 & 0 \\ 0 & e^{+i\pi Jt/2} & 0 & 0 \\ 0 & 0 & e^{+i\pi Jt/2} & 0 \\ 0 & 0 & 0 & e^{-i\pi Jt/2} \end{pmatrix}$$

where we drop the chemical shift part that is obviously refocused. Using these matrices we first calculate the result of the following operation:

$$(\pi)_y^{I_2} U_J(t)(\pi)_{-x}^{I_2} = \begin{pmatrix} 0 & 0 & -1 & 0 \\ 0 & 0 & 0 & -1 \\ 1 & 0 & 0 & 0 \\ 0 & 1 & 0 & 0 \end{pmatrix} \begin{pmatrix} e^{-i\pi Jt/2} & 0 & 0 & 0 \\ 0 & e^{+i\pi Jt/2} & 0 & 0 \\ 0 & 0 & e^{+i\pi Jt/2} & 0 \\ 0 & 0 & 0 & e^{-i\pi Jt/2} \end{pmatrix}$$

$$\times \begin{pmatrix} 0 & 0 & 1 & 0 \\ 0 & 0 & 0 & 1 \\ -1 & 0 & 0 & 0 \\ 0 & -1 & 0 & 0 \end{pmatrix}$$

$$= \begin{pmatrix} e^{+i\pi Jt/2} & 0 & 0 & 0 \\ 0 & e^{-i\pi Jt/2} & 0 & 0 \\ 0 & 0 & e^{-i\pi Jt/2} & 0 \\ 0 & 0 & 0 & e^{+i\pi Jt/2} \end{pmatrix} = U_J(-t)$$

Then, the operator corresponding to the refocusing pulse sequence becomes:

$$U_J(t)(\pi)_y^{I_2} U_J(t)(\pi)_{-y}^{I_2} = U_J(t) U_J(-t) = \begin{pmatrix} 1 & 0 & 0 & 0 \\ 0 & 1 & 0 & 0 \\ 0 & 0 & 1 & 0 \\ 0 & 0 & 0 & 1 \end{pmatrix} = \mathbf{1}$$

which shows that there is no J-coupling evolution after the application of the refocusing pulse sequence.

P4.3 – Derive the operator matrix for the SWAP and EPR generator gates for quadrupolar spins 3/2.

Solution
The matrices for the CNOT gates for quadrupolar spin 3/2 system describe in the text are:

$$CNOT_A = (\pi)_x^{23} = \begin{pmatrix} 1 & 0 & 0 & 0 \\ 0 & 1 & 0 & 0 \\ 0 & 0 & 0 & i \\ 0 & 0 & i & 0 \end{pmatrix}$$

$$CNOT_B = (\pi)_x^{12} (\pi)_x^{13} (\pi)_x^{12} = \begin{pmatrix} 1 & 0 & 0 & 0 \\ 0 & 0 & 0 & -1 \\ 0 & 0 & -1 & 0 \\ 0 & -1 & 0 & 0 \end{pmatrix}$$

Using these matrix and the corresponding Hadamard matrices of Exercise P4.1.

$$SWAP = CNOT_A CNOT_B CNOT_A = -\frac{1+i}{\sqrt{2}} \begin{pmatrix} 1 & 0 & 0 & 0 \\ 0 & 0 & 1 & 0 \\ 0 & 1 & 0 & 0 \\ 0 & 0 & 0 & 1 \end{pmatrix}$$

$$EPR^{|00\rangle,|01\rangle} = CNOT_A H_A^{|00\rangle,|01\rangle} = -\frac{1-i}{\sqrt{2}} \begin{pmatrix} 1 & 0 & 1 & 0 \\ 0 & 1 & 1 & 0 \\ 0 & 0 & 0 & 1 \\ 1 & 0 & 1 & 0 \end{pmatrix}$$

$$EPR^{|10\rangle,|11\rangle} = CNOT_A H_A^{|10\rangle,|11\rangle} = -\frac{1-i}{\sqrt{2}} \begin{pmatrix} 1 & 0 & 0 & 0 \\ 0 & 1 & 0 & 1 \\ 0 & 1 & 0 & 1 \\ 0 & 0 & 1 & 0 \end{pmatrix}$$

P4.4 – Show that the operator (4.2.18) for a EPR generator in systems of two coupled spins 1/2 produces the Bell basis states from the computational basis states.

Solution
The EPR generator operator given is Section 4.2.2 is:

$$EPR = -\frac{1+i}{\sqrt{2}} \begin{pmatrix} 1 & 0 & 1 & 0 \\ 0 & 1 & 1 & 0 \\ 0 & 1 & 0 & 1 \\ 1 & 0 & 1 & 0 \end{pmatrix}$$

The result of the application of this operation to the computation basis states becomes:

$$EPR|00\rangle = -\frac{1+i}{\sqrt{2}} \begin{pmatrix} 1 & 0 & 1 & 0 \\ 0 & 1 & 1 & 0 \\ 0 & 1 & 0 & 1 \\ 1 & 0 & 1 & 0 \end{pmatrix} \begin{pmatrix} 1 \\ 0 \\ 0 \\ 0 \end{pmatrix} = -\frac{1+i}{\sqrt{2}} \begin{pmatrix} 1 \\ 0 \\ 0 \\ 1 \end{pmatrix} = -\frac{1+i}{\sqrt{2}}(|00\rangle + |11\rangle)$$

$$EPR|01\rangle = -\frac{1+i}{\sqrt{2}} \begin{pmatrix} 1 & 0 & 1 & 0 \\ 0 & 1 & 1 & 0 \\ 0 & 1 & 0 & 1 \\ 1 & 0 & 1 & 0 \end{pmatrix} \begin{pmatrix} 0 \\ 1 \\ 0 \\ 0 \end{pmatrix} = -\frac{1+i}{\sqrt{2}} \begin{pmatrix} 0 \\ 1 \\ 1 \\ 0 \end{pmatrix} = -\frac{1+i}{\sqrt{2}}(|01\rangle + |10\rangle)$$

$$EPR|10\rangle = -\frac{1+i}{\sqrt{2}} \begin{pmatrix} 1 & 0 & 1 & 0 \\ 0 & 1 & 1 & 0 \\ 0 & 1 & 0 & 1 \\ 1 & 0 & 1 & 0 \end{pmatrix} \begin{pmatrix} 0 \\ 0 \\ 1 \\ 0 \end{pmatrix} = -\frac{1+i}{\sqrt{2}} \begin{pmatrix} 1 \\ 0 \\ 0 \\ -1 \end{pmatrix} = -\frac{1+i}{\sqrt{2}}(|00\rangle - |11\rangle)$$

$$EPR|11\rangle = -\frac{1+i}{\sqrt{2}} \begin{pmatrix} 1 & 0 & 1 & 0 \\ 0 & 1 & 1 & 0 \\ 0 & 1 & 0 & 1 \\ 1 & 0 & 1 & 0 \end{pmatrix} \begin{pmatrix} 0 \\ 0 \\ 0 \\ 1 \end{pmatrix} = -\frac{1+i}{\sqrt{2}} \begin{pmatrix} 0 \\ 1 \\ -1 \\ 0 \end{pmatrix} = -\frac{1+i}{\sqrt{2}}(|01\rangle - |10\rangle)$$

P4.5 – Exemplify the action of the three-qubit Toffoli gate of 4.2.19 in some selected computational basis states.

Solution
Taking the operator that represents the Toffoli gate with the first qubit as the target:

$$Toffoli = e^{i3\pi/8} \begin{pmatrix} 1 & 0 & 0 & 0 & 0 & 0 & 0 & 0 \\ 0 & i & 0 & 0 & 0 & 0 & 0 & 0 \\ 0 & 0 & i & 0 & 0 & 0 & 0 & 0 \\ 0 & 0 & 0 & 0 & 0 & 0 & 0 & -1 \\ 0 & 0 & 0 & 0 & i & 0 & 0 & 0 \\ 0 & 0 & 0 & 0 & 0 & -i & 0 & 0 \\ 0 & 0 & 0 & 0 & 0 & 0 & -i & 0 \\ 0 & 0 & 0 & i & 0 & 0 & 0 & 0 \end{pmatrix}$$

and applying to the three qubit states:

$$|001\rangle = \begin{pmatrix} 0 \\ 1 \\ 0 \\ 0 \\ 0 \\ 0 \\ 0 \\ 0 \end{pmatrix} ; \quad |010\rangle = \begin{pmatrix} 0 \\ 0 \\ 1 \\ 0 \\ 0 \\ 0 \\ 0 \\ 0 \end{pmatrix} ; \quad |011\rangle = \begin{pmatrix} 0 \\ 0 \\ 0 \\ 1 \\ 0 \\ 0 \\ 0 \\ 0 \end{pmatrix} ; \quad |111\rangle = \begin{pmatrix} 0 \\ 0 \\ 0 \\ 0 \\ 0 \\ 0 \\ 0 \\ 1 \end{pmatrix}$$

we obtain:

$$Toffoli|001\rangle = e^{i3\pi/8} \begin{pmatrix} 1 & 0 & 0 & 0 & 0 & 0 & 0 & 0 \\ 0 & i & 0 & 0 & 0 & 0 & 0 & 0 \\ 0 & 0 & i & 0 & 0 & 0 & 0 & 0 \\ 0 & 0 & 0 & 0 & 0 & 0 & 0 & -1 \\ 0 & 0 & 0 & 0 & i & 0 & 0 & 0 \\ 0 & 0 & 0 & 0 & 0 & -i & 0 & 0 \\ 0 & 0 & 0 & 0 & 0 & 0 & -i & 0 \\ 0 & 0 & 0 & i & 0 & 0 & 0 & 0 \end{pmatrix} \begin{pmatrix} 0 \\ 1 \\ 0 \\ 0 \\ 0 \\ 0 \\ 0 \\ 0 \end{pmatrix} = e^{i7\pi/8} \begin{pmatrix} 0 \\ 1 \\ 0 \\ 0 \\ 0 \\ 0 \\ 0 \\ 0 \end{pmatrix}$$

$$= e^{i3\pi/8}|001\rangle$$

$$Toffoli|010\rangle = e^{i3\pi/8} \begin{pmatrix} 1 & 0 & 0 & 0 & 0 & 0 & 0 & 0 \\ 0 & i & 0 & 0 & 0 & 0 & 0 & 0 \\ 0 & 0 & i & 0 & 0 & 0 & 0 & 0 \\ 0 & 0 & 0 & 0 & 0 & 0 & 0 & -1 \\ 0 & 0 & 0 & 0 & i & 0 & 0 & 0 \\ 0 & 0 & 0 & 0 & 0 & -i & 0 & 0 \\ 0 & 0 & 0 & 0 & 0 & 0 & -i & 0 \\ 0 & 0 & 0 & i & 0 & 0 & 0 & 0 \end{pmatrix} \begin{pmatrix} 0 \\ 0 \\ 1 \\ 0 \\ 0 \\ 0 \\ 0 \\ 0 \end{pmatrix} = e^{i7\pi/8} \begin{pmatrix} 0 \\ 0 \\ 1 \\ 0 \\ 0 \\ 0 \\ 0 \\ 0 \end{pmatrix}$$

$$= e^{i3\pi/8}|010\rangle$$

$$Toffoli|011\rangle = e^{i3\pi/8} \begin{pmatrix} 1 & 0 & 0 & 0 & 0 & 0 & 0 & 0 \\ 0 & i & 0 & 0 & 0 & 0 & 0 & 0 \\ 0 & 0 & i & 0 & 0 & 0 & 0 & 0 \\ 0 & 0 & 0 & 0 & 0 & 0 & 0 & -1 \\ 0 & 0 & 0 & 0 & i & 0 & 0 & 0 \\ 0 & 0 & 0 & 0 & 0 & -i & 0 & 0 \\ 0 & 0 & 0 & 0 & 0 & 0 & -i & 0 \\ 0 & 0 & 0 & i & 0 & 0 & 0 & 0 \end{pmatrix} \begin{pmatrix} 0 \\ 0 \\ 0 \\ 1 \\ 0 \\ 0 \\ 0 \\ 0 \end{pmatrix} = e^{i7\pi/8} \begin{pmatrix} 0 \\ 0 \\ 0 \\ 0 \\ 0 \\ 0 \\ 0 \\ 1 \end{pmatrix}$$

$$= e^{i3\pi/8}|111\rangle$$

$$Toffoli|111\rangle = e^{i3\pi/8} \begin{pmatrix} 1 & 0 & 0 & 0 & 0 & 0 & 0 & 0 \\ 0 & i & 0 & 0 & 0 & 0 & 0 & 0 \\ 0 & 0 & i & 0 & 0 & 0 & 0 & 0 \\ 0 & 0 & 0 & 0 & 0 & 0 & 0 & -1 \\ 0 & 0 & 0 & 0 & i & 0 & 0 & 0 \\ 0 & 0 & 0 & 0 & 0 & -i & 0 & 0 \\ 0 & 0 & 0 & 0 & 0 & 0 & -i & 0 \\ 0 & 0 & 0 & i & 0 & 0 & 0 & 0 \end{pmatrix} \begin{pmatrix} 0 \\ 0 \\ 0 \\ 0 \\ 0 \\ 0 \\ 0 \\ 1 \end{pmatrix} = e^{i11\pi/8} \begin{pmatrix} 0 \\ 0 \\ 0 \\ 1 \\ 0 \\ 0 \\ 0 \\ 0 \end{pmatrix}$$

$$= e^{i11\pi/8}|011\rangle$$

where we see that unless for a global phase the gate invert the first qubit only with the other two are in the state 1.

P4.6 – Show that the operation described in (4.3.17) applied to the equilibrium density matrix produces the density matrix corresponding to the state $|00\rangle$.

Solution
The full operation we want to show is:

$$\Delta\rho_{00} = \left[G_z(\tau)\left(\frac{\pi}{4}\right)^{I_1}_{-y} U_J\left(\frac{1}{2J}\right)\left(\frac{\pi}{4}\right)^{I_1}_{x} G_z(\tau)\left(\frac{\pi}{3}\right)^{I_2}_{x}\right]\Delta\rho_{eq}$$

$$\times \left[G_z(\tau)\left(\frac{\pi}{4}\right)^{I_1}_{-y} U_J\left(\frac{1}{2J}\right)\left(\frac{\pi}{4}\right)^{I_1}_{x} G_z(\tau)\left(\frac{\pi}{3}\right)^{I_2}_{x}\right]^\dagger$$

To show that we will need to use the pulse matrices corresponding to the pulses. For a two spin system this can be obtained from:

$$(\theta)^{I_1}_\Phi = \exp(-i\theta I_{1\Phi}); \qquad (\theta)^{I_2}_\Phi = \exp(-i\theta I_{2\Phi})$$

where $I_{1\Phi} = I_\Phi \otimes 1 \otimes 1$; $I_{2\Phi} = 1 \otimes I_\Phi \otimes 1$; $I_{3\Phi} = 1 \otimes 1 \otimes I_\Phi$ with $\Phi = x, y, z, -x, -y, -z$ and I_Φ is a single spin operator. The high temperature equilibrium deviation density matrix becomes:

$$\Delta\rho_{eq} = \frac{\hbar\omega_0}{4k_BT}(\underbrace{I_z \otimes 1}_{I_z^1} + \underbrace{1 \otimes I_z}_{I_z^2}) = \frac{\hbar\omega_0}{4kT}\begin{pmatrix} 1 & 0 & 0 & 0 \\ 0 & 0 & 0 & 0 \\ 0 & 0 & 0 & 0 \\ 0 & 0 & 0 & -1 \end{pmatrix}$$

After the application of the first $\left(\frac{\pi}{3}\right)^{I_2}_x$ operation:

$$\Delta\rho_1 = \frac{\hbar\omega_0}{4k_BT}\left[\left(\frac{\pi}{3}\right)^{I_2}_x\right]\Delta\rho_{eq}\left[\left(\frac{\pi}{3}\right)^{I_2}_x\right]^\dagger = \begin{pmatrix} 0.75 & i0.433 & 0 & 0 \\ -i0.433 & 0.25 & 0 & 0 \\ 0 & 0 & -0.25 & i0.433 \\ 0 & 0 & -i0.433 & -0.75 \end{pmatrix}$$

The application of the $G_z(\tau)$ gradient pulse kills all the off-diagonal coherences taking the deviation density matrix to:

$$\Delta\rho_2 = \frac{\hbar\omega_0}{4kT}\begin{pmatrix} 0.75 & 0 & 0 & 0 \\ 0 & 0.25 & 0 & 0 \\ 0 & 0 & -0.25 & 0 \\ 0 & 0 & 0 & -0.75 \end{pmatrix}$$

After the application of the first $\left(\frac{\pi}{4}\right)^{I_1}_x$ operation:

$$\Delta\rho_3 = \frac{\hbar\omega_0}{4k_BT}\left[\left(\frac{\pi}{4}\right)^{I_4}_x\right]\Delta\rho_2\left[\left(\frac{\pi}{4}\right)^{I_4}_x\right]^\dagger = \frac{\hbar\omega_0}{4kT}\begin{pmatrix} 0.6036 & 0 & i0.3536 & 0 \\ 0 & 0.1036 & 0 & i0.3536 \\ 0 & -i0.3536 & -0.1036 & 0 \\ 0 & 0 & -i0.3536 & -0.6036 \end{pmatrix}$$

After the application of the $U_J\left(\frac{1}{2J}\right)$ evolution:

$$\Delta\rho_4 = \frac{\hbar\omega_0}{4k_BT}\left[U_J\left(\frac{1}{2J}\right)\right]\Delta\rho_3\left[U_J\left(\frac{1}{2J}\right)\right]^\dagger = \frac{\hbar\omega_0}{4kT}\begin{pmatrix} 0.6036 & 0 & 0.3536 & 0 \\ 0 & 0.1036 & 0 & i-0.3536 \\ 0 & 0.3536 & -0.1036 & 0 \\ 0 & 0 & -0.3536 & -0.6036 \end{pmatrix}$$

After the application of the second $\left(\frac{\pi}{4}\right)^{I_1}_x$ operation:

$$\Delta\rho_5 = \frac{\hbar\omega_0}{4k_BT}\left[\left(\frac{\pi}{4}\right)^{I_4}_x\right]\Delta\rho_4\left[\left(\frac{\pi}{4}\right)^{I_4}_x\right]^\dagger = \frac{\hbar\omega_0}{4kT}\begin{pmatrix} 0.75 & 0 & 0 & 0 \\ 0 & -0.25 & 0 & -0.5 \\ 0 & 0 & -0.25 & 0 \\ 0 & -0.5 & 0 & -0.25 \end{pmatrix}$$

Then, after the application of the second $G_z(\tau)$ gradient pulse we obtain the deviation density matrix corresponding to the state $|00\rangle$:

$$\Delta\rho_{00} = \frac{\hbar\omega_0}{4kT}\begin{pmatrix} 0.75 & 0 & 0 & 0 \\ 0 & -0.25 & 0 & 0 \\ 0 & 0 & -0.25 & 0 \\ 0 & 0 & 0 & -0.25 \end{pmatrix}$$

P4.7 – Using the CNOT operations described in Section 4.3.3, derive the expression for the pseudo-pure deviation density matrix (4.3.21) starting from (4.3.19).

Solution

Let us take the high temperature deviation density matrix for this three spin system as follow:

$$\Delta\rho_{eq} = \frac{\hbar\omega_0}{8k_BT}(\underbrace{I_z \otimes \mathbf{1} \otimes \mathbf{1}}_{I_z^1} + \underbrace{\mathbf{1} \otimes I_z \otimes \mathbf{1}}_{I_z^2} + \underbrace{\mathbf{1} \otimes \mathbf{1} \otimes I_z}_{I_z^3})$$

$$= \frac{\hbar\omega_0}{16k_BT}\begin{pmatrix} 3 & 0 & 0 & 0 & 0 & 0 & 0 & 0 \\ 0 & 1 & 0 & 0 & 0 & 0 & 0 & 0 \\ 0 & 0 & 1 & 0 & 0 & 0 & 0 & 0 \\ 0 & 0 & 0 & -1 & 0 & 0 & 0 & 0 \\ 0 & 0 & 0 & 0 & 1 & 0 & 0 & 0 \\ 0 & 0 & 0 & 0 & 0 & -1 & 0 & 0 \\ 0 & 0 & 0 & 0 & 0 & 0 & -1 & 0 \\ 0 & 0 & 0 & 0 & 0 & 0 & 0 & -3 \end{pmatrix}$$

where I_z is a single spin operator and $\mathbf{1}$ is a two by two identity matrix. As discussed in the text the labeled pseudo-pure state is obtained by applying two successive CNOT gates both having the first qubit as target and control in the second and third qubits, respectively. To construct the matrix representation of such CNOT gates from the pulse operator we may find the expression for the pulse matrix for the three qubit system. For a general rotation angle this can be written as:

$$(\theta)_\Phi^{I_1} = \exp(-i\theta I_{1\Phi}); \qquad (\theta)_\Phi^{I_2} = \exp(-i\theta I_{2\Phi}); \qquad (\theta)_\Phi^{I_3} = \exp(-i\theta I_{3\Phi})$$

where $I_{1\Phi} = I_\Phi \otimes \mathbf{1} \otimes \mathbf{1}$; $I_{2\Phi} = \mathbf{1} \otimes I_\Phi \otimes \mathbf{1}$; $I_{3\Phi} = \mathbf{1} \otimes \mathbf{1} \otimes I_\Phi$ with $\Phi = x, y, -x, -y$ and I_Φ is a single spin operator. With these definitions the operator that represent these two CNOT gates can be written in terms of the sequence of pulse operator that implement two qubit CNOT gates as:

$$CNOT_{1-2} = \left(\frac{\pi}{2}\right)_z^{I_2}\left(-\frac{\pi}{2}\right)_z^{I_1}\left(\frac{\pi}{2}\right)_x^{I_1} U_{J_{12}}\left(\frac{1}{2J_{12}}\right)\left(\frac{\pi}{2}\right)_y^{I_1}$$

$$= \frac{1-i}{\sqrt{2}}\begin{pmatrix} 1 & 0 & 0 & 0 & 0 & 0 & 0 & 0 \\ 0 & 1 & 0 & 0 & 0 & 0 & 0 & 0 \\ 0 & 0 & 0 & 0 & 0 & 0 & 1 & 0 \\ 0 & 0 & 0 & 0 & 0 & 0 & 0 & 1 \\ 0 & 0 & 0 & 0 & 1 & 0 & 0 & 0 \\ 0 & 1 & 0 & 0 & 0 & 0 & 0 & 0 \\ 0 & 0 & 1 & 0 & 0 & 0 & 0 & 0 \\ 0 & 0 & 0 & 1 & 0 & 0 & 0 & 0 \end{pmatrix}$$

$$CNOT_{1-3} = \left(\frac{\pi}{2}\right)_z^{I_3}\left(-\frac{\pi}{2}\right)_z^{I_1}\left(\frac{\pi}{2}\right)_x^{I_1}U_{J_{13}}\left(\frac{1}{2J_{13}}\right)\left(\frac{\pi}{2}\right)_y^{I_1}$$

$$= \frac{1-i}{\sqrt{2}}\begin{pmatrix} 1 & 0 & 0 & 0 & 0 & 0 & 0 & 0 \\ 0 & 0 & 0 & 0 & 0 & 1 & 0 & 0 \\ 0 & 0 & 1 & 0 & 0 & 0 & 0 & 0 \\ 0 & 0 & 0 & 0 & 0 & 0 & 0 & 1 \\ 0 & 0 & 0 & 0 & 1 & 0 & 0 & 0 \\ 0 & 1 & 0 & 0 & 0 & 0 & 0 & 0 \\ 0 & 0 & 0 & 0 & 0 & 0 & 1 & 0 \\ 0 & 0 & 0 & 1 & 0 & 0 & 0 & 0 \end{pmatrix}$$

Note that the implementation of $U_{J_{12}}\left(\frac{1}{2J_{12}}\right)$ involves the application of a refocusing pulse sequence that refocus the J-coupling between the nuclei 1 and 3 and 2 and 3, but keep the coupling between 1 and 2. A similar scheme must be applied to produce $U_{J_{13}}\left(\frac{1}{2J_{13}}\right)$. Thus,

$$U_{l\text{-}pps} = CNOT_{1-2}CNOT_{1-3} = \begin{pmatrix} -i & 0 & 0 & 0 & 0 & 0 & 0 & 0 \\ 0 & 0 & 0 & 0 & 0 & -i & 0 & 0 \\ 0 & 0 & 0 & 0 & 0 & 0 & -i & 0 \\ 0 & 0 & 0 & -i & 0 & 0 & 0 & 0 \\ 0 & 0 & 0 & 0 & -i & 0 & 0 & 0 \\ 0 & -i & 0 & 0 & 0 & 0 & 0 & 0 \\ 0 & 0 & -i & 0 & 0 & 0 & 1 & 0 \\ 0 & 0 & 0 & 0 & 0 & 0 & 0 & -i \end{pmatrix}$$

Using such matrix we can obtain the labeled pseudo-pure state from above equilibrium density matrix as:

$$\Delta\rho_{l\text{-}pps} = U_{l\text{-}pps}\Delta\rho_{eq}U_{l\text{-}pps}^\dagger = \frac{\hbar\omega_0}{16k_BT}\begin{pmatrix} 3 & 0 & 0 & 0 & 0 & 0 & 0 & 0 \\ 0 & -1 & 0 & 0 & 0 & 0 & 0 & 0 \\ 0 & 0 & -1 & 0 & 0 & 0 & 0 & 0 \\ 0 & 0 & 0 & -1 & 0 & 0 & 0 & 0 \\ 0 & 0 & 0 & 0 & 1 & 0 & 0 & 0 \\ 0 & 0 & 0 & 0 & 0 & 1 & 0 & 0 \\ 0 & 0 & 0 & 0 & 0 & 0 & 1 & 0 \\ 0 & 0 & 0 & 0 & 0 & 0 & 0 & -3 \end{pmatrix}$$

P4.8 – Considering the general density matrix shown in Equation (4.4.2) show that the elements $x_{12} + iy_{12}$ and $x_{34} + iy_{34}$ can be obtained from the NMR spectrum of the I_1 nucleus while the elements $x_{13} + iy_{13}$ and $x_{24} + iy_{24}$ can be obtained from the spectrum of the I_2 nucleus.

Solution

Let us first consider that the density matrix during the acquisition periods can be represented by the general deviation density matrix of Equation (4.4.2):

$$\Delta\rho = \begin{pmatrix} x_{11} & x_{12}+iy_{12} & x_{13}+iy_{13} & x_{14}+iy_{14} \\ x_{12}-iy_{12} & x_{22} & x_{23}+iy_{23} & x_{24}+iy_{24} \\ x_{13}-iy_{13} & x_{23}-iy_{23} & x_{33} & x_{34}+iy_{34} \\ x_{14}-iy_{14} & x_{24}-iy_{24} & x_{34}-iy_{34} & x_{44} \end{pmatrix}$$

The corresponding NMR magnetization can be written as:

$$M(t) = \text{Tr}\left[U(t)\Delta\rho U^\dagger(t)I_+\right]$$

where $U(t)$ is the evolution operator during the signal detection and I_+ is the total raising operator of the two spins system. These operators are giving by:

$$U(t) = \exp\left[-i\left(\omega_0 I_z^1 + 2\pi J I_z^1 I_z^2\right)\right]$$

$$I_+ = \left(I_x^1 + i I_y^1\right) + \left(I_x^2 + i I_y^2\right) = \begin{pmatrix} 0 & 1 & 1 & 0 \\ 0 & 0 & 0 & 1 \\ 0 & 0 & 0 & 1 \\ 0 & 0 & 0 & 0 \end{pmatrix}$$

The operators I_x^i, I_y^i, I_z^i are those defined in Problem 4.6. The expression can be explicitly written as:

$$M(t) = \sum_{pq}\left(e^{-i\omega_{pq}t}[I_+]_{pq}[\Delta\rho_{pq}]^\dagger\right)$$

where ω_{pq} is the transition frequency defined by the system Hamiltonian (see operator $U(t)$). From the above sum we see that only the $\Delta\rho$ entries corresponding to the non-vanish entries of I_+ contributes to the NMR signal, i.e. the 12, 13 24 and 34 positions. Comparing to the general expression of $\Delta\rho$ we see that at these position we find the elements $x_{12}+iy_{12}$, $x_{13}+iy_{13}$, $x_{24}+iy_{24}$, $x_{34}+iy_{34}$. Since the transition 12 and 24 are associated to spin $\mathbf{I_1}$ while the transition 13 and 34 are associated to the spin $\mathbf{I_2}$ we can conclude that the elements $x_{12}+iy_{12}$ and $x_{34}+iy_{34}$ can be obtained from the NMR spectrum of the $\mathbf{I_1}$ nucleus while the elements $x_{13}+iy_{13}$ and $x_{24}+iy_{24}$ can be obtained from the spectrum of the $\mathbf{I_2}$ nucleus.

REFERENCES

[1] M.A. Nielsen, I.L. Chuang, *Quantum Computing and Quantum Information* (Cambridge University Press, Cambridge, 2000).
[2] R.R. Ernst, G. Bodenhausen, A. Wokaun, *Principles of Nuclear Magnetic Resonance in One and Two Dimensions* (Clarendon Press, Oxford, 1987).
[3] L.M.K. Vandersypen, M. Steffen, G. Breyta, C.S. Yannoni, M.H. Sherwood, I.L. Chuang, Experimental realization of Shor's quantum factoring algorithm using nuclear magnetic resonance, *Nature* **414** (2001) 883–887.
[4] A.K. Khitrin, B.M. Fung, Nuclear magnetic resonance quantum logic gates using quadrupolar nuclei, *J. Chem. Phys.* **22** (2000) 6963–6965.
[5] N. Sinha, T.S. Mahesh, K.V. Ramanathan, A. Kumar, Toward quantum information processing by nuclear magnetic resonance: Pseudopure states and logical operations using selective pulses on an oriented spin 3/2 nucleus, *J. Chem. Phys.* **14** (2001) 4415–4420.
[6] A.K. Khitrin, H. Sun, B.M. Fung, Method of multi-frequency excitation for creating pseudopure states for NMR quantum computing, *Phys. Rev. A* **63** (2) (2001) Art. No. 020301.
[7] R.S. Sarthour, E.R. de Azevedo, F.A. Bonk, E.L.G. Vidoto, T.J. Bonagamba, A.P. Gumarães, J.C.C. Freitas, I.S. Oliveira, Relaxation of coherent states in a two-qubit NMR quadrupole system, *Phys. Rev. A* **68** (2003) 022311.
[8] H. Kampermann, W.S. Veeman, Quantum computing using quadrupolar spins in solid state NMR, *Quantum Inform. Process.* **1** (5) (2002) 327–344.
[9] J. Baugh, O. Moussa, C.A. Ryan, A. Nayak, R. Laflamme, Experimental implementation of heat-bath algorithmic cooling using solid-state nuclear magnetic resonance, *Nature* **438** (2005) 470–473.
[10] R. Das, T.S. Mahesh, A. Kumar, Implementation of conditional phase-shift gate for quantum information processing by NMR, using transition-selective pulses, *J. Magn. Reson.* **159** (1) (2002) 46–54.
[11] J. Stolze, D. Suter, *Quantum Computing* (WILEY-VCH GmbH, Germany, 2004).
[12] R. Freeman, Shaped radiofrequency pulses in high resolution NMR, *Prog. Nucl. Magn. Reson. Spectrosc.* **32** (1998) 59–106.
[13] T.S. Mahesh, N. Sinha, K.V. Ramanathan, A. Kumar, Ensemble quantum-information processing by NMR: Implementation of gates and the creation of pseudopure states using dipolar coupled spins as qubits, *Phys. Rev. A* **65** (2002) Art. No. 022312.
[14] K. Doray, A. Kumar, Cascate selective pulses on connected single quantum transitions leading to selective excitation of multiple quantum coherences, *J. Magn. Reson. Ser. A* **114** (1995) 155–162.

References

[15] L.M.K. Vandersypen, I.L. Chuang, NMR techniques for quantum control and computation, *Rev. Mod. Phys.* **76** (4) (2004) 1037–1069.

[16] D. Deustch, Quantum computational networks, *Proc. R. Soc. Lond. A* **425** (1989) 73–80.

[17] M.D. Price, T.F. Havel, D.G. Cory, Multi-qubit logic gates in NMR quantum computing, *New Journal of Physics* **2** (2000) 101–109.

[18] M.D. Price, S.S. Somaroo, A.E. Dunlop, T.F. Havel, D.G. Cory, Generalized methods for the development of quantum logical gates for an NMR quantum information processor, *Phys. Rev. A* **60** (1999) 2777–2780.

[19] D.G. Cory, M.D. Price, T.F. Havel, Nuclear magnetic resonance spectroscopy: An experimentally accessible paradigm for quantum computing, *Physica D* **120** (1988) 82–101.

[20] L.M.K. Vandersypen, M. Steffen, G. Breyta, C.S. Yannoni, M.H. Sherwood, I.L. Chuang, NMR techniques for quantum control and computation, *Rev. Mod. Phys.* **76** (4) (2004) 1037–1069.

[21] E.M. Fortunato, M.A. Pravia, N. Boulant, et al., Design of strongly modulating pulses to implement precise effective Hamiltonians for quantum information processing, *J. Chem. Phys.* **116** (17) (2002) 7599–7606.

[22] H. Kampermann, W.S. Veeman, Characterization of quantum algorithms by quantum process tomography using quadrupolar spins in solid-state nuclear magnetic resonance, *J. Chem. Phys.* **122** (21) (2005) Art. No. 214108.

[23] N.A. Gershenfeld, I.L. Chuang, Bulk spin-resonance quantum computation, *Science* **275** (1997) 350–356.

[24] D.G. Cory, A.F. Fahmy, T.F. Havel, Ensemble quantum computing by NMR spectroscopy, *Proc. Nat. Acad. Sci. USA* **94** (5) (1997) 1634–1639.

[25] A. Khitrin, H. Sun, B.M. Fung, Method of multi-frequency excitation for creating pseudopure states for NMR quantum computing, *Phys. Rev. A* **63** (2) (2001) Art. No. 020301.

[26] I.L. Chuang, N. Gershenfeld, M.G. Kubinec, D.W. Leung, Bulk quantum computation with nuclear magnetic resonance: theory and experiment, *Proc. R. Soc. Lond. A* **457** (1998) 447–467.

[27] E. Knill, I.L. Chuang, R. Laflamme, Effective pure states for bulk quantum computation, *Phys. Rev. A* **57** (5) (1998) 3348–3363.

[28] G.L. Long, H.Y. Yan, Y. Sun, Analysis of density matrix reconstruction in NMR quantum computing, *J. Opt. B: Quantum Semiclass. Opt.* **3** (2001) 376–381.

[29] R. Das, T.S. Mahesh, A. Kumar, Efficient quantum-state tomography for quantum-information processing using a two-dimensional Fourier-transform technique, *Phys. Rev. A* **67** (6) (2003) Art. No. 62304-1.

[30] F.A. Bonk, R.S. Sarthour, E.R. de Azevedo, J.D. Bulnes, G.L. Mantovani, J.C.C. Freitas, T.J. Bonagamba, A.P. Guimaraes, I.S. Oliveira, Quantum-state tomography for quadrupole nuclei and its application on a two-qubit system, *Phys. Rev. A* **69** (4) (2004) Art. No. 042322.

[31] F.A. Bonk, E.R. de Azevedo, R.S. Sarthour, J.D. Bulnes, J.C.C. Freitas, A.P. Guimaraes, I.S. Oliveira, T.J. Bonagamba, Quantum logical operations for spin 3/2 quadrupolar nuclei monitored by quantum state tomography, *J. Magn. Reson.* **175** (2) (2005) 226–234.

[32] J.D. Bulnes, F.A. Bonk, R.S. Sarthour, E.R. de Azevedo, J.C.C. Freitas, T.J. Bonagamba, I.S. Oliveira, Quantum information processing through nuclear magnetic resonance, *Braz. J. Phys.* **35** (3A) (2005) 617–625.

– 5 –

Implementation of Quantum Algorithms by NMR

> *It has been proved surprisingly simple to build small NMR quantum computers, and while such computers are themselves too small for any practical use, their mere existence has brought great excitement to a field largely deprived of experimental achievements.* – J.A. Jones [Progr. in Nucl. Mag. Res. 38 (2001) 325]

Since the discovery of pseudo-pure states by Gershenfeld and Chuang [1] and Cory et al. [2] in 1997, an amazing number of papers appeared in the literature reporting the practical implementation of quantum algorithms by NMR. These include: Deutsch, Deutsch–Jozsa, Grover search and Shor factorization algorithms, besides various examples of protocols to simulate quantum systems. There are also those experiments which implemented protocols to test quantum correlations, such as entanglement. The issue of entanglement in liquid-state NMR will be discussed in separate in the next chapter. In the present chapter we will discuss various other implementations of quantum algorithms by NMR. For its importance in NMR Quantum Information Processing (QIP), we also included in this chapter some descriptions of experiments of quantum simulation experiments and discrete Wigner function measurements.

5.1 NUMERICAL SIMULATION OF NMR SPECTRA AND DENSITY MATRIX CALCULATION ALONG AN ALGORITHM IMPLEMENTATION

In the previous chapter we saw how the matrix elements of a NMR density matrix relate to spectral lines. In the context of NMR QIP, an algorithm is nothing but a radiofrequency pulse sequence which encodes quantum logic gates. Each radiofrequency pulse implements an unitary transformation, which is used to prepare the initial state, and process the information and the computation. Under a sequence of unitary operators $U(\tau_1), U(\tau_2), \ldots, U(\tau_n)$, the initial equilibrium density matrix transforms according to:

$$\rho_{final} = U(\tau_n)U(\tau_{n-1})\cdots U(\tau_1)\rho_{eq}U^{\dagger}(\tau_1)\cdots U^{\dagger}(\tau_{n-1})U^{\dagger}(\tau_n) \qquad (5.1.1)$$

Usually, in an NMR experiment, quantum state tomography is performed either on the final matrix ρ_{final}, or at each step τ_k. In the second case, it usually aims to study the information processing during the execution of the quantum algorithm. In any case, it is a good practice to follow the evolution of the experiment by calculating the resulting NMR spectrum at each step. As an example, let us calculate the density matrices along the Deutsch algorithm. The way a density matrix relates to the NMR spectrum is explained in Chapter 2.

At the initial state the qubits are in a quantum state $|\Phi_0\rangle = |0\rangle \otimes |1\rangle = |01\rangle$, and the system density matrix is then:

$$\rho_0 = |\Phi_0\rangle\langle\Phi_0| = \begin{bmatrix} 0 & 0 & 0 & 0 \\ 0 & 1 & 0 & 0 \\ 0 & 0 & 0 & 0 \\ 0 & 0 & 0 & 0 \end{bmatrix} \quad (5.1.2)$$

As a next step of the algorithm, a Hadamard logic gate is applied to both qubits leaving the system in the state $|\Phi_1\rangle = \frac{1}{2}(|0\rangle + |1\rangle)(|0\rangle - |1\rangle)$, which corresponds to the density matrix:

$$\rho_1 = |\Phi_1\rangle\langle\Phi_1| = \frac{1}{4}\begin{bmatrix} 1 & -1 & 1 & -1 \\ -1 & 1 & -1 & 1 \\ 1 & -1 & 1 & -1 \\ -1 & 1 & -1 & 1 \end{bmatrix} \quad (5.1.3)$$

One can notice that at this stage the system is in a complete superposition of states, and the next step is to perform an unitary operation U_f, which takes the two qubit system from a generic state, $|x, y\rangle$ to the state $|x, y \oplus f(x)\rangle$. This transformation $|x, y\rangle \to |x, y \oplus f(x)\rangle$ is nothing but the sum of the second qubit, with $f(x)$, that is, the computed function of the first qubit. This function, f, is the one to be verified, in order to determine if it is balanced or constant. As it can be easily verified, any binary function, when applied to the particular system state $|x, y\rangle = \frac{1}{\sqrt{2}}|x\rangle[|0\rangle - |1\rangle]$ yield the result $(-1)^{f(x)}\frac{1}{\sqrt{2}}|x\rangle[|0\rangle - |1\rangle]$. It is worth to noticing that $|0 \oplus f(x)\rangle = |0\rangle$ and $|1 \oplus f(x)\rangle = |1\rangle$ if $f(x) = 0$, or $|0 \oplus f(x)\rangle = |1\rangle$ and $|1 \oplus f(x)\rangle = |0\rangle$ if $f(x) = 1$. Therefore, the system will be in the state $|\Phi_2\rangle = \pm\frac{1}{2}[(|0\rangle + |1\rangle)(|0\rangle - |1\rangle)]$ if $f(0) = f(1)$, or $|\Phi_2\rangle = \pm\frac{1}{2}[(|0\rangle - |1\rangle)(|0\rangle - |1\rangle)]$ if $f(0) \neq f(1)$, and this will lead to one of the two possible density matrices:

$$\rho_2 = \frac{1}{4}\begin{bmatrix} 1 & -1 & 1 & -1 \\ -1 & 1 & -1 & 1 \\ 1 & -1 & 1 & -1 \\ -1 & 1 & -1 & 1 \end{bmatrix} \quad \text{for } f(0) = f(1) \quad (5.1.4)$$

$$\rho_2 = \frac{1}{4}\begin{bmatrix} 1 & -1 & -1 & 1 \\ -1 & 1 & 1 & -1 \\ -1 & 1 & 1 & -1 \\ 1 & -1 & -1 & 1 \end{bmatrix} \quad \text{for } f(0) \neq f(1) \quad (5.1.5)$$

Although both qubits are still in a superposition of states, the relative phase of the first now depends on the result of the operations $f(0)$ and $f(1)$, and will determine if the function is balanced or constant. Thus if the Hadamard gate is applied to the first qubit, the system will evolve to state described as: $|\Phi_3\rangle = \pm\frac{1}{\sqrt{2}}[|0\rangle(|0\rangle - |1\rangle)]$ or $|\Phi_3\rangle = \pm\frac{1}{\sqrt{2}}[|1\rangle(|0\rangle - |1\rangle)]$, in case $f(0) = f(1)$ or $f(0) \neq f(1)$, respectively. The density matrix for each state is:

$$\rho_3 = \frac{1}{2}\begin{bmatrix} 1 & -1 & 0 & 0 \\ -1 & 1 & 0 & 0 \\ 0 & 0 & 0 & 0 \\ 0 & 0 & 0 & 0 \end{bmatrix} \quad \text{for } f(0) = f(1) \quad (5.1.6)$$

$$\rho_3 = \frac{1}{2}\begin{bmatrix} 0 & 0 & 0 & 0 \\ 0 & 0 & 0 & 0 \\ 0 & 0 & 1 & -1 \\ 0 & 0 & -1 & 1 \end{bmatrix} \quad \text{for } f(0) \neq f(1) \tag{5.1.7}$$

Using the partial trace operation, which sums over all the state of the others qubits of the system, one may find the respective density matrices for each qubit in the system:

$$\rho_3^1 = \begin{bmatrix} 1 & 0 \\ 0 & 0 \end{bmatrix} \quad \text{and} \quad \rho_3^2 = \frac{1}{2}\begin{bmatrix} 1 & -1 \\ -1 & 1 \end{bmatrix} \quad \text{for } f(0) = f(1) \tag{5.1.8}$$

$$\rho_3^1 = \begin{bmatrix} 0 & 0 \\ 0 & 1 \end{bmatrix} \quad \text{and} \quad \rho_3^2 = \frac{1}{2}\begin{bmatrix} 1 & -1 \\ -1 & 1 \end{bmatrix} \quad \text{for } f(0) \neq f(1) \tag{5.1.9}$$

As one can see, a measurement on the first qubit is enough to find out whether the function is balanced or constant, the second qubit acting as an auxiliary bit.

5.2 NMR IMPLEMENTATION OF DEUTSCH AND DEUTSCH–JOZSA ALGORITHMS

The first experimental implementation of a quantum algorithm by NMR was reported by J.A. Jones and M. Mosca [3]. They demonstrated the Deutsch algorithm using as qubits the spins of two protons in a sample of partially deuterated cytosine. The observed J-coupling in this system is only 7.2 Hz, and the doublets were separated by 763 Hz.

The theoretical description of Deutsch algorithm was made in the Chapter 3, and the evolution of a pure-state density matrix along the algorithm calculated in the previous section. Let us only remind here that the algorithm uses quantum superposition to test a binary function, which can be of two kinds: constant or balanced. There are two possible *constant* binary functions: $f_{C1}(0) = 0$ and $f_{C1}(1) = 0$ or $f_{C2}(0) = 1$ and $f_{C2}(1) = 1$. There are also two *balanced* binary functions: $f_{B1}(0) = 0$ and $f_{B1}(1) = 1$ or $f_{B2}(0) = 1$ and $f_{B2}(1) = 0$. Therefore, the classical knowledge of whether a binary function is constant of balanced involves two bits of information, which are obtained when the given function is tested twice.[1] The Deutsch algorithm can decide whether the function is constant or balanced, by testing it only once.

The input state to the algorithm is the pseudo-pure state:

$$\rho_\epsilon = \frac{1-\epsilon}{4} I + \epsilon |01\rangle\langle 01| \tag{5.2.1}$$

which can be created by applying one of the techniques described in the previous chapter. The first qubit corresponds to the input qubit to the function, whereas the second one corresponds to the returning result. The states of the two qubits can be read directly on the NMR spectra: a doublet line pointing upwards means '0', and pointing downwards means '1'.

The algorithm involves two pairs of one-qubit Hadamard operators. The authors used instead single $(\pi/2)^{\pm y}$ pulses, which are not self-inverse as a true Hadamard, but can

---
[1] This situation is frequently compared to the testing whether a coin is fair or fake.

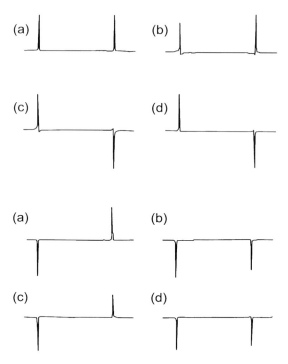

Figure 5.1 Testing binary functions in a classical algorithm. The line on the left represents the input qubit and the line on the right the output qubit. In the upper set, the input is '0', and in the lower set, it is '1'. (a) and (d) are constant, and (b) and (c) are balanced. Adapted with permission from [3].

replace it in this case, using a positive phase for the first application, $+y$, and a negative phase, $-y$, for the second one.

The four possible binary functions are implemented by four unitary operators which correspond[2] to: **1** and $CNOT_1$ for the constant functions,[3] and $CNOT_2$ and NOT_2 (i.e. applied on the second qubit) for the balanced ones. For instance, setting the first qubit to '0', the corresponding NMR lines will always point upwards. Under one of the four transformations above, the second line can point either upwards, or downwards, depending on whether the second qubit has been flipped or not by the operation. The action of these operators was tested by Jones and Mosca in the original paper [3], for the case the first spin is in either '0' or '1' state. The result is shown in Figure 5.1. Notice that this test corresponds to a classical test of binary functions.

The result of the application of the full Deutsch algorithm is shown in Figure 5.2. The lines on the right of the spectra represent the input state $|1\rangle$, which remains in $|1\rangle$ at the output. The line on the left represents the output of the calculation: the corresponding qubit always start at $|0\rangle$, but it is inverted in cases (b) and (c). These represent the balanced functions. In cases (a) and (d) it remains in $|0\rangle$ and the function is constant.

A variation of the Deutsch algorithm is the so called Deutsch–Jozsa algorithm, which uses more than one qubit binary functions [4]. A number of experimental demonstrations

---

[2] Pulse sequences to create these operators have been described in the previous chapter.

[3] $CNOT_1$ flips the second qubit when the first one is in state '1', and $CNOT_2$ flips the state of the second qubit when the first one is in '0'.

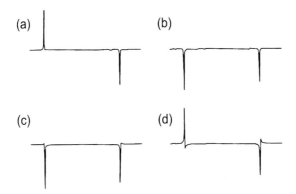

Figure 5.2 NMR implementation of the full Deutsch algorithm. The line on the right represents the eigenstate $|1\rangle$, and does not change at the output. The line on the left is inverted in cases (b) and (c) and represents balanced functions. Spectra (a) and (d) represent constant functions. Adapted with permission from [3].

of this algorithm has appeared in the literature. For instance, Mangold et al. implemented an optimized version of this algorithm in a three-qubit system [5]. They tomographed the density matrix at different stages of the implementation. Das and Kumar [6] implemented this algorithm in a quadrupole $I = 3/2$ system (see below). One interesting peculiarity of this later work is the implementation of a CNOT gate using the quadrupolar evolution, instead of the more usual selective radiofrequency pulses. Finally, an interesting implementation of the Deutsch–Jozsa algorithm using NMR pseudo-entangled states (see Chapter 6) is reported by Dorai, Arvind and Kumar [7].

5.3 GROVER SEARCH TESTED BY NMR

As discussed in the Chapter 3, the quantum search algorithm is one of the most important for quantum computation. It is used to search for one or more specific quantum states in an uniform superposition. It is often compared to a search of a name (or number) in a disordered list. The main feature of this algorithm is the operation, performed by the "oracle", which labels the state (or states) to be searched, by inverting its (their) phase. The second operation is the inversion about the mean value, i.e. the amplitude of each state in the system. These two operations must be applied to the system a certain number of times, which depends on the number of items one is looking for and the total number of elements on the system. For a two qubit system, the number of searches is only 1. Another important application is the ability to use this algorithm for searching the solution of a specific problem, which can be done by preparing the action of the "oracle" operator.

The first full implementation of Grover search algorithm by NMR was reported by Chuang, Gershenfeld and Kubinec, in 1998 [8]. The authors used hydrogen and carbon nuclear spins in chloroform as qubits. One important aspect of this work is the reconstruction of the density matrix, and its comparison with the theoretical prediction. They constructed four optimized sequences of radiofrequency pulses, one for each element labeled by the "oracle" of the quantum search algorithm (see Chapter 3). The result is shown in Figure 5.3. One observes that the deviation from the theoretical prediction increases with

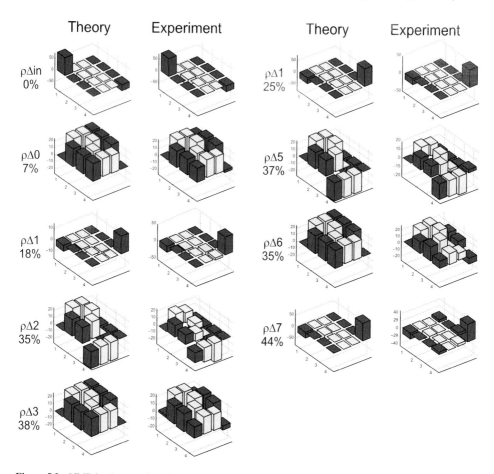

Figure 5.3 NMR implementation of the Grover search algorithm (adapted with permission from Ref. [8]). Each matrix represents a step in the algorithm implementation, and the deviation from the ideal theoretical prediction is shown as percentages.

the number of steps. This is attributed primary to inhomogeneities of the magnetic field, relaxation and imperfections of rotations.

Almost simultaneously to the publication of Chuang, Gershenfeld and Kubinec, Jones, Mosca and Hansen [9] also reported[4] an implementation of Grover search algorithm. They used the two hydrogen nuclei in partially deuterated cytosine as a quantum computer of two qubits. However, their analysis did not included tomographed density matrices.

A number of implementations of Grover search algorithm have appeared since these two original works, among them we cite the work of Xiao and Jones [10] who implemented Grover algorithm to search one or two items in a list. They used ^{1}H and ^{13}C in Na$^+$HCO$_2^-$ as qubits. Another interesting work was reported by Anwar et al. [11], in which Grover search algorithm was implemented in a highly pure state, using a pair of ^{1}H from a chemical reaction of para-hydrogen. This technique allows to achieve very high, nearly pure, states.

---

[4]The publication of Chuang et al. appeared in April 1998, and that of Jones et al. in May 1998.

5.4 QUANTUM FOURIER TRANSFORM NMR IMPLEMENTATION

Quantum Fourier Transform (QFT), as explained in Chapter 3, is a key step for quantum algorithms which exhibit exponential speed up. Its main application is in the Shor's factorization algorithm, which uses order finding and period finding [12]. These are in turn variations of the general procedure known as phase estimation [13].

An application of QFT to order finding was reported in the year 2000 by Vandersypen et al. [14]. They applied the technique to determine the order of a representative subset of 24 permutations of 4 elements. In order to implement this algorithm, they custom synthesized a molecule containing five ^{19}F spins, which served as qubits. The scheme of the molecule, as well as the chemical shifts are shown in Figure 5.4. On Figure 5.5 the NMR spectra of the equilibrium and the pseudo-pure state, used as the input of the experiment, is shown.

Another interesting implementation of the QFT in a three-qubit system (the three ^{1}C of alanine) was reported by Weinstein et al. [15]. With the technique, the authors measured the periodicity of an input state, which was followed by quantum state tomography. Their result is shown in Figure 5.6. Other interesting NMR implementation of QFT can be found in Lee et al. [16] and Weinstein et al. [17]. The first applied QFT to phase estimation and quantum counting, and the second performed the quantum process tomography of QFT.

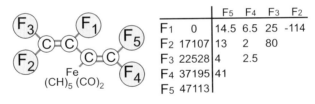

Figure 5.4 Scheme of the molecule used in Ref. [14] for the NMR implementation of a QFT protocol. The five ^{19}F nuclei form a 5-qubit system. Also shown are the respective chemical shifts (in Hz) in a field of 11.7 T, and the coupling constants (also in Hz). Adapted with permission from [14].

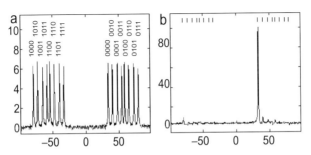

Figure 5.5 (a) Equilibrium and (b) $|00000\rangle\langle 00000|$ pseudo-pure spectra for one of the spins in Ref. [14]. The equilibrium spectrum is composed by two multiplets, each with 8 lines, corresponding to the state 0 or 1. The splitting within each multiplet is due to the different configurations of the other spins, as indicated above each line. The equilibrium positions are indicated as bars in the pseudo-pure state. Adapted with permission from [14].

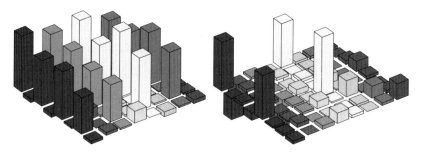

Figure 5.6 NMR implementation of QFT made by Weinstein et al. [15]. On the left the input density matrix, and on the right, the output of the QFT protocol, exhibiting the periodicity of the state. Adapted with permission from [15].

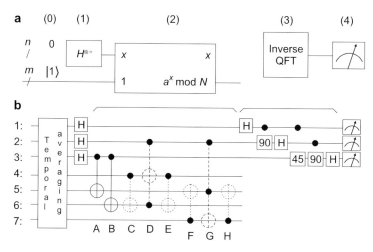

Figure 5.7 (a) General scheme of the quantum circuit implemented by Vandersypen et al. [18] to test Shor's factoring algorithm in a system containing 7 qubits. The working principles of the algorithm are explained in Chapter 3. In (b) it is shown the specific construction for $N = 15$ and $a = 7$ (see text). Numbered boxes represent rotations about the z-axis. Adapted with permission from [18].

5.5 SHOR FACTORIZATION ALGORITHM TESTED IN A 7-QUBIT MOLECULE

The discovery in 1993 by Peter Shor [12] of an efficient quantum factorization algorithm is a main breakthrough in quantum computation. Although such a discovery has dragged many people to the area of quantum information and quantum computation, due to its potential practical applications, so far only one report has appeared in the literature describing a practical implementation of the algorithm using NMR. This has been done by Vandersypen and co-workers in 2001 [18]. They used a "custom made" molecule similar to that of Figure 5.4, but with the two inner carbon nuclei ^{13}C labeled. This yields another pair of qubits, summing a total of 7 (five ^{19}F and two ^{13}C), enough to test the non-classical part of the algorithm. The generic quantum circuit and the NMR parameters for this system are shown, respectively, in Figures 5.7 and 5.8.

5.5. Shor factorization algorithm tested in a 7-qubit molecule

j	$\omega_j/2\pi$	T_{1j}	T_{2j}	J_{7i}	J_{6i}	J_{5i}	J_{4i}	J_{3i}	J_{2i}
1	−22052.0	5.0	1.3	−221.0	37.7	6.6	−114.3	14.5	25.16
2	489.5	13.7	1.8	18.6	−3.9	2.5	79.9	3.9	
3	25088.3	3.0	2.5	1.0	−13.5	41.6	12.9		
4	−4918.7	10.0	1.7	54.1	−5.7	2.1			
5	15186.6	2.8	1.8	19.4	59.5				
6	−4519.1	45.4	2.0	68.9					
7	4544.3	31.6	2.0						

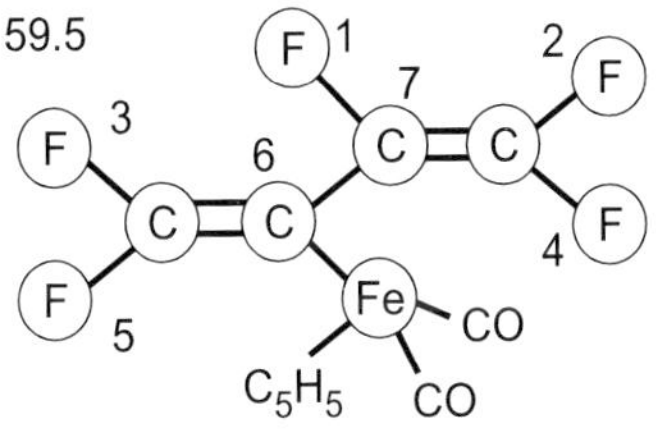

Figure 5.8 Scheme of the 7-qubit "custom made" molecule used in Ref. [18]. The difference to that shown in Figure 5.4 are the two ^{13}C labeled, C6 and C7, which adds two extra qubits to the system. Also shown are the chemical shifts and J-coupling, in Hz. Adapted with permission from [18].

The circuit was designed to test the prime factors of the integer 15 (that is, 3 and 5), and sequences of 300 pulses separated by different free evolution implemented, in which the qubits in the system interact only with each other. As explained in Chapter 3, in one of the steps of Shor algorithm it is necessary to evaluate the function $f(x) = a^x \bmod N$, where N is the number to be factorized ($N = 15$ in the present case), and a is an integer, co-prime of N, which for $N = 15$ can be 2, 4, 7, 8, 11, 13 and 14. This procedure allows the determination of the period of the function $f(x) = a^x \bmod N$, or the order, r, which is the least integer such that $a^r = 1 \bmod 15$. From the order or period it is then possible to determine at least one prime factor of N, using classical number theory techniques. If the value for a, which is randomly picked, is chosen to be $a = 2, 7, 8$, or 13, we will have $a^4 = 1 \bmod 15$, whereas $a = 4, 11$ or 14, one will have $a^2 = 1 \bmod 15$. The cases $a = 7$ and $a = 11$ were tested in the experiment.

The authors have separated the seven available qubits into two quantum registers, the first one containing 3 qubits, and the second 4. The first set is used to determine the period, and the second one is used to store the results of the controlled function $f(x) = a^x \bmod N$, conditionally to the states, $|x\rangle$, of the qubits in the first register.[5] The implementation of the algorithm starts with the system in the state $|000\rangle \otimes |0001\rangle = |0000001\rangle$, i.e., the first three qubits of the first register are in the state $|0\rangle$, and the second register is at the state $|1\rangle$. The first step of the algorithm is to put the three qubits of the first register in an uniform superposition of states. As discussed in previous chapter, this can be done by applying the Hadamard quantum gate at each qubit. In the next stage, the function $f(x)$ is computed, and the last step of the algorithm, before a measurement is done, is the application of a inverse QFT (see Figure 5.7). Figure 5.9 shows the NMR spectra of the first 3-qubits register in three different stages: thermal equilibrium, time-averaged pseudo-pure state and just after the inverse QFT and measurement. Upon measuring the state of the first 3-qubits register, from the spectra analysis, the period of $f(x)$ was determined. At the end of the order finding routine, the first three qubits of the first register will be in a mixed state

---
[5] Since the period of the function $f(x)$ is either 2 or 4, at least two qubits are necessary in the first register, to store the value of r.

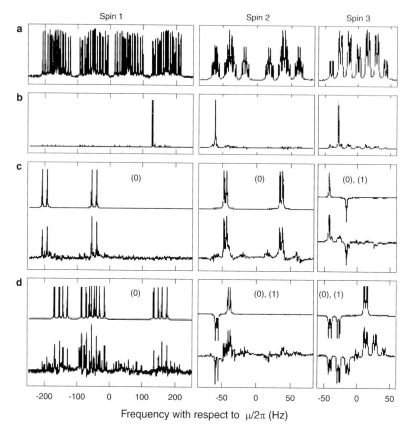

Figure 5.9 NMR spectra of the first register (3 qubits) in the implementation of the Shor' factoring algorithm [18]. (a) Equilibrium spectra; (b) Initial pseudo-pure state; (c) Output spectra for $a = 11$ (see text), and (d) Output spectra for $a = 7$ (see text).

in which the only relevant amplitudes are the ones of $|c2^n/r\rangle$, being c an integer and n the number of qubits in the first register. When using an NMR computer, it is possible to determine all the states of the computational basis that appear at the end of the computation, instead of only sampling a particular one. This feature of NMR is very useful in the Shor's algorithm, because it allows the determination of the period directly. The authors have tested two values of a: $a = 11$ and $a = 7$. For $a = 11$, one can see that the first and second qubits are in the state $|0\rangle$ while the third is in an uniform superposition of $|0\rangle$ and $|1\rangle$, as deduced directly from their NMR spectra (the lines point upwards for the state $|0\rangle$, whereas for a superposition of $|0\rangle$ and $|1\rangle$, half of them point upwards and half downwards, as may be seen on Figure 5.9(c)). It is important to notice that the information about the relative phase of each qubit was lost, since no state tomography was performed. Because the SWAP gate is not implemented at the end, the most significant qubit after the application of the inverse QFT, is the third one, and so on. Therefore, the first register is also in a uniform superposition of $|000\rangle$ and $|100\rangle$, i.e. $|0\rangle$ and $|4\rangle$ in decimal notation, which indicates that the period is $r = 2^3/4 = 2$. Replacing in the expression (see Chapter 3) $\gcd(11^{2/2} \pm 1, 15)$ yields the correct factors, 3 and 5. For the second case, $a = 7$, one can see that the first qubit is in the state $|0\rangle$ while the second and third are in an uniform superposition of $|0\rangle$ and $|1\rangle$, being this deduced directly from their NMR spectra (Figure 5.9(d)). Therefore, the first register is in an uniform superposition of $|000\rangle$, $|010\rangle$, $|100\rangle$ and $|110\rangle$, i.e. $|0\rangle$, $|2\rangle$, $|4\rangle$ and $|6\rangle$ in decimal notation, which indicates that the period is $r = 2^3/2 = 4$. Replacing in the expression $\gcd(7^{4/2} \pm 1, 15)$, this also yields the correct factors.

5.6 ALGORITHM IMPLEMENTATION IN QUADRUPOLE SYSTEMS

Quadrupole nuclei have been much less used for quantum algorithm implementations. This is partially due to the more complicated handling of quantum phases with selective pulses used in those type of experiments, and partially due to the much shorter relaxation times of quadrupole nuclei, compared to their spin 1/2 counterpart. Although various studies have been published in NMR quantum simulation, there are not many full implementations of the known quantum algorithms in nuclei with $I > 1/2$. Das and Kumar [6] report the implementation of the Deutsch–Jozsa algorithm in a $I = 3/2$ system (^{23}Na in a liquid crystal). The circuit they implemented, as well as the resulting NMR spectra, are shown in Figure 5.10. In the spectra, the same sign of the central and outer transitions indicates a constant function, whereas opposite signs indicates balanced functions.

Ermakov and Fung [19] reported an implementation of a continuous version of the Grover search algorithm in a system of $I = 3/2$ nuclei.

One interesting work is reported by Murali et al. [20], in which a half-adder and subtracter operations are implemented in a quadrupole $I = 7/2$ spin system. They used the nuclei of ^{133}Cs in a liquid crystal, for testing the half-adder and subtractor quantum circuits, that are illustrated on Figure 5.11. The operations were implemented using sequences of selective π-pulses, which invert the populations. The algorithms were tested with the sys-

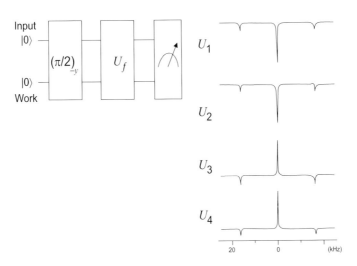

Figure 5.10 Quantum circuit and NMR spectra corresponding to the implementation of the Deutsch–Jozsa algorithm in a quadrupole $I = 3/2$ nucleus by Das and Kumar [6]. The two qubits are represented by the central and outer transitions. Transitions pointing to the same direction represent constant functions, and to opposite directions balanced ones. Adapted with permission from Ref. [6].

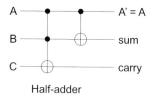

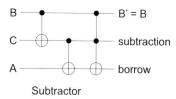

Figure 5.11 Quantum circuits that implement the half-adder and subtractor operations. Adapted with permission from Ref. [20].

tem at the thermal equilibrium state. The conclusion were drawn directly from the NMR spectra, as can be seen on Figure 5.13, which were acquired after a reading pulse followed by a gradient along the z-direction, applied to erase unwanted off-diagonal coherences.

5.7 QUANTUM SIMULATIONS

Quantum simulation is one of the most promising and interesting area of quantum information. The proposal is to map the Hamiltonian of the system to be simulated onto the Hamiltonian of the system where the simulation will take place. There has been a number of reports on quantum simulations by NMR. The idea is to apply a sequence of radiofrequency pulses which transform the natural NMR Hamiltonian to simulate a different dynamics. In this section some implementations of simulations of interesting physical systems performed using NMR quantum computers are discussed.

In 1999 Somaroo et al. [21] reported the first NMR implementation of a quantum simulation experiment: a truncated harmonic oscillator. This is a classical problem, with many applications in physics. The quantum harmonic oscillator Hamiltonian is described by:

$$\mathcal{H}_{QHO} = \sum_n \hbar\Omega\left(n + \frac{1}{2}\right)|n\rangle\langle n| \qquad (5.7.1)$$

The authors used as qubits the two protons of 2,3-dibromothiophene. The first four levels of the harmonic oscillator are mapped to the spin energy levels as:

$$|n=0\rangle \Longrightarrow |\uparrow\uparrow\rangle = |00\rangle$$
$$|n=1\rangle \Longrightarrow |\uparrow\downarrow\rangle = |01\rangle$$
$$|n=2\rangle \Longrightarrow |\downarrow\downarrow\rangle = |11\rangle$$
$$|n=3\rangle \Longrightarrow |\downarrow\uparrow\rangle = |10\rangle \qquad (5.7.2)$$

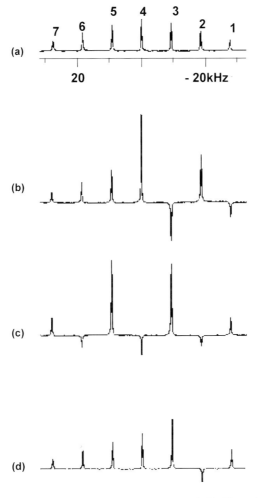

Figure 5.12 ^{133}Cs NMR spectra showing the testing, of the quantum half-adder and subtractor, at the thermal equilibrium state. (a) Equilibrium spectrum; (b) output spectrum of the half-adder circuit; (c) output spectrum of the subtractor circuit and (d) Toffoli gate. Adapted with permission from Ref. [20].

After a pulse sequence which transforms the NMR Hamiltonian into the harmonic oscillator Hamiltonian, the evolution of various coherent states was determined, as shown in Figure 5.11, where the solid lines are theoretical predictions and the points are the spectra amplitudes.

Tseng et al. [22] reported the NMR quantum simulation of the non-physical three-body interaction problem. This involves a Hamiltonian of the type $J_{123}\sigma_1^z\sigma_2^z\sigma_3^z$, which can be simulated with a pulse sequence described in [22]. The effects on the NMR spectra (they used ^{13}C labeled alanine) are extra splittings caused by the triple coupling. In a subsequent work, Tseng et al. [23] analyzed the effects of decoherence on NMR quantum simulation. They also simulated a truncated quantum harmonic oscillator, using 2,3-dibromothiophene and observed the evolution of different coherent states.

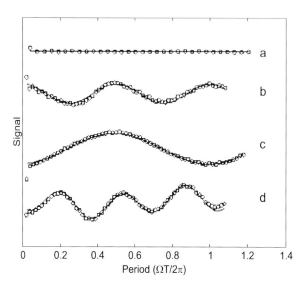

Figure 5.13 Quantum evolution of various initial coherent states of a truncated harmonic oscillator, simulated by NMR: (a) $|0\rangle$; (b) $|0\rangle + |2\rangle$; (c) and (d) different oscillations of $|0\rangle + |1\rangle + |2\rangle + |3\rangle$. Adapted with permission from [21].

A very interesting NMR simulation study of a eight-state quantum system was reported by Khitrin and Fung [24]. For that, they used the nucleus of ^{133}Cs ($I = 7/2$) in a liquid-crystal. The Hamiltonian they simulated is:

$$\mathcal{H}_{hop} = \lambda \sum_{n=0}^{n=6} \left(a_n^\dagger a_{n+1} + a_n a_{n+1}^\dagger \right) \tag{5.7.3}$$

This Hamiltonian describes a linear chain with eight sites, through which a particle jumps from one site to another. The probability to find a particle in a site is given by the population of that site. The experiments follow the time evolution of the populations, through the analysis of the spectra lines, and the results were compared with the theoretical predictions. A pseudo-pure state was created by irradiating simultaneously the sample with radiofrequency whose spectrum contained the six upper single quantum transitions, which were responsible for the exchange of the populations between the levels: $|001\rangle \Longleftrightarrow |010\rangle$, $|010\rangle \Longleftrightarrow |011\rangle$, $|011\rangle \Longleftrightarrow |100\rangle$, $|100\rangle \Longleftrightarrow |101\rangle$, $|101\rangle \Longleftrightarrow |110\rangle$ and $|110\rangle \Longleftrightarrow |111\rangle$. After that, a field gradient pulse was applied, in order to eliminate the non-diagonal coherences of the density matrix, leaving the system in a pseudo-pure state. Then, the time evolution of some coherent states, of the computational basis, were followed by applying a multi-frequency RF pulse, which simulated the Hamiltonian (5.7.3). The experimental results are in good agreement with the theoretical predictions.

There is a particular class of problems that quantum computers have difficult for simulating. Those are ones which deal with interactions of Fermionic systems. Negrevergne et al. [25] describe a NMR quantum simulation of a many-body problem, the so-called Fano–Anderson model. It consists of an n-sites ring containing an impurity atom at the center. In this system, an electron can hop either between the nearest neighbor sites, at the ring

5.7. Quantum simulations

border, or through the middle, through the impurity site. The Hamiltonian which describes this dynamics is given by:

$$\mathcal{H}_{F-A} = \sum_{l=0}^{n-1} \varepsilon_{k_l} c^\dagger_{k_l} c_{k_l} + \epsilon b^\dagger b + V\left(c^\dagger_{k_0} b + b^\dagger c_{k_0}\right) \quad (5.7.4)$$

where $c^\dagger_{k_l}$ and c_{k_l} are, respectively, the creation and annihilation operators for electrons in the ring, and $b^\dagger$ and b the equivalent operators for the impurity site. In this work, an efficient simulation of a resonant impurity scattering process in a metal is presented. The efficiency of the method was tested experimentally, through an NMR simulation experiment, implemented in a 7-qubit transcrotonic acid molecule, where the spectrum of the Fano–Anderson Hamiltonian was determined. They achieved this through a slightly different version of the scattering circuit, which uses an auxiliary qubit, in order to determine the average value of an arbitrary operator, $\langle \sigma^a_+ \rangle = \langle U(t) \rangle /2$, being $\sigma^a_+ = \sigma^a_x + i\sigma^a_y$ (the Pauli matrices of the ancilla qubit). From this, it is possible to measure $\langle e^{-i\mathcal{H}t/\hbar} \rangle$, if the operator is built to simulate $U(t) = e^{-i\mathcal{H}t/\hbar}$. The problem is therefore to rewrite the Fano–Anderson Hamiltonian (5.7.4) in terms of the Pauli matrices. This can be achieved using the Jordan–Wigner transformations [26]:

$$\begin{aligned} b &= \sigma^1_- & b^\dagger &= \sigma^1_+ \\ c_{k_0} &= -\sigma^1_z \sigma^2_- & c^\dagger_{k_0} &= -\sigma^1_z \sigma^2_- \\ &\vdots & &\vdots \\ c_{k_{n-1}} &= \left(\prod_{j-1}^n -\sigma^j_z\right)\sigma^{n+1}_- & c^\dagger_{k_{n-1}} &= \left(\prod_{j-1}^n -\sigma^j_z\right)\sigma^{n+1}_+ \end{aligned} \quad (5.7.5)$$

The transformed Hamiltonian, for a two spin system, is then described as on Equation (5.7.6), where the third term represents an interaction between the two spins through their x and y moment components. This in turn can be rewritten in terms of the usual NMR interaction $\sigma^1_z \sigma^2_z$, which appears in the NMR Hamiltonians for 1/2 spins systems, and single spins rotations [27]. The authors used three qubits to implement this Hamiltonian: one the ancilla, one the impurity and other the k_0 mode [25]. The simulation was performed for different values of ϵ, ε_{k_0} and V, for an arbitrary time, t, also a parameter of the simulation. They determined the spectrum of the Hamiltonian, by using a similar version of the scattering circuit (see Chapter 3) for calculating the function $S(t) = \langle \phi | e^{-i\mathcal{H}t} | \phi \rangle$, at different time intervals, and then applying the discrete Fourier Transform, hence obtaining the eigenvalues of $\mathcal{H}_{F-A}$.

$$\mathcal{H}_{F-A} = \frac{\epsilon}{2}\sigma^1_z + \frac{\varepsilon_{k_0}}{2}\sigma^2_z + \frac{\varepsilon_{k_0}}{2}\left(\sigma^1_x \sigma^2_x + \sigma^1_y \sigma^2_y\right) \quad (5.7.6)$$

Another interesting study is that of Yang et al. [28]. They used a NMR quantum computer to simulate the BCS superconductivity Hamiltonian. In the experiment, performed in the two-qubit chloroform dissolved in acetone-d_6, they observed the energy gap between the superconductor and normal states, directly from the NMR spectrum.

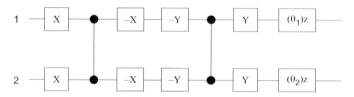

Figure 5.14 Two-qubit quantum circuit used to simulate the superconductors BCS Hamiltonian. Boxed letters indicate directions that $\pi/2$ pulses are applied. We kept the author's representation for the free-evolution period (dark circles). Adapted with permission from [28].

The reduced Hamiltonian of the BCS model is described on Equation (5.7.7), where ε_m stands for the energy required for removing an electron form the Fermi surface, $n_{\pm m}$ represent the electron number operators and $c^{\dagger}_{\pm m}$ denotes the creation ($c_{\pm m}$ – annihilation) operators. The electrons are labelled by the quantum numbers m and $-m$, since they move in pairs, called Cooper pairs, according to the BCS theory, with moment and spins in opposite directions, i.e., $m = (\mathbf{k}, \uparrow)$ and $-m = (-\mathbf{k}, \downarrow)$. The coupling coefficient, V, was taken as a constant throughout the simulation.

$$\mathcal{H}_{BCS} = \hbar \left[\sum_{m=1}^{N} \frac{\varepsilon_m}{2} (n_m + n_{-m}) + V \sum_{m,l=1}^{N} c^{\dagger}_m c^{\dagger}_{-m} c_{-l} c_l \right] \quad (5.7.7)$$

This Hamiltonian was mapped onto the a system of qubits [28], in terms of the Pauli matrices:

$$\mathcal{H}_{BCS} = \hbar \left[\sum_{m=1}^{N} \frac{\varepsilon_m}{2} \sigma_z^m + \frac{V}{2} \sum_{l>m=1}^{N} \left(\sigma_x^m \sigma_x^l + \sigma_y^m \sigma_y^l \right) \right] \quad (5.7.8)$$

After mapping the reduced Hamiltonian of the BCS model, one can easily see the similarity with the NMR Hamiltonian for spin 1/2 nuclei.

The quantum circuit used for this operation is shown in Figure 5.14. The NMR implementation of such circuit presents no particular difficulty, since it involves only simple gates.

The initial state, $|\psi_{ini}\rangle = |00\rangle + |01\rangle$, prepared using spatial averaging (Chapter 4), undergoes a Hadamard transformation on the second qubit. After the evolution under the simulated Hamiltonian, the state evolves to

$$|\psi_{fin}\rangle = \exp(-i\varepsilon\tau)|00\rangle + \cos(V\tau)|01\rangle - i\sin(V\tau)|10\rangle,$$

according to theoretical calculations. The Fourier Transform is applied to the FID and the amplitudes of the two transitions that appear in the NMR spectrum were recorded as a function of τ. Then, a second Fourier Transform was applied to the set of amplitudes and the frequency separation between the lines yielded the difference of the energy between the two eigenstates ($|01\rangle$ and $|10\rangle$), which is the information required.

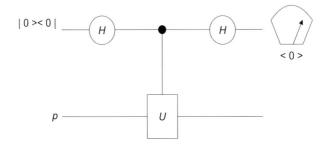

Figure 5.15 Generic scattering circuit implemented in Ref. [29] to measure the discrete Wigner function. Adapted with permission from [29].

5.8 MEASURING THE DISCRETE WIGNER FUNCTION

As a last topic in this chapter, we review the work of Miquel et al. [29] who implemented a quantum circuit to measure the discrete Wigner function of a two-qubit system through NMR. In Chapter 3 it was made a brief introduction to Wigner functions and the quantum processing of information in phase space.

The generic "scattering" circuit is shown in Figure 5.15. A probe qubit enters the upper line of the circuit and a generic multi-qubit state ρ enters the lower line. By measuring the output state of the probe, we obtain $\langle \sigma_z \rangle$, which connects to the input state ρ and the conditional transformation U through the expression: $\langle \sigma_z \rangle = \text{Re}[\text{Tr}(\rho U)]$. Therefore, if ρ is a known state, the measurement of $\langle \sigma_z \rangle$ brings information about the operator U (spectroscopy). On the opposite, the knowledge of U brings information about ρ (tomography). The authors tailored U such as the measurement of $\langle \sigma_z \rangle$ yields the Wigner function $W(q, p) = \langle \sigma_z \rangle / 2N$, where N is the dimension of the Hilbert space. The experimental result is shown in Figure 5.16, for each state of the computational basis of two qubits. They used the two carbons and the hydrogen nuclei of trichloroethylene as qubits. Each point (p, q) in phase space must be determined by a specific pulse sequence. The results are in excellent agreement with the theory, as can be seen from the calculations shown on Chapter 3.

It is interesting to discuss a little bit further the definition of the discrete Wigner function given in terms of the discrete phase-point operator, as shown on the following equation [26]:

$$A(q, p) = \frac{1}{2N} U^q R V^{-p} \exp\left(\frac{2\pi i q p}{2^n}\right) \tag{5.8.1}$$

where U and V are, respectively, the translation operators, in position ($U|q\rangle = |q + 1\rangle$) and momentum ($V|p\rangle = |p + 1\rangle$), and R, the reflection operator ($R|n\rangle = |N - n\rangle$), with $N = 2^n$, i.e. the Hilbert's space dimension. These operators can be constructed from controlled logic gates and 1-qubit transformations.

The discrete Wigner function can be written as:

$$W(q, p) = \text{Tr}\big[A(q, p)\rho\big] \tag{5.8.2}$$

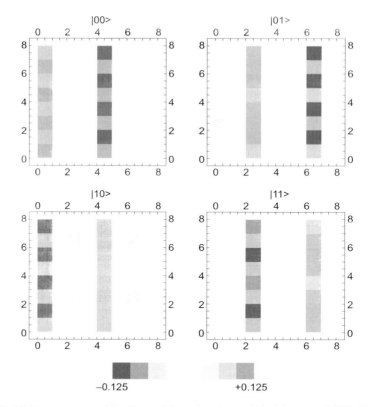

Figure 5.16 NMR measurement of the discrete Wigner function, made by Miquel et al. [29]. Colors indicate intensity (for colors see the web version of this book). Adapted with permission from [29].

which is a particularly useful expression, since it can be directly associated to the scattering circuit. The experiments were carried out using the three qubit of the trichloroethylene molecule, dissolved in chloroform. One of the carbon nuclei was used as the ancilla qubit, whereas the other carbon nucleus and hydrogen were in the quantum state, whose Wigner function was measured.

From the examples presented in this chapter, one can see the extraordinary achievement of NMR QIP. The power of the technique lies in the precise control over the radiofrequency pulses that implement the quantum logic gates, allowing the manipulation of the coherences and energy level populations. Unwanted effects, usually due to small hardware imperfections, can be corrected during and after the a protocol implementation.

Yet, there is a number of other interesting NMR QIP implementations, which have not been discussed here![6] Some of them are: Experimental quantum error correction [30], Geometric quantum computation using nuclear magnetic resonance [31], Experimental realization of quantum games on a quantum computer [34], Experimental implementation of an adiabatic quantum optimization algorithm [35], Experimental implementation of the quantum Baker's map [36], Quantum phase transition of ground-state entanglement in a Heisenberg spin chain simulated in an NMR quantum computer [37], Simulated quantum computation of molecular energies [38], Experimental implementation of heat-bath

---

[6]Experiments reporting entanglement are discussed in the next chapter.

algorithmic cooling using solid-state nuclear magnetic resonance [39], Characterization of quantum algorithms by quantum process tomography using quadrupolar spins in solid-state nuclear magnetic resonance [40] and Quantum metrology [41], Nuclear magnetic resonance implementation of a quantum clock synchronization algorithm [42].

PROBLEMS WITH SOLUTIONS

P5.1 - Consider a two-qubit quadrupole nuclear system ($I = 3/2$), with the four levels labeled according to:

$$|+3/2\rangle \equiv |00\rangle$$
$$|+1/2\rangle \equiv |01\rangle$$
$$|-1/2\rangle \equiv |10\rangle$$
$$|-3/2\rangle \equiv |11\rangle$$

(a) Using ideal selective pulses, show that a Hadamard operation on the first qubit, $|00\rangle \to |00\rangle + |10\rangle$ can be achieved with the following sequence:

$$U_H = (\pi)_1^x (\pi/2)_1^{-y}$$

where the subindex '1' means the transition $+3/2 \leftrightarrow +1/2$.

(b) Show that a CNOT operation with the control on the first qubit is generated by:

$$U_C = (\pi)_2^x (\pi)_3^x (\pi)_2^x$$

where the subindex '2' means the transition $-1/2 \leftrightarrow -3/2$.

Combine the results of (a) and (b) to create a Bell-state generator circuit for quadrupole nuclei.

Solution
(a) First, let us write the selective pulse matrices:

$$(\pi/2)_1^{-y} = \begin{pmatrix} 1/\sqrt{2} & -1/\sqrt{2} & 0 & 0 \\ 1/\sqrt{2} & 1/\sqrt{2} & 0 & 0 \\ 0 & 0 & 1 & 0 \\ 0 & 0 & 0 & 1 \end{pmatrix} \quad (\pi)_1^x = \begin{pmatrix} 0 & i & 0 & 0 \\ i & 0 & 0 & 0 \\ 0 & 0 & 1 & 0 \\ 0 & 0 & 0 & 1 \end{pmatrix}$$

$$(\pi)_2^x = \begin{pmatrix} 1 & 0 & 0 & 0 \\ 0 & 1 & 0 & 0 \\ 0 & 0 & 0 & i \\ 0 & 0 & i & 0 \end{pmatrix} \quad (\pi)_3^x = \begin{pmatrix} 1 & 0 & 0 & 0 \\ 0 & 0 & i & 0 \\ 0 & i & 0 & 0 \\ 0 & 0 & 0 & 1 \end{pmatrix}$$

From this, it is easy to see that

$$(\pi)_1^x (\pi/2)_1^{-y} |00\rangle = \frac{i}{\sqrt{2}} \begin{pmatrix} 1 \\ 1 \\ 0 \\ 0 \end{pmatrix} = i \left[\frac{|00\rangle + |01\rangle}{\sqrt{2}} \right]$$

(b) In a similar way:

$$(\pi)_2^x(\pi)_3^x(\pi)_2^x|00\rangle = |00\rangle$$
$$(\pi)_2^x(\pi)_3^x(\pi)_2^x|01\rangle = -|11\rangle$$
$$(\pi)_2^x(\pi)_3^x(\pi)_2^x|10\rangle = -|10\rangle$$
$$(\pi)_2^x(\pi)_3^x(\pi)_2^x|11\rangle = -|01\rangle$$

Therefore, the sequence of selective pulses $(\pi)_2^x(\pi)_3^x(\pi)_2^x$ implements a CNOT operator, with the control in the second qubit.

Combining the Hadamard and CNOT into the sequence:

$$\text{EPR} \equiv \underbrace{(\pi)_2^x(\pi)_3^x(\pi)_2^x}_{\text{CNOT}} \underbrace{(\pi)_1^x(\pi/2)_1^{-y}}_{\text{Hadamard}}$$

we see that:

$$\text{EPR}|00\rangle = i\left[\frac{|00\rangle - |11\rangle}{\sqrt{2}}\right]$$

and

$$\text{EPR}|01\rangle = i\left[\frac{|00\rangle + |11\rangle}{\sqrt{2}}\right]$$

P5.2 - Grover search algorithm can be implemented for any "item" in a two-qubit system using the corresponding optimized pulse sequences [8]:

$$U_0 = \bar{X}_A \bar{Y}_A \bar{X}_B \bar{Y}_B \tau \bar{X}_A \bar{Y}_A \bar{X}_B \bar{Y}_B \tau$$
$$U_1 = \bar{X}_A \bar{Y}_A \bar{X}_B \bar{Y}_B \tau \bar{X}_A \bar{Y}_A X_B \bar{Y}_B \tau$$
$$U_2 = \bar{X}_A \bar{Y}_A \bar{X}_B \bar{Y}_B \tau X_A \bar{Y}_A \bar{X}_B \bar{Y}_B \tau$$
$$U_3 = \bar{X}_A \bar{Y}_A \bar{X}_B \bar{Y}_B \tau X_A \bar{Y}_A X_B \bar{Y}_B \tau$$

where X, Y means $\pi/2$ pulses applied along the x or y-axes, respectively (barred operators represent negative direction), and τ is two-qubit free-evolution operator. Apply each of these operators to a two-qubit Hadamard state, and show that the result are the four states of the computational basis.

Solution

We start by calculating the matrices which represent the operators U_i:

$$U_0 = \frac{i}{2}\begin{pmatrix} 1 & -1 & 1 & -1 \\ 1 & -1 & -1 & 1 \\ 1 & 1 & 1 & 1 \\ -1 & -1 & 1 & 1 \end{pmatrix} \qquad U_1 = \frac{1}{2}\begin{pmatrix} -1 & -1 & -1 & -1 \\ -1 & -1 & 1 & 1 \\ -1 & 1 & -1 & 1 \\ 1 & -1 & -1 & 1 \end{pmatrix}$$

$$U_2 = \frac{1}{2}\begin{pmatrix} -1 & 1 & 1 & -1 \\ -1 & -1 & 1 & 1 \\ -1 & 1 & -1 & 1 \\ 1 & -1 & -1 & 1 \end{pmatrix} \qquad U_3 = \frac{i}{2}\begin{pmatrix} -1 & -1 & 1 & 1 \\ -1 & -1 & -1 & -1 \\ -1 & 1 & 1 & -1 \\ 1 & -1 & 1 & -1 \end{pmatrix}$$

Applying each of these operators to the two-qubit Hadamard input state:

$$|\psi_0\rangle = \frac{|00\rangle|01\rangle + |10\rangle + |11\rangle}{2} = \frac{1}{2}\begin{pmatrix}1\\1\\1\\1\end{pmatrix}$$

results in:

$$U_0|\psi_0\rangle = i|10\rangle$$
$$U_1|\psi_0\rangle = -|00\rangle$$
$$U_2|\psi_0\rangle = |11\rangle$$
$$U_3|\psi_0\rangle = i|01\rangle$$

P5.3 - A reversible half-adder circuit can be implemented in a three-qubit system, according to the quantum circuit shown in Figure 5.11 [20]: That circuit can be implemented by NMR in a quadrupole $I = 7/2$ nucleus system. Let π_{i-j} represent ideal selective π pulses applied to the transition $i - j$. Show that the sequence

$$U = \pi_6 \pi_7 \pi_5 \pi_6 \pi_7$$

implements the half-adder for any input ABC. The transitions are labeled as follows: $7 \equiv -7/2 \leftrightarrow -5/2$, $6 \equiv -5/2 \leftrightarrow -3/2$ and $5 \equiv -3/2 \leftrightarrow -1/2$.

Solution
Let us first calculate the truth-table of the reversible half-adder (Figure 5.11):

A	B	C	A'	B'	C'
0	0	0	0	0	0
0	0	1	0	0	1
0	1	0	0	1	0
0	1	1	0	1	1
1	0	0	1	1	0
1	0	1	1	1	1
1	1	0	1	0	1
1	1	1	1	0	0

Now, the $I = 7/2$ states are labeled according to:

$$|+7/2\rangle = |000\rangle$$
$$|+5/2\rangle = |001\rangle$$
$$|+3/2\rangle = |010\rangle$$
$$|+1/2\rangle = |011\rangle$$
$$|-1/2\rangle = |100\rangle$$
$$|-3/2\rangle = |101\rangle$$
$$|-5/2\rangle = |110\rangle$$
$$|-7/2\rangle = |111\rangle$$

Now, let us calculate the application of the sequence $\pi_6\pi_7\pi_5\pi_6\pi_7$ over these states, noticing that each pulse in the sequence simply inverts the population of the respective transition, and that the states $|000\rangle, |001\rangle, |010\rangle$ and $|011\rangle$ remain unaffected:

$$|111\rangle \xrightarrow{\pi_7} |110\rangle \xrightarrow{\pi_6} |101\rangle \xrightarrow{\pi_5} |100\rangle \xrightarrow{\pi_7} |100\rangle \xrightarrow{\pi_6} |100\rangle$$

$$|110\rangle \xrightarrow{\pi_7} |111\rangle \xrightarrow{\pi_6} |111\rangle \xrightarrow{\pi_5} |111\rangle \xrightarrow{\pi_7} |110\rangle \xrightarrow{\pi_6} |101\rangle$$

$$|101\rangle \xrightarrow{\pi_7} |101\rangle \xrightarrow{\pi_6} |110\rangle \xrightarrow{\pi_5} |110\rangle \xrightarrow{\pi_7} |111\rangle \xrightarrow{\pi_6} |111\rangle$$

$$|100\rangle \xrightarrow{\pi_7} |100\rangle \xrightarrow{\pi_6} |100\rangle \xrightarrow{\pi_5} |101\rangle \xrightarrow{\pi_7} |101\rangle \xrightarrow{\pi_6} |110\rangle$$

which reproduces correctly the last four lines of the truth-table.

REFERENCES

[1] N. Gershenfeld, I.L. Chuang, Bulk spin resonance quantum computation, *Science* **275** (1997) 350.
[2] D.G. Cory, A.F. Fahmy, T.F. Havel, Ensamble quantum computing by NMR spectroscopy, *Proc. Natl. Acad. Sci. USA* **94** (1997) 1634.
[3] J.A. Jones, M. Mosca, Implementation of a quantum algorithm on a nuclear magnetic resonance quantum computer, *J. Chem. Phys.* **109** (1998) 1648.
[4] D. Deutsch, R. Jozsa, Rapid solution of problems by quantum computation, *Proc. R. Soc. London A* **439** (1992) 553.
[5] O. Mangold, A. Heidebrecht, M. Mehring, NMR tomography of the three-qubit Deutsch–Jozsa algorithm, *Phys. Rev. A* **70** (2004) 042307.
[6] R. Das, A. Kumar, Use of quadrupole nuclei for quantum-information processing by nuclear magnetic resonance: implementation of a quantum algorithm, *Phys. Rev. A* **68** (2003) 032304.
[7] K. Dorai, Arvind, A. Kumar, Implementation of a Deutsch-like quantum algorithm utilizing entanglement at the two-qubit level on an NMR quantum-information processor, *Phys. Rev. A* **63** (2001) 034101.
[8] I.L. Chuang, N. Gershenfeld, M. Kubinec, Experimental implementation of fast quantum searching, *Phys. Rev. Lett.* **80** (1998) 3408.
[9] J.A. Jones, M. Mosca, R.H. Hansen, Implementation of a quantum search algorithm on a quantum computer, *Nature* **393** (1998) 344.
[10] L. Xiao, J.A. Jones, Error tolerance in an NMR implementation of Grover's fixed-point quantum search algorithm, *Phys. Rev. A* **72** (2005) 032326.
[11] M.S. Anwar, D. Blazina, H.A. Carteret, S.B. Duckett, J.A. Jones, Implementing Grover's quantum search on a para-hydrogen based pure state NMR quantum computer, *Chem. Phys. Lett.* **400** (2004) 94.
[12] P. Shor, Discrete logarithms and factoring, in: *Proc. 35th Ann. Symp. Found. Comp. Science* (1994) 124.
[13] M.A. Nielsen, I.L. Chuang, *Quantum Computation and Quantum Information* (Cambridge, 2002).
[14] L.M.K. Vandersypen, M. Steffen, G. Breyta, C.S. Yannoni, R. Cleve, I.L. Chuang, Experimental realization of an order-finding algorithm with an NMR quantum computer, *Phys. Rev. Lett.* **85** (2000) 5452.
[15] Y.S. Weinstein, M.A. Pravia, E.M. Fortunato, S. Lloyd, D.G. Cory, Implementation of the Quantum Fourier Transform, *Phys. Rev. Lett.* **86** (2001) 1889.
[16] J.-S. Lee, J. Kim, Y. Cheong, S. Lee, Implementation of phase estimation and quantum counting algorithms on an NMR quantum-information processor, *Phys. Rev. A* **66** (2002) 042316.
[17] Y.S. Weinstein, T.F. Havel, J. Emerson, N. Boulant, M. Saraceno, S. Lloyd, D.G. Cory, Quantum process tomography of the Quantum Fourier Transform, *J. Chem. Phys.* **121** (2004) 6117.
[18] L.M.K. Vandersypen, M. Steffan, G. Breyta, C.S. Yannoni, M.H. Sherwood, I.L. Chuang, Experimental realization of Shor's quantum factor in algorithm using nuclear magnetic resonance, *Nature* **414** (2001) 883.
[19] V.L. Ermakov, B.M. Fung, Experimental realization of a continuous version of the Grover algorithm, *Phys. Rev. A* **66** (2002) 042310-1.
[20] K.V.R.M. Murali, N. Sinha, T.S. Mahesh, M.H. Levitt, K.V. Ramanatham, A. Kumar, Quantum-information processing by nuclear magnetic resonance: experimental implementation of half-adder and subtractor operations using an oriented spin-7/2 system, *Phys. Rev. A* **66** (2002) 022313.

[21] S. Somaroo, C.H. Tseng, T.F. Havel, R. Laflamme, D.G. Cory, Quantum simulations on a quantum computer, *Phys. Rev. Lett.* **82** (1999) 5381.
[22] C.H. Tseng, S. Somaroo, Y. Sharf, E. Knill, R. Laflamme, T.F. Havel, D.G. Cory, Quantum simulation of a three-body-interaction Hamiltonian on an NMR quantum computer, *Phys. Rev. A* **61** (1999) 012302.
[23] C.H. Tseng, S. Somaroo, Y. Sharf, E. Knill, R. Laflamme, T.F. Havel, D.G. Cory, Quantum simulation with natural decoherence, *Phys. Rev. A* **62** (2000) 032309.
[24] A.K. Khitrin, B.M. Fung, NMR simulation of an eight-state quantum system, *Phys. Rev. A* **64** (2001) 032306.
[25] C. Negrevergne, R. Somma, G. Ortiz, E. Knill, R. Laflamme, Liquid-state NMR simulations of quantum many-body problems, *Phys. Rev. A* **71** (2005) 032344.
[26] P. Jordan, E. Wigner, Über das Paulische Äquivalenzverbot, *Z. Phys.* **47** (1928) 631.
[27] R. Somma, G. Ortiz, J. Gubernatis, E. Knill, R. Laflamme, Simulating physical phenomena by quantum networks, *Phys. Rev. A* **65** (2001) 042323.
[28] X. Yang, A.M. Wang, F. Xu, J. Du, Experimental simulation of a pairing Hamiltonian on an NMR quantum computer, *Chem. Phys. Lett.* **422** (2006) 20.
[29] C. Miquel, J.P. Paz, M. Saraceno, E. Knill, R. Laflamme, C. Negrevergne, Interpretation of tomography and spectroscopy as dual forms of quantum computation, *Nature* **418** (2002) 59.
[30] D.G. Cory, M.D. Price, W. Maas, E. Knill, R. Laflamme, W.H. Zurek, T.F. Havel, S.S. Somaroo, Experimental quantum error correction, *Phys. Rev. Lett.* **81** (1998) 2152.
[31] J.A. Jones, V. Vedral, A. Ekert, G. Castagnoli, Geometric quantum computation using nuclear magnetic resonance, *Nature* **403** (2000) 869.
[32] R.J. Nelson, Y. Weinstein, D. Cory, S. Lloyd, Experimental demonstration of fully coherent quantum feedback, *Phys. Rev. Lett.* **85** (2000) 3045.
[33] X. Fang, X. Zhu, M. Feng, X. Mao, F. Du, Experimental implementation of dense coding using nuclear magnetic resonance, *Phys. Rev. A* **61** (2000) 022307.
[34] J. Du, H. Li, X. Xu, M. Shi, J. Wu, X. Zhou, R. Han, Experimental realization of quantum games on a quantum computer, *Phys. Rev. Lett.* **88** (2002) 137902.
[35] M. Steffen, W. van Dam, T. Hogg, G. Breyta, I. Chuang, Experimental implementation of an adiabatic quantum optimization algorithm, *Phys. Rev. Lett.* **90** (2003) 067903.
[36] Y.S. Weinstein, S. Lloyd, J. Emerson, D.G. Cory, Experimental implementation of the quantum Baker's map, *Phys. Rev. Lett.* **89** (2002) 157902.
[37] X. Peng, J. Du, D. Suter, Quantum phase transition of ground-state entanglement in a Heisenberg spin chain simulated in an NMR quantum computer, *Phys. Rev. A* **71** (2005) 012307.
[38] A. Aspuru-Guzik, A.D. Dutoi, P.J. Love, M. Head-Gordon, Simulated quantum computation of molecular energies, *Science* **309** (2005) 1704.
[39] J. Baugh, O. Moussa, C.A. Ryan, A. Nayak, R. Laflamme, Experimental implementation of heat-bath algorithmic cooling using solid-state nuclear magnetic resonance, *Nature* **438** (2005) 470.
[40] H. Kampermann, W.S. Veeman, Characterization of quantum algorithms by quantum process tomography using quadrupolar spins in solid-state nuclear magnetic resonance, *J. Chem. Phys.* **122** (2005) 214108.
[41] V. Giovannetti, S. Lloyd, L. Maccone, Quantum metrology, *Phys. Rev. Lett.* **96** (2006) 010401.
[42] J.F. Zhang, G.L.Long, Z.W. Deng, W.Z. Liu, Z.H. Lu, Nuclear magnetic resonance implementation of a quantum clock synchronization algorithm, *Phys. Rev. A* **70** (2004) 062322.

– 6 –

Entanglement in Liquid-State NMR

In a beautiful example of how technology can stimulate fundamental physics, the proposals for implementing quantum computing via liquid-state NMR have sparked a debate recently on the very nature of quantum computing. – N. Linden and S. Popescu [Phys. Rev. Lett. 87 (2001) 047901-1]

The quantum circuit which implements entanglement between two qubits is quite simple, as we saw in Chapter 3: it is built only from a `Hadamard` gate, applied to the control qubit, followed by a `CNOT` gate. We also know that sequences of NMR radiofrequency pulses can implement those logic gates very easily. Now, if we have the correct tools to generate entangled states, can it be actually done in a liquid sample? How could we know that entanglement has (or has not) been produced? What are the experimental evidences for this? These matters will be addressed in this chapter.

6.1 THE PROBLEM OF LIQUID-STATE NMR ENTANGLEMENT

To address the problem of entanglement at room temperature NMR liquid-state experiments, let us start from our generic density matrix:

$$\rho_\epsilon = \frac{(1-\epsilon)}{2^n}\mathbf{1} + \epsilon\rho_1 \qquad (6.1.1)$$

Remember that this form is motivated from a high temperature approximation for the NMR equilibrium density matrix, for which $\epsilon \cong \hbar\omega_L/2^n k_B T$. But, whatever the situation, one must have $\text{Tr}(\mathbf{1}) = 2^n$ and $\text{Tr}(\rho_1) = 1$. Consequently, $\text{Tr}(\rho_\epsilon) = 1$, as it must be for density matrices. The matrix ρ_1 can represent an equilibrium mixed state, or a pseudopure state. In particular, it can represent an entangled state. For instance, for two-qubits it could be the cat-state:

$$\rho_1 = \frac{1}{2}\begin{pmatrix} 1 & 0 & 0 & 1 \\ 0 & 0 & 0 & 0 \\ 0 & 0 & 0 & 0 \\ 1 & 0 & 0 & 1 \end{pmatrix} \qquad (6.1.2)$$

In this case, we will refer to ρ_ϵ as a *pseudo-cat state*. Generally, if ρ_1 represents an entangled state, we say that ρ_ϵ is *pseudo-entangled*. Now, if ρ_1 is a cat-state, the question is whether ρ_ϵ is entangled or not. We have to keep in mind that the density matrix of the whole spin system is ρ_ϵ, and not ρ_1, but remember that NMR signals are proportional to ρ_1, and not ρ_ϵ.

So, let us assume that ρ_1 is an entangled state. It turns out that the answer whether ρ_ϵ is entangled or not, depends on the value of ϵ. It is easy to see this in the simple limit case: $\epsilon = 1$, for which $\rho_\epsilon = \rho_1$, and therefore is entangled. However, it turns out that the value of ϵ for a two-qubit room temperature NMR system is less than 1, typically of order 10^{-5}. For this value of ϵ, it has been proved that there can be no entanglement for two qubits [1].

It is not a simple matter to determine the values or, more generally, the regions of values of ϵ for which there will be (or there will be not) entanglement in the system. However, it is a simple matter to show that for $\epsilon = 10^{-5}$, ρ_ϵ cannot be entangled, even if ρ_1 is a pure cat-state. This is shown in the next section, but before we will briefly discuss the problem of *quantification* of entanglement.

Suppose that in a NMR experiment we produce an initial pseudopure state, and apply the quantum circuit that generates a cat state (see Chapter 3). Suppose also that we perform quantum state tomography on this state. We will find a matrix which will be similar to Equation (6.1.2), upon which one has to add the "background" to build the complete matrix:

$$\rho_\epsilon = \frac{1-\epsilon}{4}\mathbf{1} + \epsilon\left(\frac{|00\rangle + |11\rangle}{\sqrt{2}}\right)\left(\frac{\langle 00| + \langle 11|}{\sqrt{2}}\right) \tag{6.1.3}$$

This way of writing ρ_ϵ suggests a straightforward interpretation: a maximally mixed state added to a fraction ϵ of entangled state. However, a fraction x of entanglement can be "extracted" from the maximally mixed state itself:

$$\mathbf{1} = (1-x)\mathbf{1} + x\left\{\left(\frac{|00\rangle + |11\rangle}{\sqrt{2}}\right)\left(\frac{\langle 00| + \langle 11|}{\sqrt{2}}\right)\right.$$
$$+ \left(\frac{|01\rangle + |10\rangle}{\sqrt{2}}\right)\left(\frac{\langle 01| + \langle 10|}{\sqrt{2}}\right)$$
$$+ \left(\frac{|00\rangle - |11\rangle}{\sqrt{2}}\right)\left(\frac{\langle 00| - \langle 11|}{\sqrt{2}}\right)$$
$$+ \left.\left(\frac{|01\rangle - |10\rangle}{\sqrt{2}}\right)\left(\frac{\langle 01| - \langle 10|}{\sqrt{2}}\right)\right\} \tag{6.1.4}$$

In this way of writing, a fraction $(1-x)$ of qubits is in the maximally mixed state $\mathbf{1}$, whereas a fraction x is equally distributed in the four Bell states!

So, how to quantify entanglement? For two qubits, the elements of the Bell basis represent maximally entangled states, but as the number of qubits increases, the quantification of entanglement becomes difficult. For an arbitrary number of qubits, nobody knows how to quantify entanglement. Notice that these difficulties are present to *any physical system* where noise is present or not, and by no means is exclusive to NMR. Indeed, any quantum system in the presence of "white noise" can be written in the form (6.1.3). The difference is that in experiments of liquid-state NMR made at room temperature, that form is intrinsic. For discussions about general aspects, characterization and quantification of entanglement, see [2,3].

6.2 THE PERES CRITERIUM AND BOUNDS FOR NMR ENTANGLEMENT

There is more than one criterium to determine whether a quantum state is entangled or not. One of those was proposed by Peres, in 1996 [4]. In order to explain Peres criterium, we need to define the operation of *partial transposing* a density matrix. Let us write a density matrix ρ in the following way:

$$\rho = \sum_{i,k,j,l} \rho_{ik,jl} |i,k\rangle\langle j,l| \quad (6.2.1)$$

In the computational basis of two qubits, the kets which span ρ are $\{|i,j\rangle = |00\rangle, |01\rangle, |10\rangle, |11\rangle\}$. Usual matrix transposition would mean swapping the column labels (i,k) by line labels, (j,l). For instance, the element $\langle 01|\rho|10\rangle$ becomes $\langle 10|\rho|01\rangle$. In partial transposition, we swap only one of the indexes: $\langle 01|\rho|10\rangle$ becomes $\langle 00|\rho|11\rangle$. In general:

$$\rho^{PT} = \sum_{i,k,j,l} \rho_{ik,jl} |i,l\rangle\langle j,k| \quad (6.2.2)$$

This means the matrix element $\rho_{ik,jl}$ becomes $\rho_{il,jk}$. For instance, partial transposing the matrix representing the cat state, Equation (6.1.2) results in:

$$\rho^{PT} = \frac{1}{2}\begin{pmatrix} 1 & 0 & 0 & 0 \\ 0 & 0 & 1 & 0 \\ 0 & 1 & 0 & 0 \\ 0 & 0 & 0 & 1 \end{pmatrix} \quad (6.2.3)$$

Now, Peres criterium for entanglement states that *if a partially transposed density matrix exhibits negative eigenvalues, then the sate represented by the original matrix will be entangled*. For instance, the partially transposed density matrix for the cat state, Equation (6.2.3), has one negative eigenvalue, -0.5, and therefore the cat-state is entangled.

The idea behind Peres separability criterium is the fact that the transposed of a density matrix is another density matrix; that is, a positive operator with trace equal to one. Under partial transposition, this property is preserved for product states, but it fails for entangled states.

We can now apply Peres criterium to our pseudo-cat state:

$$\rho_\epsilon = \begin{pmatrix} (1+\epsilon)/4 & 0 & 0 & \epsilon/2 \\ 0 & (1-\epsilon)/4 & 0 & 0 \\ 0 & 0 & (1-\epsilon)/4 & 0 \\ \epsilon/2 & 0 & 0 & (1+\epsilon)/4 \end{pmatrix} \quad (6.2.4)$$

The partial transposing operation of this matrix can be easily calculated:

$$\rho_\epsilon^{PT} = \begin{pmatrix} (1+\epsilon)/4 & 0 & 0 & 0 \\ 0 & (1-\epsilon)/4 & \epsilon/2 & 0 \\ 0 & \epsilon/2 & (1-\epsilon)/4 & 0 \\ 0 & 0 & 0 & (1+\epsilon)/4 \end{pmatrix} \quad (6.2.5)$$

The eigenvalues of this matrix are: $(1+\epsilon)/4$, $(1+\epsilon)/4$, $(1+\epsilon)/4$ and $(1-3\epsilon)/4$. Thus, the smallest value of ϵ for which *all* the eigenvalues will be *non-negative* is $\epsilon = 1/3$. Below this value, according to Peres criterium, the pseudo-cat state is unentangled.

The Peres criterium is a necessary and sufficient condition for entanglement in the case of two qubits, but for larger Hilbert spaces the partial transposition operation of entangled states can take the density matrix to other positive operators, that is, with no negative eigenvalues, and the criterium fails.

The problem of determining bounds of ϵ for n qubits, is rather complicated. A general Peres criterium can be obtained for n qubits, but since its applicability is restricted to a small number of qubits, it cannot be used to analyze NMR entanglement and the scaling problem. An alternative and more general analysis was made by Braunstein and co-workers in 1999 [1]. They found that a n-qubit pseudo-entangled state will be separable for

$$\epsilon \leqslant \frac{1}{1+2^{2n-1}} \tag{6.2.6}$$

and that the non-separability region lies at

$$\epsilon > \frac{1}{1+2^{n/2}} \tag{6.2.7}$$

According to this criterium, for $n=2$, for instance, the separable region occurs for $\epsilon \leqslant 1/9$, and the entanglement region for $\epsilon > 1/5$. Nothing can be said about entanglement of states in the region: $1/9 \leqslant \epsilon < 1/5$. Notice that Peres criterium yields $\epsilon \leqslant 1/3$ for the separability region of two qubits, and therefore slightly above the entangled region of Braunstein. Of course, for liquid-state room temperature NMR, $\epsilon \cong 10^{-5}$ is inside the separable region. By increasing n (but keeping the ratio B_0/T constant), one could expect to be able to produce entanglement in liquid-state samples at room temperature. For sufficiently large n, we can approximate Equation (6.2.6) by $\epsilon \cong 4^{-n}$ and Equation (6.2.7) by $\epsilon \cong 2^{-n/2}$. However, at room temperature liquid-state NMR experiments, ϵ scales as $n2^{-n}$ [5]. Therefore, if we take $n=12$, for instance, we would have an experimental $\epsilon \approx 0.003$, a separability upper bound $\epsilon \cong 6 \times 10^{-8}$ and an entanglement lower bound $\epsilon \approx 0.0156$. Therefore, for this number of qubits, it is possible for a NMR liquid-state sample at room temperature to leave the separability region, but it will not enter the entanglement region. Recent experimental results by Negrevergne and co-workers [6] have successfully produced pseudo-cat states in a system containing 12 qubits.

The discussion about entanglement in NMR quantum computation is important not only for its conceptual and intrinsic academical interest, but also because entanglement is such an important feature of quantum mechanics and quantum information processing. However, whereas entanglement is essential in applications such as superdense coding, quantum cryptography and quantum teleportation, the situation is not so clear for quantum computation. For instance, Grover search algorithm does not use entangled states [7]. Deutsch algorithm neither. A thorough discussion on the role of entanglement and the apparent power of quantum computers was made by Laflamme and co-workers in 2001 [8]. The discussion goes about the following two questions:

1. Is it possible to have quantum information without entanglement?
2. Is entanglement responsible for the apparent power of quantum computation?

The general conclusion of the authors is that there is more to quantum information processing than entanglement and that, keeping in mind the limitations of room temperature liquid-state experiments, the NMR of these systems is an excellent test bed for the principles of quantum information and quantum computation.

The role of entanglement for quantum computation was also the subject of a report by Linden and Popescu, in 2001 [9]. Their results partially answer the second question stated above, and are obviously important to NMR quantum information processing. They analyzed quantum protocols which aim to solve exponential classical problems with polynomial resources, such as Shor factorization algorithm. They concluded that entanglement is indeed necessary to the exponential speed up of such protocols, but that it is not a sufficient condition. That is, only the existence of entanglement does not guarantee exponential efficiency, as long as the system is sufficiently noisy (such as NMR liquid-state samples with a small ϵ). In this last case, the number of necessary repetitions grows exponentially, and a polynomial efficiency is never achieved.

6.3 SOME NMR EXPERIMENTS REPORTING PSEUDO-ENTANGLEMENT

The debate about NMR liquid-state entanglement is a beautiful example of interaction between technological aspects and theoretical and experimental physics. More than 50 years of development in NMR technology, allowed the discovery of pseudopure states, and the implementation of full quantum computing protocols, including those involving entanglement, the very motivation of this chapter. We learnt from these experiments that, although current liquid-state NMR experiments present no provable quantum entanglement, the NMR implementation of quantum protocols requiring or not entanglement has been an extremely important source of results and new ideas in the field of quantum computation and quantum information. These ideas, proposals and results can lead to solutions of real obstacles for the practical implementation of quantum computers, as well as, to novel insights in fundamental quantum mechanics.[1] Since the first experiment reporting NMR entanglement, back in 1998, many others have appeared in the literature. The evidences are that, even in the absence of provable entanglement, quantum correlations persist in NMR experiments, even for liquid-state at room temperature! We finish this chapter with a brief description of some selected NMR experiments where non-separability and quantum correlations are reported. This selection is intended to be only a small sample of the activity in this area, and it does not exhaust the total number of papers.

- *NMR Greenberger–Horne–Zeillinger states* – This is a historical paper for NMR quantum information processing [10]. It was published in 1998 by Laflamme, Knill, Zurek, Catasti and Mariappan and was the first report about the implementation of a quantum circuit to entangle more than two qubits. The authors used the two ^{13}C and the ^{1}H nuclei of trichloroethylene as qubits. They describe the initial state preparation, entangling quantum circuit and quantum state tomography of the resulting density matrix. After obtaining the GHZ density matrix, a measure of fidelity $\langle \Psi_{GHZ} | \rho_\epsilon^{GHZ} | \Psi_{GHZ} \rangle = 0.95$ was obtained.

---

[1] See next chapter.

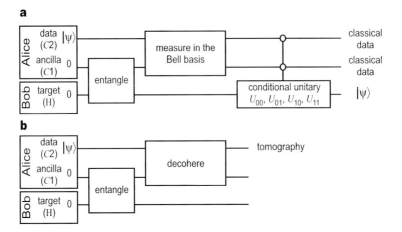

Figure 6.1 Scheme of the quantum circuit implemented by Nielsen, Knill and Laflamme for a teleportation experiment using the three qubits of trichloroethylene. The circuit in (b) is a control experiment. Adapted with permission from [11].

- *Complete quantum teleportation using nuclear magnetic resonance* – The most extraordinary application of entanglement is perhaps teleportation. In 1998 Nielsen, Knill and Laflamme reported the implementation of a complete teleportation experiment using NMR [11]. They also used the two ^{13}C and ^{1}H nuclei in trichloroethylene as qubits. The experiment aimed to teleport the state of one of the carbon nuclei to the hydrogen nucleus. A teleportation protocol based on the proposal of Brassard (Figure 6.1) and co-workers was implemented [12]. However, the original protocol requires a projective measurement of the state of ancilla and data qubits in the computational basis. To adapt the protocol to a NMR experiment, the authors replaced this step applying an idea based on the proposal made by Zurek [13] using the much faster decoherence (T_2 relaxation) of carbon nuclei, compared to hydrogen relaxation, to achieve an equivalent effect (see, however, [14] for a discussion about projective measurement in NMR experiments). The idea is that, just leaving the carbon nuclei to relax, is equivalent to a "measurement" made by the environment in the computational basis. The process was followed through the measure of entanglement fidelity, obtained from the experiment (Figure 6.2). The results showed a maximum value of 0.9 for this function, well above the value of 0.5, for perfect classical transmission.

- *Experimental demonstration of fully coherent quantum feedback* – This paper of 1999 by Nelson and co-workers describes a quantum circuit which implements a coherent quantum feedback by NMR [15]. The idea is to transmit a quantum state with its correlations from a quantum register to a target. A classical feedback can be used to control quantum states, but it involves a measurement step, which destroys quantum correlations. On the contrary, quantum feedback control is able to transmit full quantum states with correlations altogether. The experiment involves three spins (qubits) A, B, and C. The proposal is to use the quantum feedback control circuit to transmit the quantum correlations, initially between B and C, to A and C. Starting from the equilibrium state, a sequence of pulses applied over qubits B and C creates a density matrix proportional to the spin operators $-I_z^A + 2I_z^B I_z^C - 2I_y^B I_y^C$. Such a density matrix describes a thermal equilibrium for the spin A and a correlated state between B and C. Since it cannot be

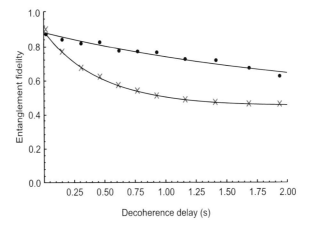

Figure 6.2 Entanglement fidelity obtained from the teleportation experiment of Nielsen, Knill and Laflamme. The top curve represents the full teleportation experiment, and the bottom one the control experiment. Adapted with permission from [11].

factorized as a product of single spin density matrices, it is a pseudo-entangled state. This state, although not provably entangled, for the reasons described in this chapter, contains quantum correlations which cannot be transmitted by classical feedback control. At the end of the process, the density matrix of A and C is found to be proportional to $-I_z^B + 2I_z^A I_z^C - 2I_y^A I_y^C$, which demonstrates the success of transmission (Figure 6.3). The efficacy of the process was evaluated by a measure of fidelity, which reached 91.5%.

- *Experimental implementation of dense coding using NMR* – This nicely simple experiment demonstrates a quantum circuit which implements quantum dense coding [16]. It was reported by Fang and co-workers in 2000. It uses the two qubits available in chloroform molecule (^{13}C and ^1H nuclei). An initial pseudo-entangled state $(|00\rangle - |11\rangle)/\sqrt{2}$ is created and then transformed to each other state of the Bell basis by the application of a one qubit operation. The resulting two-qubit state passes through a Bell analyzer and is converted back to one of the states of the computational basis. Control is made directly through the tomography of quantum state.
- *Experimental demonstration of GHZ correlations using NMR* – This experiment is a report of 2000, by Nelson and co-workers [17]. It aims to demonstrate, using NMR, the existence of quantum correlations present in the GHZ state:

$$|\psi\rangle = \frac{|000\rangle - |111\rangle}{\sqrt{2}}$$

The authors emphasize that, due to the local intrinsic nature of NMR experiments, the results cannot rule out completely the possibility of interpretation in terms of hidden variables, but it demonstrates unambiguously the quantum correlations of the state. The argument is based on the following set of expectation values for the product of spin

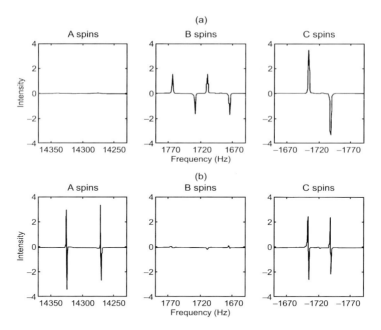

Figure 6.3 Quantum correlation feedback transfer demonstrated by Nelson and co-workers in 2000. The system of three spins starts at the equilibrium thermal state, with the deviation density matrix proportional to $I_z^A + I_z^B + I_z^C$. A sequence of pulses creates a correlated (pseudo-entangled) between B and C: $-I_z^A + 2I_z^B I_z^C - 2I_y^B I_y^C$. After the transfer protocol is implemented, the final state is $-I_z^B + 2I_z^A I_z^C - 2I_y^A I_y^C$. Adapted with permission from [15].

operators along the x and y directions:[2]

$$\langle \psi | \sigma_x^1 \sigma_y^2 \sigma_y^3 | \psi \rangle = +1$$

$$\langle \psi | \sigma_y^1 \sigma_x^2 \sigma_y^3 | \psi \rangle = +1$$

$$\langle \psi | \sigma_y^1 \sigma_y^2 \sigma_x^3 | \psi \rangle = +1$$

$$\langle \psi | \sigma_x^1 \sigma_x^2 \sigma_x^3 | \psi \rangle = -1$$

Notice that only the last value is negative. Therefore, the product of the four measurements is negative, according to the quantum mechanical prediction. This behavior cannot be explained by classical hidden variables, since each spin is measured exactly twice along the x and y directions!

The authors used the three ^{13}C nuclei in alanine as qubits for the experiment. After a sequence of pulses which prepares the ρ_{GHZ} density matrix, a measurement is performed. The GHZ correlations are displayed in the NMR spectra of spin 2, from whose lines the result for the above expectation values can be deduced (Figure 6.4). Four NMR spectra are shown, three of them corresponding to $\langle \sigma_i^1 \sigma_j^2 \sigma_{i \text{ or } j}^3 \rangle = +1$, and one corresponding to the result -1, confirming the prediction of quantum mechanics.

---

[2]Remember that $\sigma_x |0\rangle = |1\rangle$, $\sigma_x |1\rangle = |0\rangle$, $\sigma_y |0\rangle = i|1\rangle$ and $\sigma_y |1\rangle = -i|0\rangle$.

6.3. Some NMR experiments reporting pseudo-entanglement 215

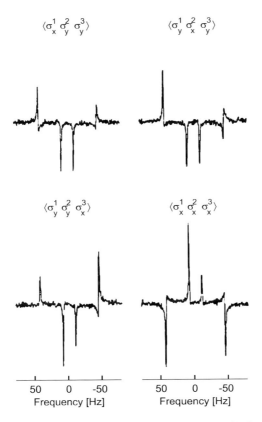

Figure 6.4 NMR report of GHZ correlations by Nelson and co-workers in 2000. Quantum mechanics predicts the expectation value of the product of spin components of three spins in a GHZ state is always positive, except for $\langle \psi_{GHZ} | \sigma_x^1 \sigma_x^2 \sigma_x^3 | \psi_{GHZ} \rangle$, which is negative and equal to -1. This correlation, which is shown in the experiment, cannot be explained by classical means. Adapted with permission from [17].

- *Entanglement transfer experiment in NMR quantum information processing* – This paper of 2002 by Boulant and co-workers [18] describes an experiment of *entanglement transfer* by NMR. The aim is to transfer an entangled state of a pair of qubits to another pair of qubits, a process which was first demonstrated using photons in 1998, by Pan and collaborators. The authors used the four ^{13}C nuclei of crotonic acid as qubits. The process was followed by quantum state tomography and the efficacy of the experiment was quantified by a measured called *attenuated correlation*. A value of 0.65 for this measure at the end of the protocol, indicated that the pseudo-entangled state was indeed transferred from one pair of qubits to the other.
- *Entanglement between an electron and a nuclear spin* $1/2$ – This interesting experiment of 2003 by Mehring, Mende and Scherer [19] describes the entanglement between the spins of a proton and an electron in the same radical •CH, produced from the irradiation of CH$_2$. Two important innovative aspects of this work are the facts that this is a solid-state experiment, performed in a single-crystal of malonic acid, and that it uses two magnetic resonance techniques combined, NMR and ESR (Electron Spin Resonance) simultaneously. In this type of experiment, entanglement is manifested through an interference pattern in the strength of the detected signal, caused by the superposi-

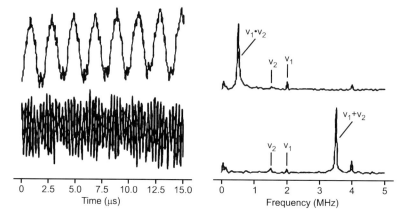

Figure 6.5 Entanglement between electron and nuclear spins, demonstrated by Mehring and co-workers in 2003. The time dependent signal is an interferogram which exhibit two patterns of oscillations, one with phase $\phi_1 + \phi_2$ and the other $\phi_1 - \phi_2$, where $\phi_{1,2}$ are rotation angles about the z-axis. On the right the Fourier transform of the signals showing the two beating frequencies. Adapted with permission from [19].

tion of the phases of the nuclear and electronic states, upon rotation about the z-axis. The pattern changes, depending on the Bell state which was created. If ϕ_1 and ϕ_2 are the phases of electronic and nuclear states, respectively, the $\Psi^\pm$ Bell state will have a phase-modulated strength given by

$$S \propto 1 \pm \cos(\phi_1 - \phi_2)$$

whereas the state $\Phi^\pm$ will show

$$S \propto 1 \pm \cos(\phi_1 + \phi_2)$$

Besides directly observing the correct phase interference in the detected signals (Figure 6.5), the authors perform quantum state tomography, from which a fidelity of 0.99 was obtained for Ψ^-.

- *Practical implementations of twirl operations* – This paper, reported in 2005 by Anwar and co-workers [20], deals with a practical implementation of a proposal made by Bennett and co-workers in 1996 for the purification of entanglement from a mixed state. The typical situation would be that in which the qubits of an initially pure entangled state $|\psi^-\rangle = (|01\rangle - |10\rangle)/\sqrt{2}$ are sent through a noisy channel. The twirl operation is a step for the purification. This operation converts an arbitrary mixed state of two qubits into a pseudo-entangled state:

$$\rho_\epsilon = \frac{1-\epsilon}{4} I + \epsilon |\psi^-\rangle\langle\psi^-|$$

The way it works is as follow: since $|\psi^-\rangle$ is invariant under the application of the same local unitary transformation (an operation called *bilateral unitary transformation*), whereas all other states get affected under such an operation, a random bilateral operation would average all other states converting them into a maximally mixed state, leaving

$|\psi^-\rangle$ unaffected, therefore resulting in the above state ρ_ϵ. Anwar and co-workers argue that in the case of a NMR experiment, a magnetic field gradient applied along the z-axis can serve as a random bilateral rotation. The authors use the two ^{1}H nuclei of cytosine dissolved in D$_2$O as qubits. The twirl operation is split in three stages, comprising different pulse sequences. Results are followed directly on NMR spectra, which, at the end of the twirl sequence, show the expected antiphase doublets, characteristic of $|\psi^-\rangle$ pseudopure state.

- *Benchmarking quantum control methods on a 12-qubit system* – This paper was reported in 2006 by Negrevergne and co-workers [6]. It describes a benchmark experiment in which a 12 qubit pseudo-cat state was successfully produced. It is the largest NMR quantum processing experiment to date. Besides, it reaches the minimum number of qubits required to leave the separability region established by Braunstein and co-workers [1] for room-temperature liquid-state NMR.

PROBLEMS WITH SOLUTIONS

P6.1 - Estimate the value of ϵ for a two-qubit C-H system in a field of 10 Tesla at room temperature. Notice that this value actually depends on the ratio B_0/T between the static magnetic field and the temperature. Repeat your estimate for $T = 100$ mK and $T = 1$ mK, in the same field. What should be the ratio B_0/T to reach $\epsilon = 0.95$?

Solution
Take

$$\epsilon \cong \frac{\hbar \omega_L}{2^N k_B T} = \frac{2\pi}{4} \frac{h}{k_B} \frac{\gamma}{2\pi} \frac{B_0}{T}$$

Let us consider ^{13}C resonance as our reference, since it will give lower values for ϵ. Replacing $\gamma/2\pi \approx 10.7$ MHz/T for ^{13}C and the values for h and k_B in the above expression, we arrive at

$$\epsilon \approx 0.75 \times 10^{-4} \frac{B}{T}$$

Now, for $B = 10$ T and $T = 300$ K, one obtains $\epsilon \approx 0.25 \times 10^{-5}$. For $T = 100$ mK, $\epsilon \approx 0.0075$ and for $T = 1$ mK, $\epsilon \approx 0.75$. In order to reach $\epsilon \approx 0.95$, we need $B/T = 0.95/0.75 \times 10^4 = 1.25 \times 10^4$.

P6.2 - Consider two qubits in a pseudopure cat-state. Calculate the resulting density matrix after a pulse of $\pi/2$ is applied over the first qubit, along the x-axis. From the density matrix, sketch the resulting NMR spectrum and calculate the transverse magnetization $M^+ = \text{Tr}(\rho_\epsilon[\sigma_x + i\sigma_y])$ for both qubits.

Solution
The spin components in the computational basis of two qubits are:

$$\sigma_1^+ = \sigma_x^1 + i\sigma_y^1 = 2\begin{pmatrix} 0 & 0 & 1 & 0 \\ 0 & 0 & 0 & 1 \\ 0 & 0 & 0 & 0 \\ 0 & 0 & 0 & 0 \end{pmatrix} \qquad \sigma_2^+ = \sigma_x^2 + i\sigma_y^2 = 2\begin{pmatrix} 0 & 1 & 0 & 0 \\ 0 & 0 & 0 & 0 \\ 0 & 0 & 0 & 1 \\ 0 & 0 & 0 & 0 \end{pmatrix}$$

Now, a $\pi/2$ pulse along x, over the first qubit is represented by

$$R_x^1(\pi/2) = \frac{1}{2}\begin{pmatrix} 1 & 0 & -i & 0 \\ 0 & 1 & 0 & -i \\ -i & 0 & 1 & 0 \\ 0 & -i & 0 & 1 \end{pmatrix}$$

Applying this matrix over the pseudo-cat state results (ignoring the identity background):

$$\rho'_\epsilon = R_x^1(\pi/2)\rho_\epsilon R_x^{1\dagger}(\pi/2) = \frac{\epsilon}{4}\begin{pmatrix} 1 & i & i & 1 \\ -i & 1 & 1 & -i \\ -i & 1 & 1 & -i \\ 1 & i & i & 1 \end{pmatrix}$$

From this, we see that:

$$M_1^+ = \text{Tr}(\rho'_\epsilon \sigma_1^+) = M_2^+ = \text{Tr}(\rho'_\epsilon \sigma_2^+) = 0$$

The NMR spectrum of the second qubit (lower energy) is given by the elements (1, 2) and (3, 4), whereas for the first qubit (higher energy) the elements are (1, 3) and (2, 4). We see that spectrum will be purely imaginary with one line pointing upward and another downward, for each qubit.

P6.3 - Consider the Hamiltonian of a N coupled homonuclear spins $1/2$ in the presence of a static magnetic field B_0 pointing along the z-axis:

$$\mathcal{H} = -\hbar\omega_L \sum_k I_z^k + \pi\hbar J \sum_{j,k} I_z^j I_z^k$$

Show that the NMR amplitude signal scales as $N2^{-N}$ for this system.

Solution
In the high temperature approximation,

$$\rho \approx \frac{1}{2^N} - \frac{\mathcal{H}}{2^N k_B T}$$

Since $\hbar J \ll \hbar\omega_L$, one can approximate

$$\rho \approx \frac{1}{2^N} + \frac{\hbar\omega_L}{k_B T} 2^{-N} \sum_{k=1}^N I_z^k$$

A population element of ρ is:

$$\langle m_1, m_2, \ldots, m_N | \rho | m_1, m_2, \ldots, m_N \rangle = \frac{I}{2^N} + \frac{\hbar\omega_L}{k_B T} 2^{-N} \sum_{k=1}^N m_k$$

where $m_k = \pm 1/2$. Since the NMR signal strength is proportional to populations difference, the sum on the right side will be proportional to N, and the signal amplitude proportional to $N2^{-N}$.

P6.4 - Apply the Peres criterium to the following two-qubit density matrices and determine the value of ϵ for which the eigenvalues of the partially transposed matrices will be negative:

$$\rho_\epsilon = \begin{pmatrix} (1+\epsilon)/4 & \epsilon/2 & 0 & 0 \\ \epsilon/2 & (1+\epsilon)/4 & 0 & 0 \\ 0 & 0 & (1-\epsilon)/4 & 0 \\ 0 & 0 & 0 & (1-\epsilon)/4 \end{pmatrix}$$

$$\rho_\epsilon = \begin{pmatrix} (1-\epsilon)/4 & 0 & 0 & 0 \\ 0 & (1+\epsilon)/4 & -\epsilon/2 & 0 \\ 0 & -\epsilon/2 & (1+\epsilon)/4 & 0 \\ 0 & 0 & 0 & (1-\epsilon)/4 \end{pmatrix}$$

Solution
The first matrix can be written as:

$$\rho_\epsilon = \frac{1+\epsilon}{4}|00\rangle\langle 00| + \frac{\epsilon}{2}|00\rangle\langle 01| + \frac{\epsilon}{2}|01\rangle\langle 00| + \frac{1+\epsilon}{4}|01\rangle\langle 01| + \frac{1-\epsilon}{4}|10\rangle\langle 10| + \frac{1-\epsilon}{4}|11\rangle\langle 11|$$

Swapping the labels of the second qubit, the second and third terms are swapped but this does not change the matrix, which can be easily verified to be a product-state $|0\rangle \otimes (|0\rangle + |1\rangle)/\sqrt{2}$. Its eigenvalues are: $(1+3\epsilon)/4$ and $(3\times)(1-\epsilon)/4$. For negative eigenvalues, this gives either negative ϵ or $\epsilon > 1$. Since ϵ must be between 0 and 1, neither solution is valid. Therefore, the state is unentangled.

On the other hand, under partial transposing, the second matrix becomes:

$$\rho_\epsilon = \begin{pmatrix} (1-\epsilon)/4 & 0 & 0 & -\epsilon/2 \\ 0 & (1+\epsilon)/4 & 0 & 0 \\ 0 & 0 & (1+\epsilon)/4 & 0 \\ -\epsilon/2 & 0 & 0 & (1-\epsilon)/4 \end{pmatrix}$$

Its eigenvalues are: $(1-3\epsilon)/4$ and $(3\times)(1+\epsilon)/4$. Therefore, the system will be entangled for $\epsilon > 1/3$.

P6.5 - Consider a pure cat-state. Suppose a selective RF pulse is applied on the first qubit, with a rotating angle θ, followed by another pulse on the second qubit with an angle ϕ. Both pulses are applied along the x direction. Show that maximum entanglement is maintained for $\theta + \phi = 2n\pi$ or $\theta + \phi = (2n+1)\pi$. Show that the amplitudes of the four lines of the corresponding NMR spectra are modulated with the same angular factor.

Solution
Take the rotating operator of qubit 1 applied on the cat-state:

$$e^{-i\theta/2\sigma_x^1}|\psi^+\rangle = \left[\cos\left(\frac{\theta}{2}\right)I - i\sin\left(\frac{\theta}{2}\right)\sigma_x^1\right]\frac{|00\rangle + |11\rangle}{\sqrt{2}}$$

$$= \cos\left(\frac{\theta}{2}\right)\frac{|00\rangle + |11\rangle}{\sqrt{2}} - i\sin\left(\frac{\theta}{2}\right)\frac{|10\rangle + |01\rangle}{\sqrt{2}}$$

Now, apply over this state a rotation of ϕ, along x on the second qubit to obtain:

$$|\psi'\rangle = \cos\left(\frac{\theta+\phi}{2}\right)\frac{|00\rangle + |11\rangle}{\sqrt{2}} - i\sin\left(\frac{\theta+\phi}{2}\right)\frac{|10\rangle + |01\rangle}{\sqrt{2}}$$

Therefore, for either $\theta + \phi = 2n\pi$ or $\theta + \phi = (2n+1)\pi$, the state will be one of the Bell eigenstates, and thus maximally entangled.

To analyze the NMR spectra of the rotated state, we have to build the deviation density matrix:

$$\epsilon |\psi'\rangle\langle\psi'| = \frac{\epsilon}{2}\begin{pmatrix} \cos^2\left(\frac{\theta+\phi}{2}\right) & +i/2\sin(\theta+\phi) & +i/2\sin(\theta+\phi) & \cos^2\left(\frac{\theta+\phi}{2}\right) \\ -i/2\sin(\theta+\phi) & +\sin^2\left(\frac{\theta+\phi}{2}\right) & +\sin^2\left(\frac{\theta+\phi}{2}\right) & -i/2\sin(\theta+\phi) \\ -i/2\sin(\theta+\phi) & +\sin^2\left(\frac{\theta+\phi}{2}\right) & +\sin^2\left(\frac{\theta+\phi}{2}\right) & -i/2\sin(\theta+\phi) \\ \cos^2\left(\frac{\theta+\phi}{2}\right) & +i/2\sin(\theta+\phi) & +i/2\sin(\theta+\phi) & \cos^2\left(\frac{\theta+\phi}{2}\right) \end{pmatrix}$$

From this wee see that the elements $(1, 2)$ and $(3, 4)$, which give rise to one qubit NMR spectrum, and the elements $(1, 3)$ and $(2, 4)$, which form the other qubit spectrum, are all the same, purely imaginary, equal to $-i/2\sin(\theta+\phi)$. However, other elements have different angular dependence. Notice that this matrix reduces correctly to the matrices representing the elements of the Bell basis, for $\theta+\phi=2n\pi$ or $\theta+\phi=(2n+1)\pi$.

REFERENCES

[1] S.L. Braunstein, C.M. Caves, R. Jozsa, N. Linden, S. Popescu, R. Schack, Separability of very noisy mixed states and implications for NMR quantum computing, *Phys. Rev. Lett.* **83** (1999) 1054.

[2] A.M. Ozorio de Almeida, Entanglement in phase space, to be published in the series *Lecture Notes in Physics*, Theoretical Foundations of Quantum Information (2007).

[3] F. Mintert, A.R.R. Carvalho, M. Kús, A. Buchleitner, Measures and dynamics of entangled states, *Phys. Rep.* **415** (2005) 207.

[4] A. Peres, Separability criterion for density matrices, *Phys. Rev. Lett.* **77** (1996) 1413.

[5] W.S. Warren, The usefulness of NMR quantum computing, *Science* **277** (1997) 1688.

[6] C. Negrevergne, T.S. Mahesh, C.A. Ryan, M. Ditty, F. Cyr-Racine, W. Power, N. Boulant, T. Havel, D.G. Cory, R. Laflamme, Benchmarking quantum control methods on a 12-qubit system, *Phys. Rev. Lett.* **96** (2006) 170501.

[7] S. Lloyd, Quantum search without entanglement, arXiv:quant-ph/9903057 v1.

[8] R. Laflamme, D.G. Cory, C. Negrevergne, L. Viola, NMR quantum information processing and entanglement, arXiv:quant-ph/0110029 v1.

[9] N. Linden, S. Popescu, Good dynamics versus bad kinematics: is entanglement needed for quantum computation?, *Phys. Rev. Lett.* **87** (2001) 047901-1.

[10] R. Laflamme, E. Knill, W.H. Zurek, P. Catasti, S.V.S. Mariappan, NMR Greenberger–Horne–Zeillinger states, *Phil. Trans. R. Soc. Lond. A* **356** (1998) 1941.

[11] M.A. Nielsen, E. Knill, R. Laflamme, Complete quantum teleportation using nuclear magnetic resonance, *Nature* **396** (1998) 52.

[12] G. Brassard, S.L. Braunstein, R. Cleve, Teleportation as a quantum computation, *Physica D* **120** (1998) 43.

[13] W.H. Zurek, Decoherence and the transition from quantum to classical, *Phys. Today* (October, 1991).

[14] J-S. Lee, A.K. Khitrin, Projective measurement in nuclear magnetic resonance, *App. Phys. Lett.* **89** (2006) 074105.

[15] R.J. Nelson, Y. Weinstein, D. Cory, S. Lloyd, Experimental demonstration of fully coherent quantum feedback, *Phys. Rev. Lett.* **85** (2000) 3045.

[16] X. Fang, X. Zhu, M. Feng, X. Mao, F. Du, Experimental implementation of dense coding using nuclear magnetic resonance, *Phys. Rev. A* **61** (2000) 022307-1.

[17] R.J. Nelson, D.G. Cory, S. Lloyd, Experimental demonstration of Greenberger–Horne–Zeillinger correlations using nuclear magnetic resonance, *Phys. Rev. A* **61** (2000) 022106.

[18] N. Boulant, E.M. Fortunato, M.A. Pravia, G. Teklemariam, D.G. Cory, T.F. Havel, Entanglement transfer experiment in NMR quantum information processing, *Phys. Rev. A* **65** (2002) 024302-1.

[19] M. Mehring J. Mende, W. Scherer, Entanglement between an electron and a nuclear spin 1/2, *Phys. Rev. Lett.* **90** (2003) 153001-1.

[20] M.S. Anwar, L. Xiao, A.J. Short, J.A. Jones, D. Blazina, S.B. Duckett, H.A. Carteret, Practical implementations of twirl operations, *Phys. Rev. A* **71** (2005) 032327-1.

Note: A.M. Souza et al. (to be published) simulated by NMR an experiment of Bell's inequalities violation. They showed that the results can be interpreted on the basis of a local model, by Memicucci et al., *Phys. Rev. Lett.* **88** (2002) 167901.

– 7 –

Perspectives for NMR Quantum Computation and Quantum Information

We propose a nuclear spin quantum computer based on magnetic resonance force microscopy (MRFM). It is shown that a MRFM single-electron spin measurement provides three essential requirements for quantum computation in solids: (a) preparation of the ground-state, (b) one- and two-qubit quantum logic gates, and (c) a measurement of the final state. – G.P. Berman, G.D. Doolen, P.C. Hammel, V.I. Tsifrinovich [Phys. Rev. B 61 (2000) 14694]

In conclusion, we have presented evidence that MRFM is now capable of detecting individual electron spin – D. Rugar, R. Budaklan, H.J. Mamin, B.W. Chui [Nature 430 (2004) 329]

Quantum information processing based on room-temperature liquid-state NMR has been extremely successful in testing quantum logic gates, algorithms, quantum system simulations, quantum correlation phenomena, etc., in small scale systems, up to a few qubits. This is basically due to the following aspects:

1. Good qubit representation;
2. Good qubit isolation;
3. Good dynamics.

However, the exponential loss of sensitivity of the NMR signal upon increasing the number of qubits, severely limits the practical applications of such systems for quantum computation, and cannot be considered for a large scale real quantum processor. Besides, as discussed in the previous chapter, it has been shown that entanglement phenomenon cannot be implemented at room-temperature liquids, in spite of the fact that NMR possesses the ideal tools for that. Just to remind the problem, suppose that a n-qubit system is in the pseudo-pure state $|00\ldots0\rangle$:

$$\rho_\epsilon = \frac{(1-\epsilon)}{2^n}\mathbf{1} + \epsilon|00\ldots0\rangle\langle0\ldots00| \tag{7.0.1}$$

Such a state is usually considered the initial state before any computation process takes place. The signal strength coming from the k-th qubit is proportional to the nuclear magnetization:

$$\gamma_n\hbar\text{Tr}(I_z^k \rho_\epsilon) = \gamma_n\hbar\frac{\epsilon}{2} \propto \frac{1}{2^n} \tag{7.0.2}$$

which means that each qubit added to the system, cuts the signal strength to half. This, in turn, means that adding new qubits to the system, say 10 qubits, would require an improvement of a factor $2^{10} = 1024$ in the sensitivity of the equipment!

So, we are faced with the following dilemma: we have an experimental technique capable of implementing a full set of universal quantum gates to perform quantum computation, we have a good representation of the qubit, but we don't have an adequate physical system to operate! This situation motivated the very meaning of the title of Linden and Popescu's paper, "Good dynamics and bad kinematics", referred in the previous chapter.

But, what would be an adequate system for a NMR large-scale quantum information processor? Basically a system containing a large number of isolated qubits, perhaps thousands, in which the interaction between any pair of them could be controlled, and the state of each single qubit could be accessed by measurement. The need of such a system may sound hopeless, since conventional NMR sensitivity is limited to about 10^{15} spins! But that only means we have to look for something unconventional, perhaps combining the best aspects and capabilities of different techniques.

Actually, along the last few years a number of innovative proposals have appeared in the literature and, against the odds, some experimental results point to a direction which may lead to a true large scale NMR-based quantum chip. These proposals will be discussed in this final chapter.

7.1 SILICON-BASED PROPOSALS: SOLUTION FOR THE SCALING PROBLEM

The first concrete proposal for a NMR large scale quantum chip was presented by B.E. Kane, and appeared in 1998 [1]. The idea was inspired in the existing semiconductor technology for conventional computers, and has various innovative ingredients.

Suppose we start with a purified silicon lattice containing only the isotope ^{28}Si. This nucleus has $I = 0$ and therefore is invisible to NMR radiofrequency pulses. Now, we insert into such a lattice a regular array of phosphorous atoms. Phosphorous acts as an electron donor to Si, and therefore acquires a charge $+e$. Besides, the nucleus ^{31}P has spin $I = 1/2$ and is 100% abundant, and so it is a good NMR qubit. The fact that the Silicon nuclei in our hypothetical lattice have spin equal to zero, means that there will be no magnetic interaction between P and Si nuclei, maximizing the relaxation time of the qubits which, at mK temperatures, can be as large as 10^{18} seconds! The scheme is shown in Figure 7.1.

Now, in order to this Si-based quantum computer work, it is necessary to control the NMR frequencies of individual ^{31}P nuclei, as well as the interaction between qubit pairs. This is done by controlling the electron density in the host lattice, through electrical gates of two types: the A-type and the J-type (Figure 7.1). To understand this idea, it is necessary to remind that nucleus–nucleus interaction can be mediated by electrons, according to:

$$\mathcal{H}_{i,j} = 2\pi J \mathbf{I}_i \cdot \mathbf{I}_j \tag{7.1.1}$$

where, J, is the constant of indirect (i.e., electron-mediated) coupling. In a semiconductor, the electronic wave-function can extend for hundreds of lattice parameters, covering the distance between ^{31}P nuclei. Therefore, if we can find a mean to control the electronic density in the host Si lattice, both, the ^{31}P hyperfine field and the ^{31}P–^{31}P interaction can be controlled. This is the purpose of the electrical gates.

Suppose that right above each ^{31}P nucleus position an electric gate of type A is deposited. The electronic density around each nucleus can be modified by a positive bias

7.1. Silicon-based proposals: solution for the scaling problem

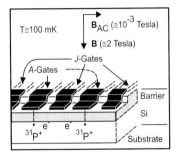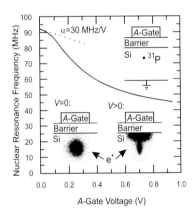

Figure 7.1 Kane's scheme for a quantum computer based in NMR of a solid-state sample. An array of ^{31}P is inserted in a silicon lattice. Electrical gates of type A control the value of the local hyperfine field, whereas gates of type J control two-qubit interactions. On the right is the estimated NMR resonance shift due to biasing of A-type gates. Adapted with permission from [1].

applied the corresponding gate. This will in turn modify the local hyperfine field, which is proportional to the local density of electrons. In the presence of a static magnetic field B pointing to the z direction, the electron–nucleus Hamiltonian is

$$\mathcal{H}_{e-n} = \gamma_e \hbar B S_z - \gamma_n \hbar B I_z + \frac{8\pi}{3} \gamma_e \gamma_n \hbar^2 |\psi(0)|^2 \mathbf{S} \cdot \mathbf{I} \quad (7.1.2)$$

where the contact field created by the electron at the nucleus site (supposed to be at $\mathbf{r} = 0$) is $\gamma_e \hbar |\psi(0)|^2 \mathbf{S}$. Upon biasing an A-type gate, $|\psi(0)|^2$ is modified and so is the local hyperfine field. With this mechanism, the NMR frequency can be controlled. This is how individual qubits are accessed and one-qubit operation is implemented in Kane's scheme. Figure 7.1 shows the variation of ^{31}P NMR frequency upon biasing of an A-type gate.

The interaction between qubits is controlled using the same idea. But now, instead of A-type gates, J-type gates are used, positioned between the nuclei. The coupling constant between two donors separated by a distance r is estimated to be:

$$4J(r) \approx 1.6 \frac{e^2}{\varepsilon a_B} \left(\frac{r}{a_B}\right)^{5/2} \exp\left(\frac{-2r}{a_B}\right) \quad (7.1.3)$$

where ε is the dielectric constant of the host material, and a_B is the semiconductor Bohr radius. This expression is obtained from a hydrogen-like model and therefore is of limited validity. This point is very important and will be discussed below. For the moment let us just accept it as a good approximation. Upon biasing a J-type gate, this coupling energy is varied, modifying the NMR frequency. In this way, two-qubit quantum operations can be implemented, in a similar fashion to what is done in conventional liquid-state NMR. Figure 7.2 shows an estimate for the variation of the NMR frequency upon biasing a J-type gate.

There are obviously many technical difficulties associated to Kane's proposal. Perhaps the most difficult one is the incorporation of the regular array of P atoms into the host

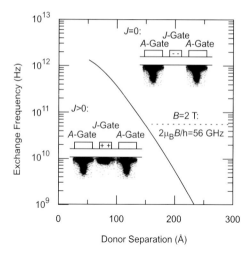

Figure 7.2 Calculated NMR frequency shift due to biasing J-type gates on Kane's proposal. Adapted with permission from [1].

lattice. Another extraordinary challenge would be depositing the gates right on the top of the atoms laying beneath the surface of the material.

But a rather more fundamental obstacle to Kane's design was raised by Koiller and co-workers in 2002 [2]. The trouble is in Equation (7.1.3) for the coupling constant. That formula is based on a hydrogenic model, a limiting approximation which was pointed out by Kane himself in the original paper. Koiller et al. investigated the detailed behavior of the donor exchange energy in Si and Ge, and found a strong oscillatory behavior for J. Depending on the crystal direction, the exchange energy can be zero at qubit sites, as shown in Figure 7.3. Since two-qubit operation must be controlled by adjusting J-type gates to modify the exchange coupling, the results of Koiller and co-workers imply that the positions of donor atoms and gates must be controlled with atomic precision, a fact which adds an enormous challenge to the practical realization of the original proposal. Nevertheless, significant experimental advances in this direction had been achieved by O'Brien and co-workers in 2001, even before the work of Koiller et al. [3]. This last work reports the use of STM (Scanning Tunneling Microscopy) lithography to fabricate an array of phosphorus atoms in a silicon lattice. Later on, in 2003, Schofield and co-workers [4] used STM H lithography to demonstrate the positioning of single P atoms in Si with an accuracy of only 1 nm! Still in the year of 2003 Skinner, Davenport, and Kane modified the original Kane's proposal to a digital approach of quantum computing in Si [5]. The new idea was to encode qubit states in the spins states of an electron and its donor nucleus. A-type gates right above the donors can switch on and off the hyperfine interaction, which is then controlled by "bit" trains of voltage pulses applied over the gates. The S-type gates positioned in the region between the atoms control two-qubit interactions. Entanglement between qubits is achieved by switching on the hyperfine interaction between the electron of one donor and the nucleus of another one. In conclusion, since the first proposal by Kane in 1998, a great deal of theoretical and experimental advances have been reported, and the problem is now much better understood. The rapid development of nanofabrication structures technology points to an optimistic future for the NMR quantum computation in silicon, based on the original idea of Kane.

7.1. Silicon-based proposals: solution for the scaling problem

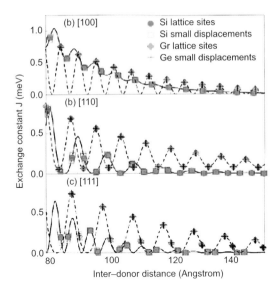

Figure 7.3 Calculated P–P exchange coupling energy in Si (solid) and Ge (dashed). Depending on crystal direction, J can be zero at qubit sites. Adapted with permission from [2].

Still concerning NMR quantum computers in silicon, Ladd and co-workers proposed in 2002 a rather different approach from that of Kane [6]. Now, instead of P donors embedded in a Si lattice, we have an "all-silicon quantum computer". The idea is based on the fact that 95.33% of natural silicon is composed by the zero nuclear spin isotopes ^{28,30}Si. On the other hand, 4.67% is made of ^{29}Si, which has $I = 1/2$ and therefore is a good qubit. In Ladd's proposal, there are no gates or impurity atoms, as in Kanes's. The starting point is a ^{28}Si(111) wafer which is cut in such a way that very regular surface steps are produced. The width of such steps is approximately 15 nm, and their length up to 2×10^4 lattice sites. Being the wafer made primary from ^{28}Si, it produces no NMR signal. Over this structure, ^{29}Si atoms are deposited. The process is such that the deposited atoms lay on the edges of the steps, forming atomic chains along each step edge. The authors argue that, in Si, nuclei can be polarized by cross relaxation with optically excited spin-polarized conduction electrons, which decay in a very short time and do not cause nuclear relaxation. This fact is pointed as an important advantage of the architecture, which is shown in Figure 7.4.

Now, if a static homogeneous magnetic field is applied in the direction of the ^{29}Si chains, the NMR frequency will be the same for all nuclei in the structure. In order to differentiate nuclei by NMR frequency, over the homogeneous field is superimposed a field gradient $\partial B/\partial z$. With this, all nuclei in a given plane perpendicular to the static field will have the same NMR frequency, but nuclei in different planes will be distinguishable by frequency. Each atomic chain along the steps is equivalent to a molecule in the liquid-state NMR, and the number of chains corresponds to the redundancy in the conventional NMR QIP.

Ladd and co-workers estimated the dimensions for the setup they proposed, including a Dy micromagnet to generate the field gradient. A dysprosium micromagnet with length of 400 μm, width 4 μm and height 10 μm generates a field gradient of $\partial B/\partial z = 1.4\,\mathrm{T}\mu\mathrm{m}^{-1}$. The distance between two neighbor ^{29}Si nuclei along the chain is $a = 1.9$ Å. Therefore, the frequency NMR resolution is $\Delta\omega = a\gamma\partial B/\partial z = 2\pi \times 2$ kHz. Overall applied RF

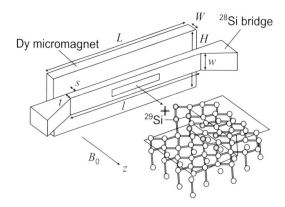

Figure 7.4 Scheme for the "all-silicon" quantum computer proposal of Ladd and co-workers. ^{29}Si isotopes ($I = 1/2$) are deposited on the steps of a heterostructure made of pure ^{28}Si ($I = 0$). A micromagnet of Dy creates a magnetic field gradient along the steps which allows differentiating the qubits by their NMR resonance frequency. The scheme is scalable and can reach thousands of qubits, depending on the initial polarization. Adapted with permission from [6].

pulses will select only a plane of nuclei, allowing single-qubit operation. For two-qubit operations, it is suggested to make use of the dipolar coupling, using specific decoupling pulse sequences to select a given pair of qubits.

In Ladd scheme, final readout of qubits is accomplished by magnetic resonance force microscopy (MRFM). This technique will be discussed in more detail in the next section, since it has become an attractive option for NMR QIP in very diluted systems. For now we just state that the observable in MRFM is the magnetic force excerpted by a field gradient on a magnetization: $F_z = M_z \partial B / \partial z$. So, the use of strong field gradients is favorable to MRFM. The force produced by the magnetization of a n-qubit pseudopure state with polarization p is

$$F_z = \frac{\hbar \Delta \omega}{2a} N \left[\left(\frac{1+p}{2} \right)^n - \left(\frac{1-p}{2} \right)^n \right] \tag{7.1.4}$$

Here, N is the number of qubit copies. To the scheme to work, this force must be comparable to the minimum force detectable by MRFM. For small polarization, the number of detectable qubits depends exponentially on p, just like in the liquid-state approach. However, for $p \approx 0.6$ and above, there is a crossover from exponential to polynomial dependence of n on p: $n \approx (1+p)/(1-p)$ (Figure 7.5). This is the main result of Ladd and co-workers proposal, for it means the system is scalable. Therefore, the usefulness of the scheme relies on the possibility to produce a large enough initial polarization, but there is no need of single spin detection and other difficulties present in the previous model.

7.2 NMR QUANTUM INFORMATION PROCESSING BASED ON MAGNETIC RESONANCE FORCE MICROSCOPY (MRFM)

Magnetic Resonance Force Microscopy (MRFM) appeared in the last few years as an important technique to implement quantum logical operations and readout of qubits states in a

7.2. NMR quantum information processing based on Magnetic Resonance Force Microscopy (MRFM)

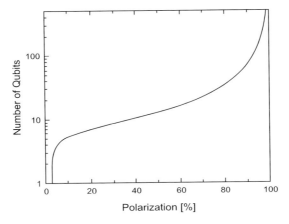

Figure 7.5 Scalability in the proposal of Ladd and co-workers. The number of available qubits depends on the initial polarization. Above $p \approx 60\%$, scaling becomes polynomial. Adapted with permission from [6].

solid state sample. The working principle of MRFM technique is quite simple: suppose we have a magnetized object with net magnetization M_z. In the presence of a magnetic field gradient $\partial B/\partial z$, the object experiences a force $F_z = \pm M_z \partial B/\partial z$. Of course, the larger is the field gradient, the larger will be the force, and the easier will be the signal detection. This is the basic idea behind some traditional magnetometry techniques [7]. Notice that if M_z points to the direction of the field gradient, the force will be positive, but if it points to the opposite direction, it will be negative. Imagine now an arrangement in which the magnetic field gradient is generated by a permanent magnet mounted as a tip on a cantilever, which is fixed at the basis and free to oscillate on the tip. If we make the magnetized object oscillate, the cantilever will experience a time-varying force, which will make it to oscillate with the sample. Now, the mechanical system made of the cantilever and the sample has – just like any other mechanical system – a resonance frequency. If the magnetized object oscillates in that frequency, then the energy absorption of the cantilever will be maximum, and so its oscillation amplitude. Now, replace the magnetized sample by a single atomic or nuclear magnetic moment, and we have the basic ingredients of an atomic-sensitive MRFM! Since the proposal of the technique is to detect the resonance of individual electron or nuclear magnetic moments, the oscillations of the cantilever can be expected to be small in the same proportion, in the range of a few angstroms above the thermal noise. To detect such a small oscillations, optical interferometry methods are used. The scheme is shown in Figure 7.6.

MRFM technique was proposed back in 1991/1992 by Sidles, Garbini and Drobny in the context of biological and molecular imaging [8,9]. The first successful experiment was performed by Rugar, Yannoni and Sidles in 1992. Electron spin resonance was detected in a sample of DPPH (diphenylpicrylhydrazil) weighting only 30 nanograms! The EPR signal appears as the enhanced amplitude of the microscope cantilever, less than 3 Å above the noise (Figure 7.7). MRFM developed to a point where the magnetic signal coming from single atomic magnetic moments can nowadays be detected. This amazing enhancement of sensitivity detection obviously opens a path towards a NMR quantum processor based on MRFM. It represents a passage from the small scale liquid-state experiments to large scale solid-sate NMR quantum processors.

228   7. Perspectives for NMR Quantum Computation and Quantum Information

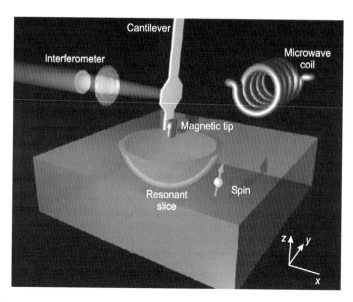

Figure 7.6 Scheme of a MRFM setup. A magnetic particle in the tip of a cantilever generates a field gradient which selects a slice below the sample surface. The NMR frequency is applied in the same frequency of the cantilever, whose oscillations are detected by optical means. Adapted with permission from [19].

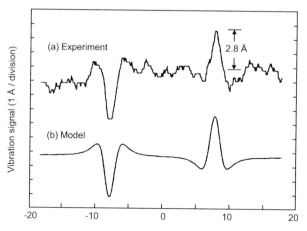

Figure 7.7 First experimental detection of a MRFM experiment, made by Rugar, Yannoni and Sidles in 1992. Electron spin resonance was detected in a sample of DPPH. Notice the maximum amplitude of vibration of the cantilever, less than 3 Å! Adapted with permission from [10].

One important parameter is the minimum force detectable in a MRFM setup. For NMR QIP applications, this will establish the effective number of qubits detectable in an experiment. So, let m_{eff} be the effective oscillating mass of the cantilever, τ its damping time-constant and B the detection bandwidth. The minimum force detectable at temperature T is [10]:

$$F_{min} = (4k_B T B)^{1/2} \sqrt{\frac{m_{eff}}{\tau}} \qquad (7.2.1)$$

7.2. NMR quantum information processing based on Magnetic Resonance Force Microscopy (MRFM)

Typical values for F_{min} are about 10^{-17} N. In the first experiment reported by Rugar and co-workers in 1992, the force on the cantilever was 10^{-14} N. It is clearly desirable to increase the sensitivity of MRFM to detect the smallest possible magnetic movement of a sample. From the above expression we see that lowering the temperature is an obvious procedure. The square root factor on the right contains parameters of the cantilever. Diminishing its effective moving mass and increasing the damping time-constant also contribute to enhance the sensitivity.

In the year of 2000, Berman and co-workers proposed a NMR quantum computer entirely based on MRFM [11]. In this approach, all three steps of a quantum computation, (a) preparation of the initial state, (b) implementation of quantum logic gates and (c) final readout, can be implemented. The idea is to use the electron–nucleus hyperfine coupling to read nuclear states through electronic states. Taking advantage of the much higher sensitivity to detect the electron magnetic moment.

To understand the idea, consider a coupled electron–nucleus system. There will be four eigenstates[1]:

$$|\uparrow 0\rangle, \quad |\uparrow 1\rangle, \quad |\downarrow 0\rangle \quad \text{and} \quad |\downarrow 1\rangle$$

The electronic transitions are

$$|\uparrow 0\rangle \longrightarrow |\downarrow 0\rangle$$

and

$$|\uparrow 1\rangle \longrightarrow |\downarrow 1\rangle$$

These two electronic transitions depend on the nuclear state and therefore have different resonance frequencies. Let ω_{e0} and ω_{e1} be the two electronic resonances, corresponding to the nuclear states $|0\rangle$ and $|1\rangle$, respectively. If a π-pulse is applied to the electron spin in the frequency ω_{e0}, the spin will rotate only if the nucleus is in the state $|0\rangle$. Now, if the cantilever resonance frequency is ω_c, applying π-pulses at frequency ω_{e0} with period $\tau_c/2 = \pi/\omega_c$, will make the cantilever oscillate only if the nucleus is in the state $|0\rangle$. In this way the nuclear state can be detected.

In order to make some numerical estimates, Berman et al. considered an array of paramagnetic moments in a non-magnetic host material, with the impurity atoms separated by $a = 50$ Å, arranged at a distance $d = 100$ Å beneath the surface of the material. This is also the distance from the cantilever tip, which possesses a ferromagnetic particle with radius $R = 50$ Å. In these conditions, the normal component of the magnetic field acting on the electronic moment is $B_z = 5.4 \times 10^{-2}$ T, which corresponds to an electronic resonance shift of 1.5 GHz, approximately (see *Problems with solutions*). Under resonance condition, the force on the cantilever, estimated as $\approx \pm 10^{-16}$ N, produces a vibration with amplitude of approximately 1.2 Å, much above the estimate of 0.3 Å due to the thermal noise, at a temperature of 1 K.

In order to operate a NMR QIP processor, besides detecting the nuclei states, it is necessary to accomplish other tasks: preparing initial state, implementing single-qubit operations and implementing the CNOT quantum gate. Initial preparation is made in the following way. Suppose the system operates at a temperature of 1 K, under a static magnetic

---
[1] The first label represents electron states, and the second label, nuclear states.

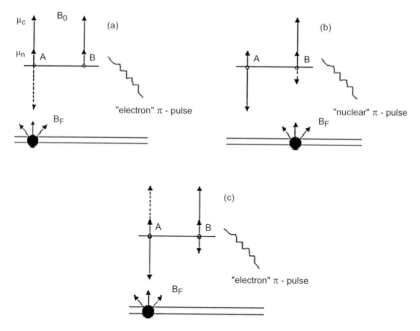

Figure 7.8 CNOT implementation in Berman MRFM scheme. The short arrows represents the nuclear spin, and the long ones the electronic spin. (a) The magnetic tip is set on the first (control) qubit. An electronic π-pulse is applied at a frequency corresponding to the nuclear spin on the ground state. This means the electronic magnetic moment will rotate only if the nuclear state is $|0\rangle$. If the nuclear state is $|1\rangle$, nothing happens. (b) The tip moves to the target qubit, and a nuclear π-pulse is applied at a frequency corresponding to the neighbor electronic spin did not rotate. But, if it did not rotate, that means the control qubit was in $|1\rangle$, and the target will be rotated. Otherwise, if it did rotate, that means the control was in $|0\rangle$ and the target will not rotate. (c) The tip moves back to control and restore the initial electronic state, accomplishing CNOT. Adapted with permission from [11].

field of $B_0 = 10$ T. In these conditions, basically all the electronic spins are in the $|\uparrow\rangle$ ground-state. Now, suppose the ferromagnetic particle is fixed (that is, not allowed to vibrate) and is displaced along the chain. If, at some position r, a nucleus is in the excited state it will be target by the ferromagnetic particle, which will create a magnetic field on that site, additional to B_0. Therefore, the resonance frequency of this particular nucleus will be $\omega_{0n}(r) = \gamma_n(B_0 + B_M)$, where B_M is the field produced by the ferromagnetic particle. Applying a π-pulse on this frequency will drive the nucleus back to its ground-state. By moving the ferromagnetic particle from site to site along the chain, 100% nuclear spin polarization can be achieved, provided the relaxation time is long enough. By the same means one-qubit quantum operation can be implemented. So, it remains to describe how to implement the two-qubit CNOT operation in this scheme.

To implement CNOT in Berman and co-workers scheme, the electronic transitions come to help. Suppose one wants to implement the CNOT operation between a nucleus at r and its neighbor at $r + a$. The nucleus at r is the target qubit and the one at $r + a$ the control. First the ferromagnetic sample is set at the control qubit position, and an electron π pulse is applied, at a frequency ω_{e0}. This pulse will take the electronic spin from the initial $|\uparrow\rangle$ state to $|\downarrow\rangle$ state only if the control qubit is in $|0\rangle$. If the control qubit is in $|1\rangle$, the electronic moment will not change upon the pulse action.

Now, the ferromagnetic particle moves to the target qubit. The field this qubit senses, depends on the direction of its neighbor electronic moments. Therefore, in the case the control electronic moment was inverted in the first step (corresponding to the nuclear moment in '0'), the target nucleus will have a NMR frequency different from that in the opposite case (that is, the nuclear moment was in '1'). So, irradiating the target qubit with a NMR frequency corresponding to the second case, the target will be inverted only if the control was initially in the state '1'. This accomplishes the CNOT operation! To operate between any pair of qubits in the chain, it is important to notice that the electronic dipolar field each qubit senses depends on its position in the chain, relative to the control qubit. The value corresponding to each nucleus can be calculated or obtained experimentally. The CNOT operation is finished by moving the ferromagnetic particle back to the control qubit position, and driving the electronic moment back to its ground-state.

In conclusion, the scheme proposed by Berman and co-workers for a MRFM-based NMR quantum processor is capable of preparing the initial state, implementing one and two-qubit quantum operations and implementing final readout of qubit states. Of course, its realization is based on the possibility of experimentally detecting single electron magnetic moments by MRFM. Such an experiment has been successfully implemented by Rugar and co-workers, as described in the next section.

7.3 SINGLE SPIN DETECTION TECHNIQUES: SOLUTION FOR THE SENSITIVITY PROBLEM

In the previous section we have described some very ingenious proposals which, if implemented in practice, could lead to a large scale quantum information processor through NMR. It is important to emphasize that those proposals circumvent the scaling problem present in liquid-state NMR QIP experiments. However, whatever the sample architecture may be, it seems unavoidable the need to detect the NMR signal of very small spin concentrations. Ideally, single spin detection should be possible. Less than two decades ago, such a strict demand could sound hopeless; conventional ESR needs a concentration of some 10^{10} spins, whereas this number increases to about 10^{15} in conventional NMR. So, we are talking of an improvement of at least 10 orders of magnitude in sensitivity!

The first indications that this could indeed be done can be traced back to the work of Manassen, Hamers, Demuth and Castellano Jr., of 1989 [12]. That work reports the observation of individual paramagnetic spins using the scanning tunneling microscopy (STM) technique. This is not a resonance technique; its observable is the quantum tunneling current between a sample surface and the tip of a STM microscope. The idea is that such a current is affected by the presence of local magnetic moments in the material surface. Therefore, under an applied static magnetic field, the Larmor precession of a local paramagnetic moment would modify the tunneling probability and modulate the current with the same frequency. The experiment was performed in surfaces of Si(111) partially oxidized. For a field of 172 G, the electronic spin precesses at 481.6 MHz, assuming an electronic g-factor equal to 2. Measurement revealed a RF component in the tunneling current exhibiting a peak around 483 MHz, in excellent agreement with the prediction. Various scans were made sweeping the surface in distance of only a few angstroms.

In 2002, Durkan and Welland also combined the techniques of STM and ESR to demonstrate the detection of single paramagnetic moments [13]. They used the organic molecule

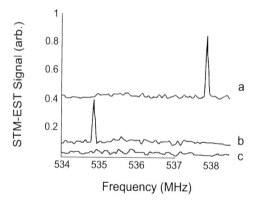

Figure 7.9 STM-ESR spectra obtained by Durkan and Welland in 2002. Peaks correspond to different regions of the sample separated by a few nanometers. Adapted with permission from [13].

known as BDPA [α, γ-bisdiphenylene β-phenylallyl], which contains free radicals which originate the magnetic signal. The substrate used was highly oriented pyrolytic graphite (HOPG). The static field was generated by permanent magnets of Sm/Co, with different sizes, in the range 190–300 G, corresponding to Larmor frequencies of 535–840 MHz, assuming an electronic g-factor equal 2. The STM-EPR signal was successfully detected from the analysis of the RF modulation on the tunneling STM current. Different spectra were obtained within an area with resolution of a few nm. The authors varied the static field and demonstrated the linear dependence of the ESR frequency upon the field, $\omega_L = g\mu_B B_0/\hbar$. From the experiment they obtain $g = 2.0 \pm 0.1$.

Different approaches to detect single electron spin states in semiconductor structures have been used with great success. Two of such experiments were reported in 2004 by Elzerman and collaborators [14] and Xiao and collaborators [15]. Such experiments can lead to a breakthrough in detection techniques for NMR QIP. In the first of these experiments, the state of a single electron spin in a quantum dot is detected. It is interesting to notice that, whereas the detection of single electron charge in quantum dots is relatively simple, the same is not true for single spin detection. One important point to note is that the detection involves a spin-to-charge conversion procedure. The dot is created in a GaAs/AlGaAs heterostructure. A magnetic field is applied to separate the up and down states of an electron. The presence or not of an electron in the dot can be controlled through gate potentials which rise or lower the energy in the dot with respect to the Fermi level of the electron bath in the heterostructure. The dependence of the energy with the magnetic field allows the tuning of the dot potential such that if the electron is in the up state it remains in the dot, but if it is in the down state, it will jump off. This jump causes a change in the electrostatic potential around the dot which is electrically detected. The detection or not of an electron is then associated to the spin state before the jump.

In the work of Xiao et al. the resonance of a single electronic spin is observed directly in a field-effect transistor (FET). After creating a paramagnetic trap, they observe the source/drain current in the FET, as a function of the ESR frequency. Under a magnetic field, the Fermi level of the channel electrons is adjusted to lye between the two electronic states of the paramagnetic trap. The idea is that, if only the lower spin state is occupied, then no electron can jump from the channel to the trap. But if only the upper spin

7.3. Single spin detection techniques: solution for the sensitivity problem

state is occupied, an electron can jump to the lower state, changing the charge in the trap, which can be sensed by the FET. Starting at low temperatures (0.4 K in the experiment) and high magnetic fields (around 16 kG), only the lower trap spin state will be occupied. The authors applied a fixed ESR frequency of 45.1 GHz and varied the magnetic field between 15,900 G and 16,200 G. The charge occupancy in the trap was observed through the source/drain current (in the range of 0.1 μA) as a function of the field. A pronounced peak was clearly observed at a field of about 16,025 G, corresponding to the electronic spin resonance with $g = 2.01$. By varying the ESR frequency, the resonance appeared at different fields, and the linear relationship between field and resonance frequency can be observed. This is done for two traps, and from the linear plot they obtain $g = 2.020 \pm 0.015$.

So far, we have described not only one, but a number of reported different experiments where techniques to detect single spin states are described, a fundamental requirement for spin-based QIP. It is also worth mentioning that various optical methods have been developed and used since 1993 to detect the magnetic resonance of single molecular spins (see Kohler et al. [16], Wrachtrup et al. [17], and, more recently, Jelezko et al. [18]). Optics is a fast developing area, and optical methods are extremely promising in the context of quantum information and quantum computation. However, the experiment that best approaches the original proposal of Berman and co-workers for a NMR QIP processor described in the previous section, was implemented in 2004 by Rugar, Budaklan, Mamin and Chui [19]. The main motivation of the authors was the improvement of resolution of magnetic resonance imaging below 1 μm. They used MRFM to detect a single electronic spin ESR signal in vitreous silica. The paramagnetic centers were produced irradiating the sample with a 2-Gy dose of ^{60}Co γ-rays. An estimative for the spin concentration in the sample after irra-

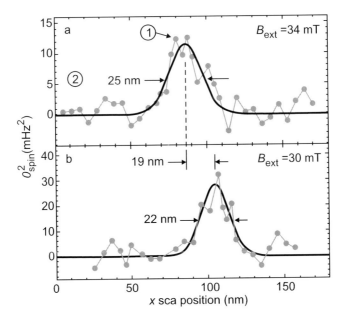

Figure 7.10 Single spin detection by MRFM. The result is from Rugar et al. (2004). The two plots correspond to different values of the external field. Changing the external field modifies the resonant slice, which in turn causes a shift in the peak. The average distance between spins in the sample is 300 Å. Adapted with permission from [19].

diation showed to be about $10^{13}/cm^3$. The field gradient was produced by a 150 nm-wide SmCo particle and the experiment done at 1.6 K. These conditions sets up a resonant slice in the sample of approximately 250 nm below the cantilever tip, according to the authors. The ESR frequency was set at 2.96 GHz. Scanning the sample surface, resonance peaks were detected for different values of the external field (34 and 30 mT), with peak centers 19 nm apart, and peak widths of about 23 nm, as shown in Figure 7.10. This shift in peak centers is not caused by different spins in the sample, since that, from the estimated spin concentration, the average distance between spins is of about 300 Å. The shift is due to the reduction of the external field, which causes a reduction in the radius of the resonant slice, moving the region of detection. In fact, the much larger distance between spins guarantees that the detected signal is due to a single spin!

7.4 NMR ON A CHIP: TOWARDS THE NMR QUANTUM CHIP INTEGRATION

In the previous sections we revised different proposals for a large scale NMR quantum processor and various ways to detect single spin states in different structures, including the use of magnetic resonance force microscopy (MRFM), a technique capable of not only to detecting single spin states, but also to prepare initial states and implement one and two-qubits quantum operation. Such a high sensitivity is obtained through the combination of resonant, optical, mechanical, and electrical methods. In the same way of classical computing technology, in which discrete electronic elements of circuits were replaced by millions of components integrated into single chips, it would be desirable to have some of the NMR quantum computing hardware integrated into a single quantum chip. It appears that the first step towards this direction has already been given by Yusa and co-workers [20] and co-workers in the year of 2005. They described the observation of multiple quantum coherences of nuclear spins in a semiconductor structure which had part of a NMR hardware integrated to it. So, the work of Yusa et al. is neither about NMR QIP scalability, nor about single spin detection techniques; it is about two other important aspects of NMR QIP: chip integration and direct observation of NMR multiple quantum coherences. We saw that the complete characterization of NMR qubits must be done by measuring all the density matrix elements through quantum state tomography (see Chapter 4). However, conventional NMR detects the signal coming only from first order coherences, directly linked to the transverse magnetization, $M_x \pm iM_y$. Therefore, to observe higher order coherences, long sequences of RF pulses must be used. The approach of Yusa et al. is totally different, and allows the direct measurement of higher order coherences through electric methods.

The architecture of their device, shown in Figure 7.11 is based on GaAs/AlGaAs heterostructure which contains a RF antenna gate and a micrometer point contact region through which electrical resistance is measured. In this tiny region, the electronic system is a two-dimensional gas. Under some conditions, the hyperfine interaction between the electrons and local nuclei, leads to nuclear spins polarization. The idea is that the electrical resistivity of the current through the contact depends on the nuclear spin state. Therefore, changes in the spin directions caused by NMR leads to a change in the resistivity, ΔR. So, it is basically an electrical method to detect NMR. The experiment was performed with three different NMR isotopes: ^{69}Ga, ^{71}Ga, and ^{75}As. All three have $I = 3/2$ and therefore present an unequally spaced manifold of energy levels, due to the quadrupole

7.4. NMR on a chip: towards the NMR quantum chip integration

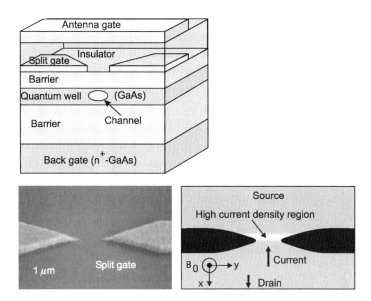

Figure 7.11 NMR on a chip: the scheme shows the architecture of the heterojunction built by Yusa and co-workers in 2005, containing part of a NMR hardware integrated to it. In a conventional NMR experiment only first-order coherences are observed. In Yusa et al. experiment, higher order coherences are directly measured through the resistance variation measured across the split gate. The resistance depends on the nuclear magnetic state. Adapted with permission from [20].

interaction (Chapter 2). Conventional NMR is capable to detect the transitions only of those pair of levels satisfying the selection rule $\Delta m = 1$. However, such a restriction does not exist for the proposed electrical detection method. Single- ($|+3/2\rangle \longleftrightarrow |+1/2\rangle$, ($|+1/2\rangle \longleftrightarrow |-1/2\rangle$) and ($|-1/2\rangle \longleftrightarrow |-3/2\rangle$), double- ($|+3/2\rangle \longleftrightarrow |\pm -1/2\rangle$ and ($|+1/2\rangle \longleftrightarrow |-3/2\rangle$)), and triple-quantum transitions ($|+3/2\rangle \longleftrightarrow |-3/2\rangle$) are observed by adjusting the intensity of the RF field. Such transitions appear as fast oscillating signals in the resistivity ΔR.

Finally, we would like to mention that during the writing of the book we could hardly keep track of new results coming out everyday in the literature. To finish, we mention the papers of Kitchen et al. [21] describing an Mn *atom-by-atom substitution* in GaAs using STM, and the work of Savukov, Lee and Romalis, *Optical detection of liquid-state NMR* [22]. Although not directly related to quantum computation, these works represent new development techniques towards large-scale NMR QIP: the former to a direction of quantum chip manufacturing, and the later to the increase in resolution and sensitivity.

In conclusion, this chapter reviewed a number of ideas and experiments which, combined together, may lead to the construction of a large scale NMR quantum processor. It is interesting to notice that all these proposals have in common the dependence on the precise manipulation of matter at molecular and atomic scale. Therefore, we can conclude that nanofabrication is the promising technology for NMR quantum computing science.

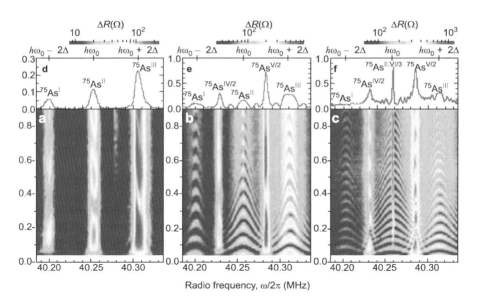

Figure 7.12 Direct detection of multiple quantum coherences in Yusa et al. experiment. Varying the intensity of the radiofrequency field induces multiple quantum transitions which appear as the oscillations observed in the resistance measured across the split gate. Adapted with permission from [20].

PROBLEMS WITH SOLUTIONS

P7.1 - Consider the Hamiltonian of a $I = 1/2$ nucleus coupled to an electron, in the presence of a static field B in the z-direction:

$$\mathcal{H}_{e-n} = \gamma_e \hbar B S_z - \gamma_n \hbar B I_z + A \mathbf{S} \cdot \mathbf{I} \tag{7.4.1}$$

Write explicitly the matrix $\mathcal{H}_{e-n}$ in the basis $|I_z, S_z\rangle$, and find its eigenvalues and the NMR frequencies for the case $A \ll \hbar \gamma_e B$.

Solution

Let us write the basis as $|0\uparrow\rangle, |0\downarrow\rangle, |1\uparrow\rangle, |1\downarrow\rangle$. In this basis, the nucleus and electron spin matrices are:

$$I_x = \frac{1}{2}\begin{pmatrix} 0 & 0 & 1 & 0 \\ 0 & 0 & 0 & 1 \\ 1 & 0 & 0 & 0 \\ 0 & 1 & 0 & 0 \end{pmatrix}; \quad S_x = \frac{1}{2}\begin{pmatrix} 0 & 1 & 0 & 0 \\ 1 & 0 & 0 & 0 \\ 0 & 0 & 0 & 1 \\ 0 & 0 & 1 & 0 \end{pmatrix}$$

$$I_y = \frac{i}{2}\begin{pmatrix} 0 & 0 & -1 & 0 \\ 0 & 0 & 0 & -1 \\ 1 & 0 & 0 & 0 \\ 0 & 1 & 0 & 0 \end{pmatrix}; \quad S_y = \frac{i}{2}\begin{pmatrix} 0 & -1 & 0 & 0 \\ 1 & 0 & 0 & 0 \\ 0 & 0 & 0 & -1 \\ 0 & 0 & 1 & 0 \end{pmatrix}$$

$$I_z = \frac{1}{2}\begin{pmatrix} 1 & 0 & 0 & 0 \\ 0 & 1 & 0 & 0 \\ 0 & 0 & -1 & 0 \\ 0 & 0 & 0 & -1 \end{pmatrix}; \quad S_z = \frac{1}{2}\begin{pmatrix} 1 & 0 & 0 & 0 \\ 0 & -1 & 0 & 0 \\ 0 & 0 & 1 & 0 \\ 0 & 0 & 0 & -1 \end{pmatrix}$$

Performing the matrices products and replacing in the Hamiltonian, we obtain:

$$\mathcal{H}_{e-n} = \begin{pmatrix} \gamma_e \hbar B - \gamma_n \hbar B + A/4 & 0 & 0 & 0 \\ 0 & -\gamma_e \hbar B - \gamma_n \hbar B - A/4 & A/2 & 0 \\ 0 & A/2 & \gamma_e \hbar B + \gamma_n \hbar B - A/4 & 0 \\ 0 & 0 & 0 & -\gamma_e \hbar B + \gamma_n \hbar B + A/4 \end{pmatrix}$$

The eigenvalues of this Hamiltonian are:

$$E_1 = \hbar B(\gamma_e - \gamma_n) + \frac{A}{4}$$

$$E_2 = \hbar B(-\gamma_e + \gamma_n) + \frac{A}{4}$$

$$E_3 = -\frac{A}{4} + \frac{1}{2}\sqrt{A^2 + 4\hbar^2 B^2(\gamma_e + \gamma_n)^2}$$

$$E_4 = -\frac{A}{4} - \frac{1}{2}\sqrt{A^2 + 4\hbar^2 B^2(\gamma_e + \gamma_n)^2}$$

Considering $A \ll \hbar \gamma_e B$, the energies $E_{3,4}$ can be written as

$$E_{3,4} \approx -\frac{A}{4} \pm \frac{1}{2} 2(\gamma_e + \gamma_n)\hbar B \left[1 + \frac{A^2}{4\hbar^2 B^2(\gamma_e + \gamma_n)^2}\right]^{1/2}$$

$$E_{3,4} \approx -\frac{A}{4} \pm \hbar(\gamma_e + \gamma_n)B \pm \frac{A^2}{8(\gamma_e + \gamma_n)\hbar B}$$

For the electron up-state, the NMR frequency will be:

$$\frac{E_3 - E_1}{\hbar} = \omega_{n\uparrow} = 2\gamma_n B - \frac{A}{2\hbar} + \frac{A^2}{8(\gamma_e + \gamma_n)B}$$

And for the electron down-state:

$$\frac{E_4 - E_2}{\hbar} = \omega_{n\downarrow} = 2\gamma_n B - \frac{A}{2\hbar} - \frac{A^2}{8(\gamma_e + \gamma_n)B}$$

P7.2 - Calculate the NMR frequency shift for ^{31}P nuclei in an array where neighbors nuclei are separated by a distance of $a = 1.9$ Å, under a static field of 7 T and a field gradient applied along the chain of 1.4 T/μm. Repeat the calculation for ^{29}Si.

Solution

Take $\gamma/2\pi \approx 17.2$ MHz/T for ^{31}P. Under a field of 7 T, the NMR frequency will be $17.2 \times 7 = 120.4$ MHz. Now, the field shift in the array, along the gradient direction and neighboring sites, is 1.4 T/μm $\times 1.9 \times 10^{-4}/\mu$m $= 2.66 \times 10^{-4}$ T. Therefore, the NMR frequency shift will be 45.7×10^{-4} MHz or 4.57 kHz. For silicon, we have $\gamma/2\pi \approx 8.4$ MHz/T and $\Delta \nu = 8.4 \times 2.66 \times 10^{-4}$ MHz $= 2.2$ kHz.

P7.3 - Estimate the probabilities $p(\uparrow)$ and $p(\downarrow)$, of an electronic spin to be in the up or down state in a field of 10 T at the temperature of 1 K. Repeat the calculation for a $I = 1/2$ nuclear spin.

Solution
First, let us calculate the factor

$$\frac{\mu_B B}{k_B T} = \frac{5.8 \times 10^{-5} \text{ eV/T} \times 10 \text{ T}}{8.6 \times 10^{-5} \text{ eV/K} \times 1 \text{ K}} \approx 6$$

Now, the canonical partition function is

$$\mathcal{Z} = e^{\mu_B B/2k_B T} + e^{-\mu_B B/2k_B T} = e^3 + e^{-3} = 20.15$$

The probability $p(\uparrow)$ will be:

$$p(\uparrow) = \frac{e^{\mu_B B/2k_B T}}{\mathcal{Z}} = \frac{e^3}{20.15} = 99.75\%$$

Obviously, $p(\downarrow) = 0.25\%$. Notice that this produces a net magnetic moment of $(0.9975 - 0.0025)\mu_B = 0.995 \mu_B$.

In the nuclear case, we know that $\mu_n \approx 0.001 \mu_B$. For the same field and temperature, we have $\mu_n B/k_B T \approx 0.006$, $\mathcal{Z} \approx 1.99$ and $p(\uparrow) \approx 50.25\%$ and $p(\downarrow) \approx 49.75$. This produces a net magnetic moment (in units of Bohr magneton) of $5 \times 10^{-4} \mu_B$.

P7.4 - Consider a two-level system composed by N non-interacting spin $1/2$ nuclei. Calculate the force excerpted over this system when exposed to a magnetic field gradient $\partial B/\partial z = B_0/a$, where a is the distance between the spins, and show that it is given by $\mu_N B_0 N p/a$.

Solution
The magnetic force is $F = M \partial B/\partial z = M B_0/a$. To calculate M, let $n_\uparrow = p_\uparrow N$ and $n_\downarrow = p_\downarrow N$ be the number of up and down spins, respectively. Here, $p_\uparrow$ and $p_\downarrow$ are the probabilities of occupancy of the two levels. The population difference between the levels is therefore $n_\uparrow - n_\downarrow = N(p_\uparrow - p_\downarrow) = Np$, where $p = (p_\uparrow - p_\downarrow)$ is the polarization. But, on the other hand, $n_\uparrow + n_\downarrow = N$ is the total number of spins. Thus,

$$n_\uparrow = N\frac{1+p}{2} \quad \text{and} \quad n_\downarrow = N\frac{1-p}{2}$$

The magnetization will be:

$$M = \mu_N (n_\uparrow - n_\downarrow) = \mu_N N \left[\left(\frac{1+p}{2}\right) - \left(\frac{1-p}{2}\right) \right]$$

and the magnetic force:

$$F = \frac{\mu_N B_0}{a} N \left[\frac{1+p}{2} - \frac{1-p}{2} \right] = \frac{\mu_N B_0}{a} N p$$

P7.5 - The minimum detectable force on a cantilever of a MRFM apparatus can be written as

$$F_{min} = \sqrt{\frac{4 k k_B T B}{\omega_0 Q}}$$

where k is the spring constant of the cantilever, ω_0 its resonance frequency, B the bandwidth and Q the quality factor. Estimate this force taking $k = 0.0042$ N/m, $\omega_0 = 2\pi \times 23$ rd/s, $Q = 10^4$ and $B = 0.6$ kHz. Make the calculation at 4.2 K and at room temperature.

Solution

$$F_{min} = \sqrt{\frac{4 \times 0.0042 \times 1.38 \times 10^{-23} \times 0.6 \times 10^3}{2 \times \pi \times 23 \times 10^4}} \times T^{1/2}$$

$$F_{min} = 9.8 \times 10^{-10} T^{1/2}$$

Therefore, for $T = 4.2$ K, $F_{min} = 2.0 \times 10^{-9}$ N. For 300 K, $F_{min} = 1.6 \times 10^{-7}$ N.

P7.6 - The force on the cantilever of a MRFM device, excerpted by n qubits with initial polarization p is,

$$F_z = \frac{\hbar \Delta \omega}{2a} N \left[\left(\frac{1+p}{2} \right)^n - \left(\frac{1-p}{2} \right)^n \right]$$

Setting $F_z = F_{min} = 6.2 \times 10^{-16}$ N, $a = 1.9$ Å, $N = 10^5$ and $\Delta \omega = 2\pi \times 2$ kHz, calculate n for $p = 1, 5, 10, 20, 50, 80$ and 90%. Make a numerical calculation to obtain the necessary initial polarization for 2, 3, 6, 8, 10, 20, 50, 100 and 200 qubits. Make a plot of n vs. p and notice how the number of qubits quickly increases as $p \to 100\%$.

Solution
Replacing the given numerical values we find

$$\frac{\hbar \Delta \omega}{2a} N = 2.2 \times 10^{-15}$$

Therefore,

$$\left[\left(\frac{1+p}{2} \right)^n - \left(\frac{1-p}{2} \right)^n \right] = 0.28$$

A numerical solution for this equation yields the following set of values:

Number of qubits	Initial polarization
2	0.28
3	0.36
6	0.62
8	0.71
10	0.76
20	0.88
50	0.95
100	0.97
200	0.99

Figure 7.13 shows a plot of n vs. $100 \times p$. The number of available qubits increases very fast as $p \to 100\%$.

P7.7 - Estimate the magnetic field in the z-direction of a magnetized sphere of radius $R = 50$ Å right over an electronic moment at a distance of $d = 150$ Å from the center of the sphere. For that, consider the field of a dipole with magnetic moment given by

$$\mathbf{m} = \frac{4}{3} \pi R^3 M \mathbf{k}$$

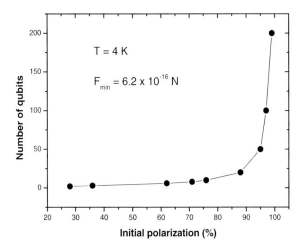

Figure 7.13 Problem (P7.6): increase in the number of available qubits in the "all-silicon" quantum computer proposal of Ladd and co-workers, as a function of the initial polarization. The calculation is made for $T = 4$ K.

where M is the saturation magnetization of the sphere. Use $\mu_0 M = 2.2$ T, and calculate the ESR frequency in this field.

Solution
The starting point is to write the field produced by a magnetic dipole **m** at a position **r**:

$$\mathbf{B} = \frac{\mu_0}{4\pi} \frac{3(\mathbf{m} \cdot \mathbf{e}_r)\mathbf{e}_r - \mathbf{m}}{r^3}$$

where $\mathbf{e}_r$ is the unit vector pointing to the direction of **r**. Replacing the expression for **m**, one obtains

$$B = \frac{\mu_0}{4\pi} \frac{1}{d^3} \left[3\frac{4}{3}\pi R^3 M - \frac{4}{3}\pi R^3 M \right] = \frac{2}{3}\mu_0 M \left(\frac{R}{d}\right)^3$$

Replacing numerical values:

$$B = \frac{2}{3} \times 2.2 \times \frac{1}{3^3} = 0.054 \text{ T}$$

Using $\gamma_e/2\pi = 28$ GHz/T for the electronic gyromagnetic ratio, one obtains $\Delta\omega_e = 28 \times 0.054 = 1.5$ GHz for the resonance shift.

P7.8 - From the results of the previous exercise, obtain the field gradient and the magnetic force over an electronic magnetic moment of 1 μ_B.

Solution
To calculate the force over a magnetic moment, one must obtain the field gradient. In the case of the field calculated in the previous exercise, we simply calculate the derivative of B with respect to the moment position, d:

$$\frac{\partial B}{\partial d} = -\frac{3}{d} \times \frac{2}{3}\mu_0 M \left(\frac{R}{d}\right)^3$$

The magnetic force will be

$$F = \mp \mu_B \frac{3}{d} \times \frac{2}{3}\mu_0 M \left(\frac{R}{d}\right)^3$$

Replacing numbers,

$$F = \mp 9.3 \times 10^{-24} \frac{3}{150 \times 10^{-10}} \times 0.054 = \mp 0.01 \times 10^{-14} = \mp 10^{-16} \text{ N}$$

REFERENCES

[1] B.E. Kane, A silicon-based nuclear spin quantum computer, *Nature* **393** (1998) 133.
[2] B. Koiller, X. Hu, S. Das Sarma, Exchange in silicon-based quantum computer architecture, *Phys. Rev. Lett.* **88** (2002) 027903-1.
[3] J.L. O'Brien, S.R. Schoefield, M.Y. Simmons, R.G. Clark, A.S. Dzurak, N.J. Curson, B.E. Kane, N.S. McAlpine, M.E. Hawley, G.W. Brown, Towards the fabrication of phosphorous qubits for a silicon quantum computer, *Phys. Rev. B* **64** (2001) 161401-1.
[4] S.R. Schoefield, N.J. Curson, M.Y. Simmons, F.J. Rueß, T. Hallam, L. Oberbeck, R.G. Clark, Atomically precise placement of single dopants in Si, *Phys. Rev. Lett.* **91** (2003) 136104-1.
[5] A.J. Skinner, M.E. Davenport, B.E. Kane, Hydrogenic spin quantum computing in silicon: a digital approach, *Phys. Rev. Lett.* **90** (2003) 087901-1.
[6] T.D. Ladd, J.R. Goldman, F. Yamaguchi, Y. Yamamoto, E. Abe, K.M. Ytoh, All-silicon quantum computer, *Phys. Rev. Lett.* **89** (2002) 017901-1; see also T.D. Ladd, D. Maryenko, Y. Yamamoto, E. Abe, K.M. Itoh, *Phys. Rev. B* **71** (2005) 104401.
[7] J. Crangle, *Solid State Magnetism* (Edward Arnold, London, 1991).
[8] J.A. Sidles, Noninductive detection of single proton magnetic resonance, *App. Phys. Lett.* **58** (1991) 2854.
[9] J.A. Sidles, J.L. Garbini, G.P. Drobny, The theory of oscillator-coupled with potential applications to molecular imaging, *Rev. Sci. Instrum.* **63** (1992) 3881.
[10] J.A. Sidles, J.L. Garbini, K.J. Bruland, D. Rugar, O. Züger, S. Hoen, C.S. Yannoni, Magnetic resonance force microscopy, *Rev. Mod. Phys.* **67** (1995) 249.
[11] G.P. Berman, G.D. Doolen, P.C. Hammel, V.Y. Tsifrinovich, Solid-state nuclear spin quantum computer based on magnetic resonance force microscopy, *Phys. Rev. B* **61** (2000) 14694.
[12] Y. Manassen, R.J. Hamers, J.E. Demuth, A.J. Castellano Jr., Direct observation of the precession of individual paramagnetic spins on oxidized silicon surfaces, *Phys. Rev. Lett.* **62** (1989) 2531.
[13] C. Durkan, M.E. Welland, Electronic spin detection in molecules using scanning-tunneling-microscopy-assisted electron-spin resonance, *App. Phys. Lett.* **80** (2002) 458.
[14] J.M. Elzerman, R. Hanson, L.H. Willems van Beveren, B. Witkamp, L.M.K. Vandersypen, L.P. Kouwenhoven, Single-shot read-out of an individual electron spin in a quantum dot, *Letters to Nature* **430** (2004) 431.
[15] M. Xiao, I. Martin, E. Yablonovitch, H.W. Jiang, Electrical detection of the spin resonance of a single electron in a silicon field-effect transistor, *Letters to Nature* **430** (2004) 435.
[16] J. Kohler, J.A.J.M. Disselhorst, M.C.J.M. Donckers, E.J.J. Groenen, J. Schmidt, W.E. Moerner, Magnetic resonance of a single molecular spin, *Letters to Nature* **363** (1993) 242.
[17] J. Wrachtrup, C. von Borczyskowski, J. Bernard, M. Orrit, R. Brown, Optical detection of magnetic resonance in single molecule, *Letters to Nature* **363** (1993) 244.
[18] F. Jelezko, I. Popa, A. Gruber, C. Tietz, J. Wrachtrup, A. Nizovtsev, S. Kilin, Single spin states in a defect center resolved by optical spectroscopy, *App. Phys. Lett.* **81** (2002) 2160.
[19] D. Rugar, R. Budakian, H.J. Mamin, B.W. Chui, Single spin detection by magnetic resonance force microscopy, *Letters to Nature* **430** (2004) 329.
[20] G. Yusa, K. Muraki, K. Takashina, K. Hashimoto, Y. Hirayama, Controlled multiple quantum coherences of nuclear spins in a nanometer-scale device, *Letters to Nature* **434** (2005) 1001.
[21] D. Kitchen, A. Richardella, J.-Ming Tang, M.E. Flatté, A. Yazdani, Atom-by-atom substitution of Mn in GaAs, visualization of their hole-mediated interactions, *Letters to Nature* **442** (2006) 436.
[22] I.M. Savukov, S.-K. Lee, M.V. Romalis, Optical detection of liquid-state NMR, *Letters to Nature* **442** (2006) 1021.

Index

^{28}Si(111) wafer, 225
12 qubit pseudo-cat state, 217
$\pi/8$ (T), 102
$\pi/8$ (T) gate, 98
$\pi/8$, 140, 141
ρ_{GHZ} density matrix, 214
"scattering" circuit, 199

A
AB system, 63
adiabatic quantum optimization algorithm, 200
Alain Aspect, 109
Alan Turing, 1, 10
alanine, 214
Albert Einstein, 1, 108
all-silicon quantum computer, 225
Alonzo Church, 1
alphabet, 10
amplitude of the NMR signal, 5
analog-to-digital converter, 81
ancilla, 212
AND, 2, 12
antiferromagnetic materials, 54
antiphase doublets, 217
Anwar, 5, 216
Arthur Eckert, 3
artifacts, 81
Aspect, 4
asymmetry parameter, 58
atom-by-atom substitution in GaAs using STM, 235
atomic-sensitive MRFM, 227
attenuated correlation, 215
AX system, 62

B
Barret, 4
basis, 96, 97
BB84, 94
BCS model, 198
BDPA [α, γ-bisdiphenylene β-phenylallyl], 232
Bell analyzer, 213
Bell's basis, 107, 129
Bell's inequality, 4, 109
benchmarking quantum control methods on a 12-qubit system, 217
Berman, 6, 229

bilateral unitary transformation, 216
binary digit, 1, 96
black holes, 24
Bloch sphere, 147, 168–171
Boltzmann factor, 18, 37
Boris Podolsky, 108
bosons, 126
Boulant, 215
Bouwmeester, 4
bra, 95
Brassard, 212
Braunstein, 5
Budaklan, 233
bulk semiconductor materials, 9

C
cantilever resonance frequency, 229
carry bit, 13, 28
Castellano Jr., 231
cat state, 107–110
Catasti, 211
central transition, 59, 70
charge occupancy, 233
Charles Bennett, 2, 94
chemical shielding tensor, 54
chemical shift anisotropy (CSA), 55
chemical shifts, 51, 54, 55
chemical-shift interaction, 54
chip integration, 234
Chuang, 4, 183, 187, 188
Chui, 233
circuit representations, 12
classical communication channel, 111
classical information, 109, 110
classical transmission, 212
Claude Shannon, 1
CNOT, 101, 102, 140, 146, 148–151, 154–157, 161, 162, 168, 174
CNOT gate, 99, 207
CNOT$_a$, 101
CNOT$_b$, 101
Cohendet, 126
coherence transfer pathways, 73, 82
coherences, 45
communication channel, 1, 109
complete quantum teleportation using nuclear magnetic resonance, 212

complete set of logic gates, 12
complete set of universal quantum gates, vii
complete teleportation experiment, 212
completeness relation, 96
components integrated into single chips, 234
computational basis, 96
computational resource, 106
computer memory, 10
computer program, 10, 21
computer technology, 14
conditional NOT operation, 12
conservation of the scalar product, 95
construction of a large scale NMR quantum
    processor, 235
contact field, 223
control bit, 13
control qubits, 101
controlled swap, 20
controlled-NOT gate, 102
controlled-swap gate, 30
conventional NMR sensitivity, 222
Cory, 4, 183
coupled spin 1/2 systems, 163
coupling constant, 224
creation of coherence, 50
cross relaxation, 225
cross-polarization (CP), 77
crossover from exponential to polynomial
    dependence, 226
crotonic acid, 215
CYCLOPS, 81
cytosine, 217

D

damping time-constant, 229
Das, 193
Davenport, 6, 224
David Deutsch, 2, 94, 112
David Hilbert, 1
decoherence, 212
Demuth, 231
density matrix, 45, 104, 105
density matrix formalism, 44
density matrix tomography, 163, 168
detecting single electron magnetic moments by
    MRFM, 231
detection of single paramagnetic moments, 231
determining bounds of ϵ, 210
determining eigenvalues and eigenvectors, 130
Deutsch algorithm, 2, 183, 186
Deutsch–Jozsa algorithm, 113, 185, 186
deviation density matrix, 47
diamagnetic substances, 53
dielectric constant, 223
digital approach of quantum computing, 224

digital signal processor, 81
dipolar coupling, 55, 226
dipolar decoupling, 76
dipolar interaction, 54
direct measurement of higher order coherences
    through electric methods, 234
direct observation of NMR multiple quantum
    coherences, 234
direct tensorial product, 63
discrete Wigner function, 183, 199
disorder, 17
double rotation (DOR), 77
double-resonance, 65, 76, 77
DPPH (diphenylpicrylhydrazil), 227
Drobny, 227
Durkan, 231
dynamic angle spinning (DAS), 77
dynamics, 5

E

effective magnetic field, 39
effective moving mass, 229
effective number of qubits detectable, 228
effective nutation frequency, 51
eigenstates, 4, 96
Einstein, 4
electric field gradient, 58
electric field gradient tensor, 57
electric quadrupole moment, 35
electrical method to detect NMR, 234
electron density, 222
electron donor, 222
electron magnetic moment, 229
electron spin in a quantum dot, 232
electron–nucleus Hamiltonian, 223
electronic density, 222
electronic dipolar field, 231
electronic resonance shift, 229
electronic wave-function, 222
electrostatic potential, 232
Elzerman, 232
energy eigenvalues, 23
energy levels, 17
energy spectrum, 18
ensemble averages, 44
entangled systems, 109
entanglement, 2, 4, 94, 97, 106, 183
entanglement between an electron and a nuclear
    spin 1/2, 215
entanglement transfer experiment in NMR
    quantum information processing, 215
entropy, 2, 3, 17, 107
entropy growth, 12
entscheidungsproblem, 1
EPR, 149, 153, 174
EPR paper, 4

equalization of the populations, 50
erasure, 18, 20
Erhenfest theorem, 31
Ermakov, 193
Erwin Schrödinger, 1
ESR (Electron Spin Resonance), 215
ESR frequency, 232
evolution operator, 45
exclusive-OR gate, 12
experimental demonstration of fully coherent quantum feedback, 212
experimental demonstration of GHZ correlations using NMR, 213
experimental implementation of dense coding using NMR, 213
experimental quantum error correction, 200
exponential efficiency, 211
exponential loss of sensitivity, 221
exponential speedup, 4
external interactions, 53

F
Fahmy, 4
Fang, 213
Fano–Anderson, 196, 197
Fast Fourier Transform (FFT), 102
Fermi golden rule, 39
Fermi level, 232
Fermionic systems, 196
fermions, 126
ferromagnetic materials, 54
ferromagnetic particle, 230
field gradient, 225
field-effect transistor (FET), 232
final readout, 226
first order coherences, 234
Fourier transform, 43
Fredkin gate, 20, 30
free induction decay (FID), 41
free precession, 52
frequency NMR resolution, 225
full adder, 14
Fung, 193

G
GaAs/AlGaAs heterostructure, 232, 234
Garbini, 227
gate T, 100
geometric quantum computation, 200
George Boole, 1
Gershenfeld, 4, 183, 187, 188
GHZ correlations, 214
GHZ density matrix, 211
Gilles Brassard, 94
global phase, 141, 149, 173, 176
good dynamics and bad kinematics, 222

Gordon Moore, 24
gradient pulse, 159, 161, 177, 178
Grangier, 4
graphical representation of gates, 100
Grover, 113, 183, 187, 188
Grover search algorithm, 188
gyromagnetic ratio, 35

H
Hadamard, 98, 102, 103, 140–142, 146, 149, 150, 169, 191, 198
Hadamard gate, 98, 100, 146, 207
Hadamard transformations, 146
half-adder, 14
half-integer spin nuclei, 59
Hamers, 231
Hamiltonian, 22, 93, 130
Hansen, 188
hard or non-selective pulses, 70
hard pulses, 52, 70, 143
Hartmann–Hahn condition, 78
Havel, 4
Hawking radiation, 24
heat-bath algorithmic cooling, 201
Heisenberg Hamiltonian, 4
Heisenberg spin chain, 200
Helmholtz free-energy, 18
Hermitians, 98
heteronuclear decoupling, 76
high temperature approximation, 207
high-resolution solid-state NMR spectroscopy, 78
high-temperature limit, 46
higher order coherences, 234
Hilbert space, 95, 96, 99, 125, 130
homonuclear decoupling, 76
homonuclear spins, 218
host lattice, 222
hyperfine field, 6
hyperfine magnetic fields, 54

I
ideal gas, 29
identity matrix, 5
implementation of quantum algorithms by NMR, 183
implementing a full set of universal quantum gates, 222
implementing single-qubit operations, 229
implementing the CNOT quantum gate, 229
improvement of resolution of magnetic resonance imaging below 1 μm, 233
impurity atoms, 229
indirect or scalar coupling, 56
information content, 29
interaction representation, 74
interference pattern, 215

internal energy, 18
internal interactions, 53
internal state, 10, 11
inversion of the populations, 50
irreducible spherical tensorial representations, 61
irreversibly, 12
Isaac Chuang, 93, 94
isolated quantum states, 26
isolated systems, 2
isotropical chemical shift, 54

J

J-coupling, 56
J.A. Jones, 185
J.W. Turkey, 1
James Clerk Maxwell, 10
Jelezko, 233
John Bardeen, 24
John Bell, 4, 109
Jones, 5, 186, 188
Jordan–Wigner, 197
Jozsa, 113, 183

K

Kane, 6, 224
ket, 95
Keyes, 25
kinetic energy, 29
Kitchen, 6, 235
Knight shift, 54
Knill, 4, 211, 212
knowledge, statistics and thermodynamics, 15
Kohler, 233
Koiller, 6, 224
Kubinec, 187, 188
Kumar, 193

L

Ladd, 6, 225
Laflamme, 4, 211, 212
Landauer's principle and the Maxwell demon, 20
large scale NMR-based quantum chip, 222
large scale quantum information processor through NMR, 231
large scale solid-sate NMR quantum processors, 227
Larmor frequency, 36
Larmor precession, 36, 231
Lee, 6, 189, 235
Leonhardt, 126
Linden, 5, 211
Liouville–von Neumann equation, 44
liquid-state NMR, vii
local hyperfine field, 223
logic states, 26
logical gates, 66, 150

logical labeling, 137, 162
logical processing, 21
longitudinal or spin-lattice relaxation time, 43
longitudinal relaxation, 41, 74
Louis de Broglie, 1
Lov Grover, 3, 94

M

M. Mosca, 185
magic-angle, 77
magic-angle spinning (MAS), 76
magnetic dipole moment, 34
magnetic field, 17
magnetic field gradient, 217
magnetic force, 226
magnetic hysteresis, 17
magnetic interaction, 222
magnetic materials, 9
magnetic moment, 22
magnetic resonance force microscopy (MRFM), vii, 6, 226
magnetization, 17
Mamin, 233
Manassen, 231
manipulation of matter at molecular and atomic scale, 235
Mariappan, 211
master equation, 75
matrix elements of a NMR density matrix, 183
Max Planck, 1
maximum memory, 23
maximum number of operations per bit per second, 23
maximum speed, 23
Maxwell demon, 10, 20
measure of entanglement fidelity, 212
measure of fidelity, 211
measurement operators, 95
measurements, 4, 95, 108
measuring the discrete Wigner function, 199
measuring the Wigner function, 127
Mehring, 215
Mende, 215
Michael Nielsen, 93
micrometer point contact, 234
minimum force detectable by MRFM, 226
minimum force detectable in a MRFM setup, 228
mixed state, 153
Moore's law, 24
Moore's law. Quantum computation, 24
Mosca, 5, 186, 188
MRFM-based NMR quantum processor, 231
multi-frequency pulses, 161
multiple-quantum coherences, 74

multiple-quantum magic-angle spinning
(MQ-MAS), 77
multiple-quantum (MQ) coherences, 73
multiplicity, 29
Murali, 193

N

n-qubit pseudo-entangled state, 210
n-qubits density matrix, 5
NAND, 102
NAND (NOT-AND), 12
nanofabrication, 6, 224, 235
nanoscience, 9
nanotechnology, 9
Nathan Rose, 108
natural limits for computation, 22
natural phenomena as computing processes. The physical limits of computation, 21
natural resources, 27
Negrevergne, 5, 196, 210, 217
Neil Gershenfeld, 94
Nelson, 212
new development techniques towards large-scale NMR QIP, 235
Niels Bohr, 1
Nielsen, 4, 212
NMR, 4, 10, 33, 183, 185, 187
NMR density matrices, 5
NMR Greenberger–Horne–Zeillinger states, 211
NMR Hamiltonian, 195
NMR implementation, 185
NMR on a chip: towards the NMR quantum chip integration, 234
NMR QIP, 200
NMR quantum computation, 94
NMR quantum computer, 200
NMR quantum computer entirely based on MRFM, 229
NMR quantum computing science, 235
NMR quantum information processing based on magnetic resonance force microscopy (MRFM), 226
NMR radiofrequency pulses, 207
NMR sensitivity, vii
NMR spectra, 185
NMR spectrometer, 79
noisy channel, 216
non-equilibrium states, 4
non-locality, 108, 109
non-selective irradiation, 52
non-separability region, 210
NOR (NOT-OR), 12
NOT, 2, 12, 140, 141
NOT gate, 141
nuclear gyromagnetic ratio, 30
nuclear magnetic resonance, 3, 10
nuclear paramagnetism, 38
nuclear spin Hamiltonian, 53
nuclear spin quantum number, 33
nuclear spins, vii, 2, 33
nucleons, 34
nucleus–nucleus interaction, 222
nutation angle, 40
nutation frequency, 40
nutation NMR spectroscopy, 71

O

O'Brien, 224
observables, 107
observation, 4
observation of individual paramagnetic spins using the scanning tunneling microscopy (STM) technique, 231
occupation number, 18
off-resonance pulses, 51
optical detection of liquid-state NMR, 235
optical interferometry methods, 227
optical methods, 233
OR, 2, 12
order, 119
oscillatory behavior for J, 224
output, 12

P

Pan, 215
paramagnetic centers, 233
paramagnetic moments, 229
paramagnetic shift, 54
paramagnetic substances, 54
paramagnetic trap, 232
partial trace, 105
partial transposing, 209
partition function, 18, 45
Paul Benioff, 2
Pauli matrices, 22, 47, 97
Peres criterium and bounds for NMR entanglement, 209
permanent magnet, 227
Peter Shor, 2, 94, 112, 190
phase correction, 142, 150, 169
phase cycling, 81
phase estimation, 117
phase gate (S), 98
phase of a RF pulse, 49
phase shift, 140
phase-sensitive detection, 80
phases of electronic and nuclear states, 216
Podolsky, 4
polynomial resources, 211
Popescu, 5, 211
populations, 18, 45
postulates of quantum mechanics, 95

practical implementation of quantum algorithms by NMR, 183
practical implementations of twirl operations, 216
preparing initial state, 229
principal axis system (PAS), 55
principle of maximization of the entropy, 18
probability distribution, 17
problem of liquid-state NMR entanglement, 207
product operators, 68
projective measurement, 212
propagator, 45
pseudo-cat state, 207
pseudo-entangled, 207
pseudo-pure states, vii, 5, 153, 155–162, 169, 178, 179
pulse duration, 40
pure state, 153–155
purification, 216
purified silicon lattice, 222
pyrolytic graphite (HOPG), 232

Q
QC, 10
QFT, 103, 117, 189, 191
QIP, vii, 10
QST, 163
quadrature detection, 81
Quadrupolar coupling, 57
quadrupolar interaction, 54
quadrupolar spins, 201
quadrupole systems, 193
quantification of entanglement, 208
quantum algorithms, vii, 2, 111, 112
quantum Baker's map, 200
quantum bit of information, 26
quantum bits, 96
quantum circuits, 100, 207
quantum clock synchronization algorithm, 201
quantum communication, vii
quantum computation, 2, 26, 93, 94, 96, 111
quantum computing, 94, 102
quantum correlations, 183, 211
quantum cryptography, vii, 4, 210
quantum error correction codes, 4
quantum factorizing algorithm, 116
quantum feedback, 212
Quantum Fourier Transform NMR implementation, 189
Quantum Fourier Transform (QFT), 102, 189
quantum games, 200
quantum hardware, 2
quantum harmonic oscillator, 194
quantum information, 2, 93, 94, 96
quantum information in phase space, 125
quantum information processing, 137, 140, 146, 163

quantum logic gates, viii, 15, 97
quantum logic operations, 5
quantum mechanics, 1, 93–95, 108
quantum metrology, 201
quantum NOT operation, 23
quantum order-finding, 119
quantum phase transition, 200
quantum process tomography, 201
quantum protocols, viii, 211
quantum resources, 2, 3
quantum resources for information processing, vii
quantum search algorithm, 113
Quantum simulations, vii, 124, 194
quantum state tomography, viii, 5, 104, 106, 156, 162, 163, 166, 183, 211
quantum states in phase space, 127
quantum superposition, 4
quantum systems simulations, 2
quantum teleportation, 4, 210
quantum tunneling current, 231
qubits, vii, 3, 26, 96, 137

R
R. Landauer, 2
R_k, 103
radiofrequency, 38
radiofrequency pulses, vii, 140
read/write head, 10
Redfield theory, 75
refocusing, 143, 147, 149, 174
refocusing similar, 150
registers, 117
relative shift, 55
relaxation, 41
relaxation time, 222
resonance, 33
resonance of a single electronic spin, 232
resonant slice, 234
reversible gate, 2
reversible versus irreversible computation, 18
RF antenna gate, 234
RF Hamiltonian, 38
RF pulses, 40, 140–143, 151, 155, 159, 163, 164, 171
Richard Feynman, 2
Riebe, 4
Roger, 4
Roger Penrose, 10
Rolf Landauer, 10
Romalis, 6, 235
Rosen, 4
rotating frame, 39
rotation operator, 48, 140
RSA protocol, 3
Rugar, 6, 227, 233

S

S gate, 100
saturation, 43, 50
Savukov, 6, 235
scalable physical system, vii
scalar coupling, 56
scaling factor, 5
Scherer, 215
Schofield, 224
Schrödinger equation, 2, 26
second law of thermodynamics, 17
second order corrections, 59
secular approximation, 54, 56–58
selection rule, 235
selective excitation, 52
selective pulses, 52, 143, 145–147, 149, 151, 152, 157, 161, 165–168, 171
self-reversibility, 98
semiconductor Bohr radius, 223
semiconductor physics, 1
semiconductor technology, 222
set of universal logic gates, 102
Seth Lloyd, 4, 10, 21
Shannon or information entropy, 17
Shor, 183, 189
Shor factorization algorithm, 2, 117, 190
Sidles, 227
signal routing, 81
signal-to-noise (S/N) ratio, 43
silicon lattice, 6
silicon scalable quantum computer, 6
silicon-based proposals: solution for the scaling problem, 222
single spin detection, 231
single spin detection techniques: solution for the sensitivity problem, 231
singlet spin state, 4
Skinner, 6, 224
small flip-angle pulses, 72
SMP, 151–153
soft or selective RF pulses, 70
soft pulses, 52, 70, 145
solid-state experiment, 215
Somaroo, 194
some NMR experiments reporting pseudo-entanglement, 211
source/drain current, 232, 233
spatial averaging, 159, 161, 162
spectral density functions, 75
spin, 17, 96
spin 1/2 systems, 137, 139, 140, 147, 149, 159
spin 3/2 systems, 138–140, 146, 158, 168
spin configuration, 29
spin eigenstates, 22
spin selective pulse, 143
spin-lattice relaxation, 43
spin-locking, 78
spin-spin relaxation, 41
spin-to-charge conversion, 232
spinning sidebands, 77
spread in energy, 23
state evolution, 23
state labeling, 161
statistical, 16
statistical ensembles, 4
statistical entropy, 18
Stirling's formula, 30
STM H lithography, 224
STM (Scanning Tunneling Microscopy) lithography, 224
Strongly Modulated Pulses (SMP), 151, 152
subsystems, 96
superdense coding, 2, 4, 109, 210
superposition, 4
superposition principle, 27
SWAP, 101, 140, 149, 156, 157, 174
SWAP circuit, 101

T

T, 141
T_2 relaxation, 212
target qubits, 101
techniques to detect single spin states, 233
teleport, 110, 111
teleportation, 94, 212
temporal averaging, 154, 156, 157, 159
tensor product, 99
tensorial product, 99
theory of information, 1
thermal equilibrium density matrix, 46
thermal noise, 229
thermal reservoir, 17
Toffoli, 150, 175
Toffoli gate, 19
torque, 23
transverse magnetization, 234
transverse or spin-spin relaxation time, 42
transverse relaxation, 41, 74
trichloroethylene, 211, 212
truth table, 12
Tseng, 195
Turing Machine, 1, 9, 10
Turing Machines, logic gates and computers, 9
twirl operation, 216
two qubit operation, 100
two-qubit system, 137, 139, 153, 161

U

ultimate laptop, 22
uncertainty principle, 22
unitary operations, 140, 147, 157, 159
unitary operator, 26

unitary propagator, 2
unitary transformations, 2, 95, 97, 183

V
Vandersypen, 189, 190
Von Neumann's entropy, 107

W
Walter Houser Brattain, 24
Warren, 5
Welland, 231
Werner Heisenberg, 1
Wigner, 125, 199
Wigner function, 125, 126, 200
Wigner functions for discrete systems, 125
Wigner–Eckart theorem, 35
Williams and Clearwater, 24
Willian Bradford Shockley, 24
Wolfgang Pauli, 1
Wooters, 126

Wrachtrup, 233

X
Xiao, 232
XOR, 12

Y
Y. Jack Ng, 9
Y. Lecerf, 19
Yang, 197
Yannoni, 227
Yusa, 6, 234

Z
Z, 141
z-rotation, 142
Zeeman Hamiltonian, 36
Zeeman interaction, 35
zero-field NMR, 79
Zurek, 211, 212